Springer Series in Computational Mathematics

21

R. Hammer M. Hocks
U. Kulisch D. Ratz

Numerical Toolbox for Verified Computing I

Basic Numerical Problems

Theory, Algorithms, and Pascal-XSC Programs

With 28 Figures

Springer-Verlag
Berlin Heidelberg New York
London Paris Tokyo
Hong Kong Barcelona
Budapest

Prof. Dr. Ulrich Kulisch
Dr. Rolf Hammer
Dr. Dietmar Ratz
Dipl-Math. oec. Matthias Hocks
Institut für Angewandte Mathematik
Universität Karlsruhe
Postfach 6980
D-76128 Karlsruhe

Mathematics Subject Classification (1991): 65-01, 65G10, 65F, 65H, 65K

ISBN-13:978-3-642-78425-5 e-ISBN-13:978-3-642-78423-1
DOI: 10.1007/978-3-642-78423-1

CIP data applied for

Typesetting: Camera ready by the authors using TEX
41/3140 – 5 4 3 2 1 0 – Printed on acid-free paper

Preface

As suggested by the title of this book *Numerical Toolbox for Verified Computing*, we present an extensive set of sophisticated tools to solve basic numerical problems with a *verification of the results*. We use the features of the scientific computer language PASCAL–XSC to offer modules that can be combined by the reader to his/her individual needs. Our overriding concern is reliability – the *automatic verification of the result* a computer returns for a given problem. All algorithms we present are influenced by this central concern. We must point out that there is no relationship between our methods of *numerical result verification* and the methods of *program verification* to prove the correctness of an implementation for a given algorithm.

This book is the first to offer a general discussion on

- arithmetic and computational reliability,
- analytical mathematics and verification techniques,
- algorithms, and
- (most importantly) actual implementations in the form of working computer routines.

Our task has been to find the right balance among these ingredients for each topic. For some topics, we have placed a little more emphasis on the algorithms. For other topics, where the mathematical prerequisites are universally held, we have tended towards more in-depth discussion of the nature of the computational algorithms, or towards practical questions of implementation. For all topics, we present examples, exercises, and numerical results demonstrating the application of the routines presented.

The different chapters of this volume require different levels of knowledge in numerical analysis. Most numerical toolboxes have, after all, tools at varying levels of complexity. Chapters 2, 3, 4, 5, 6, and 10 are suitable for an advanced undergraduate course on numerical computation for science or engineering majors. Other chapters range from the level of a graduate course to that of a professional reference. An attractive feature of this approach is that you can use the book at increasing levels of sophistication as your experience grows. Even inexperienced readers can use our most advanced routines as black boxes. Having done so, these readers can go back and learn what secrets are inside.

The central theme in this book is that practical methods of numerical computation can be simultaneously efficient, clever, clear, and (most importantly) reliable.

We firmly reject the alternative viewpoint that such computational methods must necessarily be so obscure and complex as to be useful only in "black box" form where you have to believe in any calculated result.

This book introduces many computational verification techniques. We want to teach you to take apart these black boxes and to put them back together again, modifying them to suit your specific needs. We assume that you are mathematically literate, i.e. that you have the normal mathematical preparation associated with an undergraduate degree in a mathematical, computational, or physical science, or engineering, or economics, or a quantitative social science. We assume that you know how to program a computer and that you have some knowledge of scientific computation, numerical analysis, or numerical methods. We do not assume that you have any prior formal knowledge of numerical verification or any familiarity with interval analysis. The necessary concepts are introduced.

Volume 1 of *Numerical Toolbox for Verified Computing* provides algorithms and programs to solve basic numerical problems using automatic result verification techniques.

Part I contains two introductory chapters on the features of the scientific computer language PASCAL–XSC and on the basics and terminology of interval arithmetic. Within these chapters, the important correlation between the arithmetic capability and computational accuracy and mathematical fixed-point theory is also discussed.

Part II addresses one-dimensional problems: evaluation of polynomials and general arithmetic expressions, nonlinear root-finding, automatic differentiation, and optimization. Even though only one-dimensional problems treated in this part, the verification methods sometimes require multi-dimensional features like vector or matrix operations.

In Part III, we present routines to solve multi-dimensional problems such as linear and nonlinear systems of equations, linear and global optimization, and automatic differentiation for gradients, Hessians, and Jacobians.

Further volumes of *Numerical Toolbox for Verified Computing* are in preparation covering computational methods in the field of linear systems of equations for complex, interval, and complex interval coefficients, sparse linear systems, eigenvalue problems, matrix exponential, quadrature, automatic differentiation for Taylor series, initial value, boundary value and eigenvalue problems of ordinary differential equations, and integral equations. Editions of the program source code of this volume in the C++ computer language are also in preparation.

Some of the subjects that we cover in detail are not usually found in standard numerical analysis texts. Although this book is intended primarily as a reference text for anyone wishing to apply, modify, or develop routines to obtain mathematically certain and reliable results, it could also be used as a textbook for an advanced course in scientific computation with automatic result verification.

We express our appreciation to all our colleagues whose comments on our book were constructive and encouraging, and we thank our students for their help in testing our routines, modules, and programs. We are very grateful to Prof. Dr. George Corliss (Marquette University, Milwaukee, USA) who helped to polish the text and the contents. His comments and advice based on his numerical and computational experience greatly improved the presentation of our tools for Verified Computing.

Karlsruhe, September 1993 The Authors

The computer programs in this book

are available in several machine-readable formats. To purchase diskettes in IBM-PC compatible format, use the order form at the end of the book. The source code is also available by anonymous ftp from

`iamk4515.mathematik.uni-karlsruhe.de` (129.13.129.15)

in subdirectory

`pub/toolbox/pxsc.`

Technical questions, corrections, and requests for information on other available formats and software products should be directed to *Numerical Toolbox Software, Institut für Angewandte Mathematik, Universität Karlsruhe, D-76128 Karlsruhe, Germany*.

Table of Contents

List of Figures

Chapter 1

Introduction

This is a reference book for numerical methods with automatic result verification. The methods presented here are practical, reliable, and elegant. We provide theory, algorithmic descriptions, and implementations for methods to solve some basic numerical problems in a reliable way. Also, this book can help you learn how to develop such methods and how to proceed for other problems beyond the scope of the book.

We warn of potential computational problems of standard numerical routines, but we do not claim that our advice is infallible! Therefore, we offer our practical judgment on solving many basic numerical problems with automatic result verification. As you gain experience, you will form your own opinion of how reliable our advice is.

We presume that the reader is able to read computer programs in PASCAL and/or PASCAL–XSC (i.e. PASCAL eXtension for Scientific Computation), the language in which all the "tools" in this edition of *Numerical Toolbox for Verified Computing* are implemented. Its wide range of concepts for scientific computation makes PASCAL–XSC especially well suited as a specification language for programming with automatic result verification. We discuss the features of PASCAL–XSC in Chapter 2.

In the following sections, we first give some hints on the methodology and structure of this book. Then we explain the textual and algorithmic notation we use. We also describe the basic concepts and form of our implementations, and we make some comments on the computational environment. We conclude with a section on the motivation and the necessity of methods with numerical result verification.

1.1 Advice for Quick Readers

This book is organized in three parts:

1. the preliminaries,
2. the one-dimensional problems, and
3. the multi-dimensional problems.

The preliminaries give an overview on the extensions of the programming language PASCAL–XSC and on the properties and advantages of interval arithmetic. The reader with little experience with numerical verification and interval techniques should read the section on interval arithmetic with care and interest. Its introductory character shows what arithmetic improvements can do for the reliability of computed results.

We discuss only basic one- and multi-dimensional problems in parts II and III, respectively. We especially direct the attention of readers unfamiliar with numerical result verification to the following chapters:

- Linear Systems of Equations (Chapter 10),
- Evaluation of Polynomials (Chapter 4),
- Automatic Differentiation (Chapter 5), and
- Nonlinear Equations in One Variable (Chapter 6).

Some other chapters, e.g. optimization, require a little more background knowledge on numerical verification and interval techniques. These algorithms sometimes use the basic algorithms mentioned above (e.g. the linear system solver).

1.2 Structure of the Book

In order to make this *Numerical Toolbox* easy and efficient to use, we have chosen some general standards to keep the different chapters of this book uniform. This will help you find the important information on each topic, e.g. the theoretical background or the implementation of a desired algorithm, without necessarily reading the whole chapter.

Each chapter starts with an introduction to describe the numerical and mathematical problem and to motivate a strategy for its solution. A comprehensive discussion of the theoretical basis necessary to solve the problem and to verify the computational result is given in the section "*Theoretical Background*". Under the heading "*Algorithmic Description*", we present the complete algorithm in a very short, clear, and consistent way.

A very important part of each chapter is the section "*Implementation and Examples*" where an actual implementation for the topic discussed in the chapter is presented. As promised in the title, all routines are designed as "tools" that can be combined in several ways to satisfy your needs in solving a specific numerical problem. The routines are built of different modules so they may be integrated in an existing program, e.g. to overcome a numerical trap at a certain point of your own complex problem-solving algorithm. All algorithms presented in this volume are available on a floppy disk and via ftp (see Page vii) to help you use the verification algorithms immediately for testing or for integrating them into existing applications. To demonstrate the effectiveness of each algorithm, we present sample programs that solve practical problems. We point out the superiority of verification methods and precisely defined machine arithmetic over most standard numerical problem-solving strategies.

Most chapters include some "*Exercises*" to encourage you to study the behavior or to improve the algorithms we present. The section "*References and Further*

Reading" at the end of each chapter points out to some similar or more sophisticated and specialized algorithms. The appendix contains basic modules that are used by certain verification algorithms.

1.3 Typography

We use the following type faces for certain words, names, or paragraphs:

italics emphasizes certain words in the text.

boldface marks reserved words of PASCAL–XSC (e.g. **begin**, **module**) in the text or in excerpts of programs.

slanted characterizes predefined identifiers of PASCAL–XSC (e.g. *integer*, *real*) and identifiers from programming examples when they appear in the body of the text.

`typewriter` distinguishes listings and run-time results of programs, including some kind of pretty-printing to improve the readability (see Section 1.5).

References are always indicated as [*nr*], where the number *nr* corresponds to an entry in the bibliography.

1.4 Algorithmic Notation

The algorithms presented in Parts II and III of this book are specified in a pseudo-code similar to PASCAL code, including mathematical notation (cf. Chapter 3) wherever this is possible. Algorithm 1.1 is a simple example of our algorithmic notation.

Algorithm 1.1: DoSomething (x, y, z, Err) {Procedure}

1. $a := x + y;$ $b := x - y;$ {Initializations}
2. {Do something special}
   if $(a > 0)$ **and** $(b > 1)$ **then**
      $c := \sqrt{a};$ $d := \ln(a);$
      $a := -c;$ $b := -d;$
   else if $(b = 1)$ **then**
      return Err := "Error number 1 occurred!";
   else
      return Err := "Another error occurred!";
3. **return** $z := a \cdot b \cdot c/d;$

For clarity, the steps of the algorithms are numbered, and the single statements are separated by semicolons. We use indentation and numbering to mark blocks of statements and to avoid "**begin** – **end**" notation, whenever this is possible. Within the algorithmic steps, subalgorithms may be used in the form of procedures or functions

ProcedureName (*ArgumentList*)

Variable := FunctionName (*ArgumentList*).

The statements

return or **return** *ArgumentList*,

cause an immediate return from the current algorithm, where the second form returns new values for the arguments of *ArgumentList*. Loops have three different forms:

for *Index* := *InitialValue* **to** *FinalValue* **do** *Statement(s)*

repeat *Statement(s)* **until** *LogicalExpression*

while *LogicalExpression* **do** *Statement(s)*.

The statements

$\textbf{exit}_{\textit{Index}\text{-loop}}$, $\textbf{exit}_{\textit{while}\text{-loop}}$, and $\textbf{exit}_{\textit{repeat}\text{-loop}}$

terminate a loop. The statement

$\textbf{next}_{\textit{Index}}$

terminates the current iteration of the loop by transferring control to the end of the loop specified by *Index*. There the next iteration of the loop may be initiated, depending on the nature of the loop, its associated increments and tests. When using indexed expressions in mathematical notation, the loop is characterized by a sequence of indices enclosed in parentheses, e.g. $(k = 1, \ldots, n)$.

Conditional statements are

if *LogicalExpression* **then** *Statement(s)* **else** *Statement(s)*

with optional **else**-branch.

All comments are enclosed in braces.

1.5 Implementation

The authors feel very strongly that you should be able to count on the programs, modules, and code fragments in this book to compile and run as we claim. Therefore, we have compiled and run each of the code fragments (as parts of entire running programs), modules, and programs. Similarly, the results we give were actually produced by the example programs, except that we show both the printed output and the user-supplied input exactly as it will appear on your screen. The code and results have been included directly, without modification, in this manuscript by a pretty-print program, so you are seeing exactly the code that we ran. We include modules or programs in the text in this typographical style:

```
{ Short description of the global routines ... }
module dosome;

const
  NoError      = 0;
  ErrorOne     = 1;
  AnotherError = 2;

global function DoSomethingErrMsg( ErrCode: integer ) : string;
var
  Msg : string;
begin
  case ErrCode of
    NoError      : Msg := '';
    ErrorOne     : Msg := 'Error number 1 occurred';
    AnotherError: Msg := 'Another error occurred';
    else         : Msg := 'Code not defined';
  end;
  if (ErrCode <> NoError) then Msg := 'Error: ' + Msg + '!';
  DoSomethingErrMsg := Msg;
end;

global procedure DoSomething( x, y: real; var z: real; var Err: integer);
var
  a, b, c, d : real;
begin
  a := x + y;  b := x - y;  Err := NoError;        { Initializations }

  if (a > 0) and (b > 1) then                      { Do something special }
    begin
      c := sqrt(a);  d := ln(b);  a := -c;  b := -d;
    end
  else if (b = 1) then
    Err := ErrorOne
  else
    Err := AnotherError;

  if (Err = NoError) then  z := a * b * c / d;    { Else z is undefined }
end;

{ Module initialization }
begin
  { Nothing to initialize }
end.
```

We handle all errors and exceptional cases with a function *DoSomethingErrMsg*. It associates an error message with the error code returned by the routine *DoSomething*.

1.6 Computational Environment

We intend to make the programs in this book as generally useful as possible, not just in terms of the subjects covered, but also in terms of their degree of portability among very different computers. Specifically, we intend that all the programs should work on mainframes, workstations, and on personal computers. We selected PASCAL–XSC (language reference [44] and [45]) as the implementation environment for our algorithms, because it was designed with this type of portability in mind. The PASCAL–XSC system is available from Numerik Software GmbH [65].

All our programs have been run and tested with the PASCAL–XSC Compiler version 2.02 on PCs 386/486, on Sun SPARCstations, and on HP9000 workstations. Our programs contain no known bugs, but we make no claim that they are bug-free or that they are immediately suitable for any particular application. This book has the benefit of about ten years of developments in the field of scientific computation with result verification, and we have been very careful in implementing and testing our routines. Comments for improvements of text and programs are welcome.

1.7 Why Numerical Result Verification?

Floating-point arithmetic is the fast way to perform scientific and engineering calculations. Today, individual floating-point operations on most computers are of maximum accuracy, in the sense that the rounded result differs at most by 1 unit in the last place from the exact result. However, after two or more operations, the result can be completely wrong. Computers now carry out up to 10^{11} floating-point operations in a second. Thus, particular attention must be paid to the reliability of the computed results. In recent years, techniques have been developed in numerical analysis which make it possible for the computer itself to verify the correctness of computed results for numerous problems and applications. Moreover, the computer frequently establishes the existence and uniqueness of the solution in this way. For example, a verified solution of a system of ordinary differential equations is just as valid as a solution obtained by a computer algebra system, which still requires a valid formula evaluation. Furthermore, the numerical routine remains applicable even if the problem does not have a closed-form solution.

In the rest of this section, we give some historical remarks, some details on the prerequisite knowledge, and some motivation for the methods presented in this book.

1.7.1 A Brief History of Computing

Methods of calculation have put their stamp on mathematics throughout history. Progress in connection with numbers has meant progress in mathematics also. The invention of mechanical calculating machines by B. Pascal, G. W. Leibniz, and others, of logarithms and the slide rule, and of many other mechanical and analog computational aids, represent significant milestones in scientific computing.

A great leap in development took place about 50 years ago with the development of the first electronic computers. This technology immediately made it possible to perform arithmetic operations faster than its mechanical or electromechanical predecessors by a factor of about 1000. The great technological gains of this century would have been impossible without modern computation. Today's automobile, airplane, space travel, modern radio and television, and not least the rapid further development of the computer itself were enabled by enormous computational capacity. On the other hand, advances in computer hardware gave massive stimulation to further development of algorithmic and numerical mathematics.

Further development of circuit technology, which again would not have been possible without the computer, has increased the computational capacity of a processor by a factor of about 10^9, compared to the first electronic computers of the 1950's. Comparison of the numbers 10^3 and 10^9 shows that the real computer revolution took place *after* the development of the first electronic computers. Remarkably, there was nothing equivalent to this development on the arithmetical-mathematical side. The enormous advances in computer technology really suggested an attempt to make the computer also more powerful arithmetically. On this, however, mathematicians exerted hardly any influence. In the area of scientific and engineering applications of concern here, the computer is used today in essentially the same way it was in the middle 1950's. Only the four floating-point operations of addition, subtraction, multiplication, and division are still being performed, albeit much faster.

As more and more computations can be done faster and faster, larger scientific problems can be addressed, and it becomes increasingly critical to make scientific computation more trustworthy than the exclusive use of ordinary floating-point arithmetic. This development began about 25 years ago and has gained a considerable following (see, for example, the extensive bibliographies in [64] or [28]). However, methods for validated computation have yet to achieve universal acceptance.

1.7.2 Arithmetic on Computers

Usually, electronic computers now have two different number systems built into their hardware: integers and floating-point numbers. Integer arithmetic operates over a limited range of integers. To the extent that the hardware (and also the software) are intact, integer arithmetic and its software extensions operate without error. For example, a rational arithmetic or a multiple-precision arithmetic can be constructed using integer arithmetic. In this way, the computer carries along as many digits as necessary to represent the result of an operation exactly. In number theory or computer algebra, this is the preferred method for arithmetic. Also, most applications in computer science can be carried out error-free with integer arithmetic. Examples are compilation or other forms of translation, and algorithms for searching or sorting.

On the other hand, in numerical or scientific analysis, one works with real numbers. A real number is represented by an infinite decimal or binary fraction. For numerical or scientific computation, real numbers must be approximated by a finite fraction in a computer, ordinarily, by floating-point numbers. Floating-point arithmetic is built into the hardware, and is consequently very fast. However, each floating-point operation is subject to error. Allthough, each floating-point operation on most modern computers is of maximum accuracy, the result after only a few operations can be completely wrong. We consider two simple examples.

Example 1.1 Let the two real vectors $x = (10^{20}, 1223, 10^{18}, 10^{15}, 3, -10^{12})$ and $y = (10^{20}, 2, -10^{22}, 10^{13}, 2111, 10^{16})$ be given. Let the scalar product be denoted by $x \cdot y$ (the corresponding components are multiplied, and all the products are summed). This gives (using exact integer arithmetic)

$$x \cdot y = 10^{40} + 2446 - 10^{40} + 10^{28} + 6333 - 10^{28} = 8779.$$

In contrast, floating-point arithmetic on *every* computer (including those with IEEE-arithmetic) gives the value zero for this scalar product. The reason for this is that the summands are of such different orders of magnitude that they cannot be processed correctly in ordinary floating-point format. This catastrophic error occurs even though the data (the components of the vectors) use up less than 5% of the exponent range of small and medium size computers!

Example 1.2 For the second example, we consider a floating-point system with base 10 and a mantissa length of 5 digits. We want to compute the difference of the two numbers $x = 0.10005 \cdot 10^5$ and $y = 0.99973 \cdot 10^4$ in the presented floating-point arithmetic. This time, both operands are of the same order of magnitude. The computer gives the completely correct result $x - y = 0.77000 \cdot 10^1$. Now, suppose that the two numbers x and y are the results of two previous multiplications. Since we are dealing with 5-digit arithmetic, these products of course have 10 digits. Suppose the unrounded products are

$$x_1 \cdot x_2 = 0.1000548241 \cdot 10^5 \quad \text{and} \quad y_1 \cdot y_2 = 0.9997342213 \cdot 10^4.$$

If we subtract these two numbers, normalize the result, and round it to 5 places, we get $0.81402 \cdot 10^1$, which differs in every digit from the result computed in floating-point arithmetic. In the second case, the value of the expression $x_1 \cdot x_2 - y_1 \cdot y_2$ was computed to the closest 5-digit floating-point number. In contrast, pure floating-point arithmetic with rounding after each individual operation gave a completely wrong result.

One should be very, very careful with the argument, "nothing like that happens in *my* calculations." It is very difficult to know all the data that enter as intermediate results into an hour-long computation on a workstation or a supercomputer. Many computer users appear to be like a wood carver who is forced to carve a work of art with a dull knife, even though it is so easy to sharpen a knife!

In classical error analysis of numerical algorithms, the error of each individual floating-point operation is estimated. It is evident that this is no longer practically possible for an algorithm for which 10^{14} floating-point operations are performed in an hour. Thus, an error analysis is not usually performed. Indeed, the fact that the computed result could be wrong is often not taken into consideration.

From the mathematical point of view, the problem of correctness of computed results is of central importance because of the high computational speeds attainable today. The determination of the correctness of computed results is essential in many applications such as simulation or mathematical modeling. One needs an assurance of correct computational results to distinguish between inaccuracies in computation and actual properties of the model. A mathematical model can be developed systematically only if errors entering into the computation can be essentially excluded.

1.7.3 Extensions of Ordinary Floating-Point Arithmetic

In the past, scientific computations were carried out using a slide rule or tables of logarithms of numbers and standard functions. The use of logarithms avoids the necessity of keeping track of exponents, as when using a slide rule, but precision is limited in either case. A modern mathematical coprocessor chip can be viewed as automating computation using logarithm tables with a large (but fixed) number of places, including addition and subtraction logarithms. It is so fast, however, that there is no way to observe the cumulative effect of the neglect of small quantities or of cancellation errors.

Early scientific computation can be characterized as multiplication-division intensive. In contrast, accountants and bookkeepers labored with addition and subtraction of long columns of numbers of various magnitudes. Errors in these addition-subtraction intensive computations were not tolerated, and various methods of validation (such as double-entry bookkeeping) were developed to ensure that answers are correct to the penny. In order to handle potentially millions of additions and subtractions accurately, it is evident that the "slide-rule whiz" capabilities of ordinary floating-point arithmetic have to be augmented by the abilities of a patient accountant. Given the state of the art, there is no reason why this cannot be done on a single chip.

Computers were invented to take on complicated work for people. The evident discrepancy between computational power and control of computational errors suggests also turning over the process of error estimation itself to the computer. This has been done successfully for practically all the basic problems of numerical analysis and many applications. To achieve this, the computer arithmetic has to be made more powerful than is ordinarily the case. One thus proceeds from the above observations: In floating-point arithmetic, most errors occur in accumulations, that is, by execution of a sequence of additions and subtractions. On the other hand, multiplication and division in floating-point arithmetic are relatively stable operations. In fixed-point arithmetic, however, accumulation is performed without errors. One only has to provide a fixed-point register in the arithmetic unit which covers the entire floating-point range. If such a hardware register is not available, then it can be simulated in the main memory by software. The resulting loss of speed in many cases is outweighed by the gain in certainty.

If this register is made twice as long as the total dynamic range of the floating-point number system, then dot products of vectors of any finite dimension can always be computed exactly. The products (the summands in the dot product) have double mantissa lengths, and the exponent range is also doubled. For the IEEE arithmetic standard data format, about four times as many bits or bytes are necessary because of its substantially greater exponent range. Even this relatively long fixed-point register presents no problem to modern technology. The possibility of exactly computing the dot product of floating-point vectors of any finite dimension opens a new dimension in numerical analysis. In particular, the optimal dot product proves itself to be an essential instrument in attaining higher computational accuracy.

With the dot product, all operations between floating-point numbers, operations for complex numbers,[1] and in particular, all operations on vectors or matrices with real or complex components can be carried out with maximum accuracy, that is, with only a single rounding in each component by performing the operations in the fixed-point register and then rounding once to store the result back into the appropriate floating-point format. Only a few years ago, just the error-free computation of the product of two floating-point matrices was considered to be harder than the calculation of the eigenvalues of a symmetric matrix! We view the optimal dot product as a basic arithmetic operation. In addition, higher precision floating-point arithmetic can be based on the optimal dot product.

In order to adapt the computer also for automatic error control, its arithmetic must be extended with still another element. All operations on floating-point numbers (addition, subtraction, multiplication, division, and the dot product of floating-point vectors) must be supplied with directed roundings, that is, rounding to the nearest smaller and larger floating-point numbers. An interval arithmetic for real and complex floating-point numbers, as well as for vectors and matrices with real and complex floating-point components can be built with these operations. Intervals bring the continuum into the computer and open a completely new dimension in numerical analysis. An interval is represented in the computer by a pair of floating-point numbers. It describes the continuum of real numbers bounded between these two floating-point numbers. Operations on two intervals in the computer result from operations on two appropriately chosen endpoints of the operand intervals. In this, the computation of the lower endpoint of the result interval is rounded downward, and the computation of the upper endpoint is rounded upward. The result is certain to contain all results of the operation considered applied individually to elements of the first and second operand intervals. The interval evaluation of an arithmetic expression such as a polynomial costs about twice as much as the evaluation of the expression in simple floating-point arithmetic. It provides a superset of the range of the expression or function in the given interval of arguments.

[1] For complex division, an additional consideration is necessary.

1.7.4 Scientific Computation with Automatic Result Verification

With these two arithmetic aids (optimal dot product and interval arithmetic), we are able to perform scientific computations with automatic result verification. We will illustrate this by two simple examples.

Example 1.3 We first address the question of whether a given real function defined by an arithmetic expression has zeros in a given interval $[x]$. This question cannot be answered with mathematical certainty if only floating-point arithmetic is available. One can evaluate the function say 1000 times in the interval $[x]$. If all computed function values turn out to be positive, then the probability is high that the function does not have any zeros in $[x]$. However, this conclusion is certainly not reliable. Because of roundoff errors, a positive result could be computed for a negative function value. The function could also descend to a negative value which was not detected due to the choice of the evaluation points. On the other hand, a single interval evaluation of the function may suffice to solve this problem with complete mathematical certainty. If the computed interval result does not contain zero, then the range of values also cannot contain zero, because the computed interval result is a superset of the range. As a consequence, we conclude that the function does not have any zeros in the interval $[x]$. An interval evaluation of the range of the function may fail to validate that the function has no root either because the function really *does* have a root or because of overestimation of the range. In many cases, however, a single interval function evaluation (costing the same as about two floating-point evaluations) provides a guarantee not available from hundreds, thousands, or millions of floating-point evaluations (see Figure 1.1).

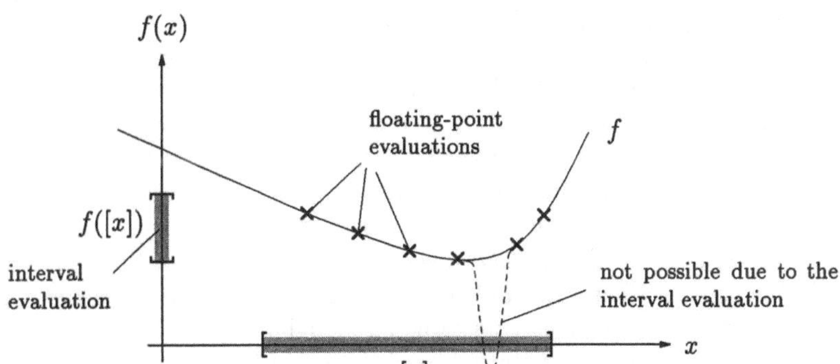

Figure 1.1: A single interval evaluation provides the guarantee that f does not descend to negative values (dashed graph)

Example 1.4 As a second example, we sketch a method by which one can verify the correctness of the computed solution for a linear system of equations $A \cdot x = b$. First, an approximation for the solution is computed, for example by Gaussian elimination. This approximation is expanded by augmenting it with a small number $\pm\varepsilon$ in each component direction, giving an n-dimensional cube $[x]$ with the computed approximation as its midpoint. This cube is just an interval vector consisting of intervals in each coordinate direction. Now, one transforms the system of equations in an appropriate way into a fixed-point form $x = B \cdot x + c$, where one naturally makes use of the fact that one has already computed a presumed approximate solution. For example, one might get B and c from the residual form of $A \cdot x = b$ as discussed in Chapter 10. Then the image $[y] := B \cdot [x] + c$ is computed. If $[y]$ is contained in $[x]$, which can be verified simply by comparison of endpoints, then $[y]$ contains a fixed-point of the equation $x = B \cdot x + c$ by Brouwer's fixed-point theorem (cf. Theorem 3.1). If $[y]$ is strictly contained in $[x]$, so that the endpoints do not touch, then the fixed-point is unique. That is, the computer program has verified computationally that the original matrix A of the system of equations is nonsingular. By iteration using the optimal dot product, the fixed-point can then be determined to arbitrary accuracy in practice. This example illustrates that validation ($Ax = b$ has a unique solution in $[y]$) is distinct from achieving high accuracy (iterating $[x] := B \cdot [x] + c$ until the width of the interval components of $[x]$ are small). In this way, the user obtains a mathematically certain and reliable result.

Examples 1.3 and 1.4 of numerical computation with automatic result verification described here leave open the question of what to do if the verification step fails. For rootfinding (Example 1.3), solving linear systems (Example 1.4), and many other problems, refined methods have been developed which allow valid assertions. A great advantage of automatic result verification is that the computer itself can quickly establish that the computation performed has not led to a correct and usable result. In this case, the program can choose an alternative algorithm or repeat the computation using higher precision.

Similar techniques of automatic result verification can be applied to many other algebraic problem areas, such as the solution of nonlinear systems of equations, the computation of zeros, the calculation of eigenvalues and eigenvectors of matrices, optimization problems, etc. In particular, valid and highly accurate evaluation of arithmetic expressions and functions on the computer is included. These routines also work for problems with uncertain or interval data.

The technique of enclosing function values with high accuracy is an essential aid for scientific computation. Newton's method for the computation of the roots of a system of equations is frequently unstable in a neighborhood of a root, for example when a root results from the mutual cancellation of positive and negative terms. In the immediate vicinity of a root, these terms are only approximately equal and cancellation occurs in ordinary floating-point arithmetic. However, in the case of insufficiently accurate function evaluation, Newton's method can easily overshoot its goal, possibly into the domain of attraction of another root. An iteration process which converges to a solution x in the real numbers thus may not converge to the value x in floating-point arithmetic. The iteration can swing over to an entirely different root. A concluding verification step is thus also indicated for iteration processes.

The techniques sketched here for numerics with automatic result verification use the optimal dot product and interval arithmetic as essential tools in addition to floating-point arithmetic. Interval arithmetic permits computation of guaranteed bounds for the solutions of a problem. One obtains high accuracy by means of the optimal dot product. The combination of both instruments is indeed a breakthrough in numerical analysis. Naive application of interval arithmetic alone always leads to reliable bounds. However, these can be so wide that they are practically useless. This effect was observed 20 years ago by many numerical analysts, leading frequently to the rejection of this useful, necessary and blameless tool. But there are applications, e.g. global optimization methods, which use interval arithmetic with only low accuracy as an essential tool. On the other hand, high accuracy alone is a waste. Validation by interval arithmetic is necessary to finally get results of *high and guaranteed* accuracy.

As already mentioned, interval arithmetic allows us to bound the range of a function over a continuum of points, including those points that are not finitely representable. Hence, interval analysis supplies the prerequisite for solving global problems like the problem of finding *all* zeros of a nonlinear function or the global optimization problem with guarantee, that the global minimum points and the global minimum values have been found.

With automatic differentiation as an essential tool, researchers also have developed methods with automatic result verification for problems of numerical analysis including numerical quadrature or cubature, integral equations, ordinary and partial differential equations. Experimentation with verified solutions of differential equations proves to be very interesting. For example, a "solution" computed by conventional methods for estimation can follow completely different trajectories if only two places of accuracy are lacking at a critical location, which cannot be recognized by an ordinary approximate method. In the case of a verified method, the width of the inclusion explodes at such a location, indicating that the accuracy must be increased.

In recent years, reliable solutions have been computed for a number of engineering and scientific problems with problem-solving routines with automatic result verification. Frequently, problems for which ordinary solvers and solution methods fail are handled. GAMM (Gesellschaft für Angewandte Mathematik und Mechanik) and IMACS (International Association for Mathematics and Computers in Simulation) have held symposia on the theme "Computer Arithmetic and Scientific Computation with Automatic Result Verification" regularly since 1980. Proceedings of many of these meetings have appeared ([6], [37], [57], [55], [56], [53], [63], [86], [87]). Applications belong to a wide variety of scientific fields including high-speed turbines, filter calculations, plasma flows in a rectangular channel, gear-drive vibrations, high-temperature superconductivity, infiltration of pollution into ground water, inductive drives for buses, periodic solutions of the oregonator (chemical kinetics), geometric modeling, dimensioning of sewers, verified quadrature in chemical transport engineering, nonlinear pile driving, refutation of certain "chaotic" solutions in the three-body problem of celestial mechanics, development of solutions of the Schrödinger equation in wave functions, calculation of magnetohydrodynamic flows for large Hartmann numbers, optical properties of liquid crystal cells, simulation of semiconductor diodes, optimization of VLSI circuits, rotor oscillations, geometric methods for CAD systems, centered beam fracture, and robotics. A documentation of these applications and an extensive bibliography is found in [1].

1.7.5 Program Verification versus Numerical Verification

In conclusion, it does not follow in general that if an algorithm has been verified to be correct in the sense of computer science, then a numerical result computed with it will be correct. In this connection, recall the examples above. Here, numerics with automatic result verification is an important additional, simple tool. Naturally, with it alone, one still is not absolutely certain, because the verification step could still lead to a positive result because of a programming error. Verification of the algorithm, compiler, the operating system, and also the computer hardware can never be made superfluous by interval arithmetic. Another useful situation occurs when interval techniques yield a guarantee that the wrong answer has been computed. This demands further debugging. In case of success, numerical result verification at least implies with high probability that all these components have furnished a correct result. Additionally, this result is certainly generally independent of whether a program verification has been carried out or not. On the other hand, when the computed result cannot be verified to be correct, this also has a positive value. This result is established quickly and automatically by the computer, without anything having to be done externally. The user can provide program alternatives in this case, for example, choice of different algorithms or methods, or a change to higher precision.

Part I

Preliminaries

Chapter 2

The Features of PASCAL–XSC

In this chapter, we give a short overview of the new concepts of the programming language PASCAL–XSC, a universal PASCAL eXtension for Scientific Computation with extensive predefined modules for scientific computation. For a complete language reference and examples, we refer to [44] and [45].

PASCAL–XSC is available for personal computers, workstations, mainframes and supercomputers. Its modern language concepts make PASCAL–XSC a powerful tool for solving scientific problems. The mathematical definition of the arithmetic is an intrinsic part of the language, including optimal arithmetic operations with directed roundings that are directly accessible in the language. Further arithmetic operations for intervals and complex numbers and even for vector/matrix operations provided by precompiled arithmetical modules are defined with maximum accuracy according to the general principle of semimorphism (see Sections 3.5 and 3.6).

PASCAL–XSC contains the following features:

- ISO Standard PASCAL
- Universal operator concept (user-defined operators)
- Functions and operators with arbitrary result type
- Overloading of procedures, functions, and operators
- Overloading of the assignment operator
- Module concept
- Dynamic arrays
- Access to subarrays
- String concept
- Controlled rounding
- Optimal (exact) scalar product
- Predefined type *dotprecision* (a fixed-point format to cover the entire range of floating-point products)
- Additional predefined arithmetic types such as *complex, interval, rvector, rmatrix* etc.

- Highly accurate arithmetic for all predefined types
- Highly accurate mathematical functions
- Exact evaluation of expressions (#-expressions)

These new language features are discussed in the following sections.

2.1 Predefined Data Types, Operators, and Functions

PASCAL–XSC adds the following numerical data types to those available in Standard PASCAL:

	complex	*interval*	*cinterval*
rvector	*cvector*	*ivector*	*civector*
rmatrix	*cmatrix*	*imatrix*	*cimatrix*

Here, the prefix letters *r*, *i*, and *c* are abbreviations for *real*, *interval*, and *complex*. Hence, *cinterval* means *complex interval*, *cimatrix* denotes complex interval matrices, and *rvector* specifies real vectors. The vector and matrix types are defined as dynamic arrays and can be used with arbitrary index ranges.

PASCAL–XSC also supplies the data type *dotprecision* representing a fixed-point format covering the entire range of floating-point products. The type *dotprecision* allows scalar results — especially sums of floating-point products — to be stored exactly. It is used in connection with *accurate expressions* (see Section 2.7).

Many operators are predefined for these types in the arithmetic modules (see Section 2.9). All of these operators, as well as the operators for type *real*, deliver results of maximum accuracy.

PASCAL–XSC provides 11 new operator symbols beyond those provided by Standard PASCAL. These are the operators $\circ<$ and $\circ>$, with $\circ \in \{+,-,*,/\}$ for operations with downwardly and upwardly directed rounding, and the operators $**$, $+*$, $><$ needed in interval computations for the intersection, the interval hull, and the test for disjointness, respectively.

Tables 2.1, 2.2, and 2.3 show all predefined operators in connection with the possible combinations of operand types. In Tables 2.1 and 2.3, the symbol $\triangleleft$ is used as an abbreviation: $\triangleleft \in \{=, <>, <, <=, >, >=\}$. The operators of the first row of Table 2.2 are monadic, i.e. there is no left operand. Let $\circ \in \{+,-,*,/\}$, and $\bullet \in \{+,-,*\}$. For vector and matrix types, $*$ denotes the scalar or matrix product.

All usual operations, even those in the higher mathematical spaces, have been realized as operators and can be used in conventional mathematical notation to make programs more easily readable.

Table 2.1: Predefined basic operators $(\triangleleft \in \{=, <>, <, <=, >, >=\})$

left operand \ right operand	integer	boolean	char	string	**set**
monadic	$+, -$	**not**			
integer	$+, -, *, /,$ **div**, **mod**, $\triangleleft$				**in**
boolean		**or**, **and**, $=, <>, <=, >=$			**in**
char			$+$ $\triangleleft$	$+$ $\triangleleft$, **in**	**in**
string			$+$ $\triangleleft$	$+$ $\triangleleft$, **in**	
set					$+, -, *, =, <>, <=, >=$
enumeration type					**in**

Table 2.2: Predefined arithmetical operators $(\circ \in \{+, -, *, /\}, \bullet \in \{+, -, *\})$

left operand \ right operand	integer real complex	interval cinterval	rvector cvector	ivector civector	rmatrix cmatrix	imatrix cimatrix
monadic	$+, -$	$+, -$	$+, -$	$+, -$	$+, -$	$+, -$
integer real complex	$\circ, \circ<, \circ>, +*$	$+, -, *, /, +*$	$*, *<, *>$	$*$	$*, *<, *>$	$*$
interval cinterval	$+, -, *, /, +*$	$+, -, *, /, +*, **$	$*$	$*$	$*$	$*$
rvector cvector	$*, *<, *>, /, /<, />$	$*, /$	$\bullet, \bullet<, \bullet>, +*$	$+, -, *, +*$		
ivector civector	$*, /$	$*, /$	$+, -, *, +*$	$+, -, *, +*, **$		
rmatrix cmatrix	$*, *<, *>, /, /<, />$	$*, /$	$*, *<, *>$	$*$	$\bullet, \bullet<, \bullet>, +*$	$+, -, *, +*$
imatrix cimatrix	$*, /$	$*, /$	$*$	$*$	$+, -, *, +*$	$+, -, *, +*, **$

In Table 2.3, the operators $<=$ and $<$ denote the subset relations, whereas $>=$ and $>$ denote the superset relations if the operands are interval data types. As already mentioned, $><$ denotes the test for disjointness for interval types. The operator **in** tests for membership of a point in an interval or for strict enclosure of an interval in the interior of another interval. We also call this the inner inclusion relation (see Section 3.1 for details).

Table 2.3: Predefined relational operators (◁ ∈ {=, <>, <, <=, >, >=})

right operand / left operand	integer real complex	interval cinterval	rvector cvector	ivector civector	rmatrix cmatrix	imatrix cimatrix
integer real complex	◁	**in** =, <>				
interval cinterval	=, <>	**in**, ><, ◁				
rvector cvector			◁	**in** =, <>		
ivector civector			=, <>	**in**, ><, ◁		
rmatrix cmatrix					◁	**in** =, <>
imatrix cimatrix					=, <>	**in**, ><, ◁

Compared with Standard PASCAL, PASCAL–XSC provides an extended set of mathematical functions (see Table 2.4). These functions are available for the types *integer*, *real*, *complex*, *interval*, and *cinterval* with generic names and deliver results of maximum accuracy.

Table 2.4: Predefined mathematical functions

Function	Generic Name	Function	Generic Name
Absolute Value	abs		
Square	sqr	Square Root	sqrt
Exponential Function	exp	Natural Logarithm (Base e)	ln
Power Function (Base 2)	exp2	Logarithm (Base 2)	log2
Power Function (Base 10)	exp10	Logarithm (Base 10)	log10
Sine	sin	Arc Cosine	arccos
Cosine	cos	Arc Sine	arcsin
Tangent	tan	Arc Tangent	arctan
Cotangent	cot	Arc Cotangent	arccot
Hyperbolic Sine	sinh	Inverse Hyperbolic Sine	arsinh
Hyperbolic Cosine	cosh	Inverse Hyperbolic Cosine	arcosh
Hyperbolic Tangent	tanh	Inverse Hyperbolic Tangent	artanh
Hyperbolic Cotangent	coth	Inverse Hyperbolic Cotangent	arcoth

PASCAL–XSC provides type transfer functions *intval*, *inf*, *sup*, *compl*, *re*, and *im* for conversion between and access to the components of the numerical data types (for scalar, vector and matrix types). Also, some additional functions like *diam*, *mid*, or *transp* for computing diameter and midpoint of an interval or the transpose of a matrix are provided.

2.2 The Universal Operator Concept

PASCAL–XSC makes programming easier by allowing the programmer to define functions and operators with arbitrary result type. The advantages of these concepts are illustrated by the simple example of polynomial addition. If we define the type *polynomial* by

```
const max_degree = 20;
type  polynomial = array [0..max_degree] of real;
```

in Standard PASCAL, then the addition of two polynomials is implemented as a *procedure*

```
procedure add ( a, b : polynomial; var c : polynomial );
{ Computes c = a + b for polynomials }
var
  i : integer;
begin
  for i := 0 to max_degree do c[i] := a[i] + b[i];
end;
```

Several calls of *add* have to be used to compute the expression $z = a + b + c + d$:

```
add(a,b,z);
add(z,c,z);
add(z,d,z);
```

In PASCAL–XSC, we define a *function* with the result type *polynomial*

```
function add ( a, b : polynomial ) : polynomial;
{ Delivers the sum a + b for polynomials }
var
  i : integer;
begin
  for i := 0 to max_degree do add[i] := a[i] + b[i];
end;
```

Now, the expression $z = a + b + c + d$ may be computed as

```
z := add(a,add(b,add(c,d)));
```

Even clearer is the *operator* in PASCAL–XSC

```
operator + ( a, b : polynomial ) result_polynomial : polynomial;
{ Delivers the sum a + b for polynomials }
var
  i : integer;
begin
  for i := 0 to max_degree do result_polynomial[i] := a[i] + b[i];
end;
```

Now, the expression may be written in the common mathematical notation

```
z := a + b + c + d;
```

A programmer may also define a new name as an operator. A priority is assigned in a priority declaration.

2.3 Overloading of Procedures, Functions, and Operators

PASCAL–XSC permits the overloading of function and procedure identifiers. A generic name concept allows the programmer to apply the identifiers *sin, cos, exp, ln, arctan*, and *sqrt* not only for *real* numbers but also for intervals, complex numbers, or elements of other mathematical spaces. Overloaded functions and procedures are distinguished by number, order, and type of their parameters. The result type is *not* used for distinction.

As illustrated above, operators also may be overloaded. Even the assignment operator (:=) may be overloaded so that the mathematical notation may be used for assignments:

```
operator := ( var p : polynomial; r : real );
var
  i : integer;
begin
  p[0] := r;
  for i := 1 to max_degree do p[i] := 0;
end;

var
  x : real;
  q : polynomial;
begin
  x := 1.5;
  q := x;    { Polynomial with constant value 1.5 }
end.
```

The overloading concept also applies to the predefined procedures *read* and *write* in a slightly modified way. The first parameter of a newly declared input/output procedure must be a **var**-parameter of file type. The second parameter represents the quantity that is to be input or output. All following parameters are interpreted as format specifications. One could provide an overloaded output facility for polynomials as

```
procedure write ( var t : text; p : polynomial; w : integer );
var
  i          : integer;
  PolyIsZero : boolean;   { Signals 'p' is a zero polynomial }
begin
  PolyIsZero := true;
  for i := 0 to max_degree do
    if (p[i] <> 0) then
    begin
      if PolyIsZero then write(t,'  ')  else  write(t,'+ ');
      writeln(t,p[i]:w,' * x↑',i:1);
      PolyIsZero := false;
    end;
  if PolyIsZero then writeln(t,'   0 (= zero polynomial)');
end;
```

The file parameter is omitted from the calling of an overloaded input/output procedure if the standard file *input* or *output* is assumed. The format parameters must be introduced and separated by colons. Moreover, several input or output statements can be combined to a single statement as in Standard PASCAL.

```
var
  r : real;
  p : polynomial;
begin
  r := 2;
  p := 5;
  write(p : 7, r : 10, r/5);
end.
```

2.4 Module Concept

The module concept allows the programmer to separate large programs into modules and to develop and compile them independently of each other. The control of syntax and semantics may be carried out beyond the bounds of the modules. Modules are introduced by the reserved word **module** followed by a name and a semicolon. The body of a module is built up quite similarly to that of a PASCAL program. The significant exception is that the objects to be exported from the module are identified by the reserved word **global** directly in front of the reserved words **const, type, var, procedure, function**, and **operator** and directly after **use**. Moreover, if **global** is placed after the equality sign in a type declaration, then the module exports both the type identifier and the internal structure (e.g. names of components, component types) of the type, which is then called a *non-private* type. Without this second **global**, types are called *private*.

Modules are *imported* into other modules or programs via a **use**-clause. The semantics of the **use**-clause are that all objects declared **global** in the imported module are also known in the importing module or program.

The example of a polynomial arithmetic module illustrates the structure of a module:

```
module poly;

use { Other modules ... }
  utility;

{ Local declarations }
{--------------------}
const
  max_degree = 20;

var
  LocalVariable : integer;

function LocalFunction : real;
var
  Value : real;
begin
  { Do some computations ... }
  LocalFunction := Value;
end;
```

```
procedure LocalProcedure;
begin { Do something ... } end;

{ Other local declarations ... }

{ Global declarations }
{---------------------}
global type polynomial = array [0..max_degree] of real;

global procedure read ( var t : text; var p : polynomial );
begin { Input statements ... } end;

global procedure write ( var t : text; p : polynomial );
begin { Output statements ... } end;

global operator + ( a, b : polynomial ) result_polynomial : polynomial;
{ Delivers the sum a + b for polynomials }
var
  i : integer;
begin
  for i := 0 to max_degree do result_polynomial[i] := a[i] + b[i];
end;

{ Other global declarations ... }

{ Initialization part of the module }
{-----------------------------------}
begin
  LocalVariable := 10;
end. { module poly }
```

2.5 Dynamic Arrays and Subarrays

The concept of dynamic arrays enables the programmer to implement algorithms independently of the length of the arrays used. The index ranges of dynamic arrays are not defined until run-time. Procedures, functions, and operators may be programmed in a fully dynamic manner, since allocation and release of local dynamic variables are executed automatically. Hence, the memory is used optimally.

For example, a dynamic type *dyn_poly* may be declared:

```
type dyn_poly = dynamic array [*] of real;
```

When declaring variables of this dynamic type, the index bounds have to be specified:

```
var p, q : dyn_poly[0..2*n];
```

where the values of the expressions for the index range are computed during program execution. The two functions *lbound(...)* and *ubound(...)* and their abbreviations *lb(...)* and *ub(...)* access the bounds of dynamic arrays which are specified only during execution of the program. The multiplication of two polynomials may be realized dynamically as follows:

```
operator * ( a, b : dyn_poly ) product : dyn_poly[0..ub(a)+ub(b)];
{ Delivers the product a * b of two dynamic polynomials a, b }
var
  i, j   : integer;
```

```
    result : dyn_poly[0..ub(a)+ub(b)];
  begin
    for i := 0 to ub(a)+ub(b) do
      result[i] := 0;
    for i := 0 to ub(a) do
      for j := 0 to ub(b) do
        result[i+j] := result[i+j] + a[i] * b[j];
    product := result;
  end;
```

A PASCAL–XSC program using dynamic arrays for polynomials follows the template

```
program dynatest (input, output);
type dyn_poly = dynamic array [*] of real;

operator * ( a, b : dyn_poly ) product : dyn_poly[0..ub(a)+ub(b)];
var
  i, j   : integer;
  result : dyn_poly[0..ub(a)+ub(b)];
begin
  { Statements for the computation of the product ... }
  product := result;
end;

procedure read ( var f : text; p : dyn_poly );
begin  { Input statements ... }  end;

procedure write ( var f : text; p : dyn_poly );
begin  { Output statements ... }  end;

procedure dyn_work (degree: integer);
var
  p, q : dyn_poly[0..degree];
  r    : dyn_poly[0..2*degree];
begin
  writeln('Enter p: '); read(p);
  writeln('Enter q: '); read(q);
  r := p * q;
  writeln('p*q = ', r);
end;

var
  actual_degree : integer;

begin  { Main program }
  write('Enter actual degree: ');  read(actual_degree);
  dyn_work(actual_degree);
end.
```

The following example demonstrates that it is possible to access a row or a column of dynamic arrays as a single object. This is called *slice* notation.

```
program slice;
var
  v      : rvector[1..5];
  A      : rmatrix[1..5,1..5];
begin
  v       := A[2];    { 2nd row of A    }
  A[*,3]  := v;       { 3rd column of A }
end.
```

2.6 Data Conversion

This section is *critical* to the appropriate use of every routine in this book. The concept is simple: People work in a decimal notation, while computers use a binary representation for numbers. The most common error made by even the most experienced interval guru is to forget that numbers such as 0.1 in input data or as literals in the code are *not* what they seem!

Numerical computations nearly always are executed on non-decimal floating-point systems. The floating-point format usually used is a binary format, so it is inevitable that literal *real* constants must be converted into that data format. This conversion can cause an error. For example, the literal constant 1.1 is not exactly representable in a binary format (see Section 3.7 for details). To execute this conversion in a controlled way, an additional notation for real literal constants is necessary. While the usual PASCAL notation of *real* numbers implies the conversion with rounding to the nearest floating-point (machine) number, it is possible to specify *real* constants that are converted with rounding to the next-smaller or the next-larger floating-point number by the notations

$$(< \pm \textit{Mantissa}\ \mathrm{E}\ \textit{Exponent}\ ) \quad \text{and} \quad (> \pm \textit{Mantissa}\ \mathrm{E}\ \textit{Exponent}\ )\ ,$$

respectively. The E and *Exponent* may be omitted as usual, in which case *Mantissa* must contain a decimal point. The parentheses are mandatory. For example, we can program

```
program round_input;
use i_ari;
var
  x, y : real;
  z    : interval;
begin
  x := (< 1.1);                          { Round 1.1 downwardly        }
  y := (> 1.0E-1);                       { Round 0.1 upwardly          }
  z := intval( (< 0.1), (> 0.1) );  { Enclose 0.1 in an interval }
end.
```

To realize a controlled rounding when entering *real* data from the console or from a text file, the procedures *read* (or *readln*) provide an additional format control parameter r. This *integer* parameter specifies the rounding mode of the *real* value during the input process. The value of a variable x of type *real* can be entered by

```
read(x : r);
```

which causes the value to be rounded (cf. Section 3.5):

$r < 0$	round to the next-smaller representable number,
$r = 0$ (or absent)	round to the nearest representable number, or
$r > 0$	round to the next-larger representable number.

A rounding parameter can also be used to convert a string representing a literal *real* constant into a *real* value. Moreover, similar rules apply to the output of *real* values and to their conversion into strings.

2.7 Accurate Expressions (#-Expressions)

The implementation of algorithms with automatic result verification or validation in this book makes extensive use of the accurate evaluation of dot product expressions, i.e. expressions which can be reduced to a single dot product. This includes matrix and vector expressions, where the computation of each component can be reduced to such a dot product form.

To evaluate this kind of expression, the new data type *dotprecision* was introduced. This data type accommodates the full floating-point range with double exponents (see [54] or [53]). *Accurate expressions* (#-expressions) can be formed based on this data type by an accurate symbol (#, #*, #<, #>, or ##) followed by an *exact expression* enclosed in parentheses. The exact expression must have the form of a dot product expression in scalar, vector, or matrix structure and is evaluated without any rounding error. Because of this, the result of an accurate expression is of maximum accuracy in the sense that in every component of the result there is no floating-point number between the exact value and the computed one. That is, the rounded and the exact result differ at most by 1 unit in the last place of the mantissa.

To obtain the unrounded or correctly rounded result of a dot product expression, you need to parenthesize the expression and precede it by the symbol #, optionally followed by a symbol for the rounding mode. Table 2.5 shows the possible rounding modes for the dot product expression form.

Table 2.5: Rounding modes for accurate expressions

Symbol	Expression Form	Rounding Mode
#*	scalar, vector or matrix	nearest
#<	scalar, vector or matrix	downwards
#>	scalar, vector or matrix	upwards
##	scalar, vector or matrix	smallest enclosing interval
#	scalar only	exact, no rounding

In practice, dot product expressions may contain a large number of terms making an explicit notation very cumbersome. To alleviate this difficulty in mathematics, the symbol $\sum$ is used. For instance, if A and B are n-dimensional matrices, then the evaluation of

$$d = \sum_{k=1}^{n} A_{i,k} \cdot B_{k,j}$$

represents a dot product expression. PASCAL–XSC provides the equivalent shorthand notation **sum** for this purpose. The corresponding PASCAL–XSC statement for this expression is

```
d := #( for k:=1 to n sum( A[i,k] * B[k,j] ) );
```

where d is a *dotprecision* variable.

Accurate or dot product expressions are used mainly in computing a residual. In the case of a linear system $Ax = b$, $A \in I\!R^{n \times n}$, $x, b \in I\!R^n$, with $Ay \approx b$, an enclosure of the residual $b - Ay$ can be computed as

```
##( b - A * y );
```

We emphasize that there is only one interval rounding operation per component. To get verified enclosures for linear systems of equations, it may be necessary to evaluate an enclosure of the residual $I - RA$ where $R \approx A^{-1}$ and I is the identity matrix. This can be programmed as

```
##( id(A) - R * A );
```

where an interval matrix is computed with only one rounding operation per component. The function *id(...)* generates an identity matrix of appropriate dimension according to the shape of A (see Section 2.9).

2.8 The String Concept

The tools provided for handling strings in Standard PASCAL do not allow convenient text processing. For this reason, PASCAL–XSC includes a string concept for the convenient handling of textual information and symbolic computation. With this new data type *string*, you can work with strings of up to *maxint* characters, by specifying a maximum string length less than *maxint* in the declaration part. Thus, a string *s* declared by

```
var s : string[40];
```

can be up to 40 characters long. The following string operations are available:

- concatenation of strings (operator +)
- actual length of a string (function *length*)
- conversions *string* → *real*, *string* → *integer*, *real* → *string*, and *integer* → *string* (functions *rval*, *ival*, and *image*)
- extraction of substrings (function *substring*)
- position of first appearance of a string in another string (function *pos*)
- comparisons (relational operators <=, <, >=, >, <>, =, and **in**)

2.9 Predefined Arithmetic Modules

The following predefined arithmetic modules are available:

- interval arithmetic (*i_ari*)
- complex arithmetic (*c_ari*)

- complex interval arithmetic (*ci_ari*)
- real matrix/vector arithmetic (*mv_ari*)
- interval matrix/vector arithmetic (*mvi_ari*)
- complex matrix/vector arithmetic (*mvc_ari*)
- complex interval matrix/vector arithmetic (*mvci_ari*)

These modules may be imported via the **use**-statement described in Section 2.4. As an example, Table 2.6 shows the operators provided by the module for interval matrix/vector arithmetic.

Table 2.6: Predefined arithmetic and relational operators from module *mvi_ari*

left operand \ right operand	integer real	interval	rvector	ivector	rmatrix	imatrix
monadic				+, −		+, −
integer real				*		*
interval			*	*	*	*
rvector		*, /	+*	+*, +, −, *, in, =, <>		
ivector	*, /	*, /	+*, +, −, *, =, <>	+*, **, +, −, *, in, =, <>, ><, <=, <, >=, >		
rmatrix		*, /		*	+*	+*, +, −, *, in, =, <>
imatrix	*, /	*, /	*	*	+*, +, −, *, =, <>	+*, **, +, −, *, in, =, <>, ><, <=, <, >=, >

In addition to these operators, the module *mvi_ari* provides the following generically named standard operators, functions, and procedures

intval, inf, sup, diam, mid, blow, transp, null, id, read, and *write*.

The function *intval* is used to generate interval vectors and matrices, whereas *inf* and *sup* are selection functions for the infimum and supremum of an interval object. The diameter and the midpoint of interval vectors and matrices are determined via *diam* and *mid*. *Blow* yields an interval inflation, and *transp* is used to get the transpose of a matrix (refer to Chapter 3 for details on the mathematical meaning of these terms).

Zero vectors and matrices are generated by the function *null*, while *id* returns an identity matrix of appropriate shape. Finally, there are the generic input/output procedures *read* and *write*, which may be used in connection with all matrix/vector data types defined in the modules mentioned above.

2.10 Why PASCAL–XSC?

As already said in the introduction of this book, PASCAL–XSC is the language in which all our "tools" are implemented. The reason for this is that its wide range of modern language concepts for scientific computation makes PASCAL–XSC a powerful tool for solving scientific problems and especially well suited as a specification language for programming with automatic result verification. Moreover, PASCAL–XSC is available for personal computers, workstations, mainframes and supercomputers, and so it supports a high degree of portability among very different computers for all our routines, modules, and programs.

Chapter 3

Mathematical Preliminaries

Interval arithmetic is the basic mathematical tool in verified numerical computing. This chapter summarizes the mathematical terms used in this book. After an introduction to real, complex, and extended interval arithmetic, we make some comments on the realization of arithmetics on a digital computer. We also touch on the problem of data conversion. Finally, we point out how to use fixed-point theorems to derive algorithms for verified numerical computing.

Let us start with some remarks on our notation. If M is an arbitrary set, then M^n and $M^{n \times m}$ denote the sets of n-dimensional vectors and $n \times m$ matrices over M. Vectors are defined to be column vectors. The identity matrix is denoted by I. The maximum norm of vectors and matrices is denoted by $\| \cdot \|_\infty$. To specify the index of iterates, we use raised indices enclosed in round brackets: $x^{(k)}$.

3.1 Real Interval Arithmetic

Detailed treatments of interval arithmetic are given in the standard textbooks of Moore [61], Alefeld and Herzberger [2], [3], or Neumaier[64]. Here, we give only an introduction and some examples.

A *real interval*, or just an interval, is a closed and bounded subset of the real numbers $I\!R$

$$[x] := [\underline{x}, \overline{x}] := \{x \in I\!R \mid \underline{x} \leq x \leq \overline{x}\},$$

where $\underline{x}$ and $\overline{x}$ denote the *lower* and *upper bounds* of the interval $[x]$, respectively. A real interval covers the range of real numbers between its two bounds. Sometimes, lower and upper bounds are also called the *infimum* and the *supremum*, respectively. The set of real intervals is denoted by $I\!I\!R$. An interval is called *thin* or *a point interval* if $\underline{x} = \overline{x}$. It is called *thick* if $\underline{x} < \overline{x}$. For a thin interval, we may write x instead of $[x]$. That is, a thin interval is just another notation for a real number.

Since intervals are sets, the terms equality ($=$), membership ($\in$), subset ($\subseteq$), proper subset ($\subset$), superset ($\supseteq$), proper superset ($\supset$), and intersection ($\cap$) are defined in the usual sense of set theory. For instance, the proper subset relation $[x] \subset [y]$ is defined as $[x] \subseteq [y]$ and $[x] \neq [y]$. An interval $[x]$ is said to be *contained in the interior* of $[y]$ if $\underline{y} < \underline{x}$ and $\overline{x} < \overline{y}$. In this case, we write $[x] \overset{\circ}{\subset} [y]$. We also call this relation the *inner inclusion relation*. Sometimes, we need the *hull* of two intervals defined by

$$[x] \,\underline{\cup}\, [y] := [\min\{\underline{x}, \underline{y}\}, \max\{\overline{x}, \overline{y}\}].$$

Example 3.1 Let $[x] = [1,3]$, and $[y] = [1,\pi]$. Then the following relations are true: $\frac{4}{3} \in [x]$, $[x] \subseteq [y]$, and $[y] \supset [x]$. But $[x] \not\overset{\circ}{\subset} [y]$ since $\underline{x} = \underline{y}$. Moreover, we have $[x] \,\underline{\cup}\, [4,5] = [1,5]$, and $[x] \cap [y] = [x]$.

The terms *diameter*, *radius*, and *midpoint* of an interval $[x]$ are defined as

$$d([x]) := \mathrm{diam}([x]) := \overline{x} - \underline{x},$$

$$r([x]) := \mathrm{rad}([x]) := \frac{\overline{x} - \underline{x}}{2}, \text{ and}$$

$$m([x]) := \mathrm{mid}([x]) := \frac{\underline{x} + \overline{x}}{2}.$$

Thus, the membership relation $x \in [x]$ may be expressed as $|x - m([x])| \leq r([x])$ using the radius and midpoint notation. In this sense, the radius of $[x]$ is an upper bound for the absolute error of the midpoint $m([x])$ considered as an approximation of an unknown number $x \in [x]$.

The *smallest* and the *greatest absolute value* of an interval $[x]$ are denoted by

$$\begin{aligned} \langle[x]\rangle &:= \min\{|x| \mid x \in [x]\}, \text{ and} \\ |[x]| &:= \max\{|x| \mid x \in [x]\} = \max\{|\underline{x}|, |\overline{x}|\}. \end{aligned} \tag{3.1}$$

Thus, $\langle[x]\rangle = 0$ if $0 \in [x]$. We note that $\langle[x]\rangle$ and $|[x]|$ are real numbers. However, the absolute value of an interval is an interval and is denoted by $\mathrm{abs}([x])$. Hence, $\mathrm{abs}([x])$ is an elementary interval function, as discussed below. The attributes of an interval are shown graphically in Figure 3.1.

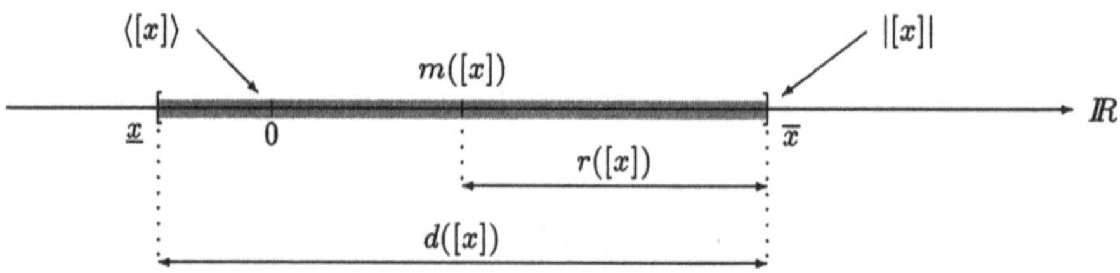

Figure 3.1: Attributes of an interval $[x]$

Suppose $[x]$ is an enclosure of a real number x, the exact value of which is unknown. To get a measure for the quality of the enclosure $[x]$, we define the *relative diameter* of an interval $[x]$ by

$$d_{\mathrm{rel}}([x]) := \begin{cases} \dfrac{d([x])}{\langle[x]\rangle} & \text{if } 0 \notin [x] \\ d([x]) & \text{otherwise.} \end{cases} \tag{3.2}$$

The relative diameter is an upper bound of the relative error of x with respect to an arbitrary element of $[x]$.

The *distance* $q([x],[y])$ between two intervals $[x]$ and $[y]$ is defined by

$$q([x],[y]) := \max\{|\underline{x}-\underline{y}|, |\overline{x}-\overline{y}|\}. \tag{3.3}$$

The distance is nonnegative and vanishes if and only if $[x]=[y]$. It does not depend on the order of its arguments, and the triangle inequality holds. Thus, q is a metric, and the set $I\mathbb{R}$ provided with the metric q is a *metric space*. Moreover, $(I\mathbb{R}, q)$ is a complete metric space. See Figure 3.2 for a graphical interpretation of the distance between two intervals.

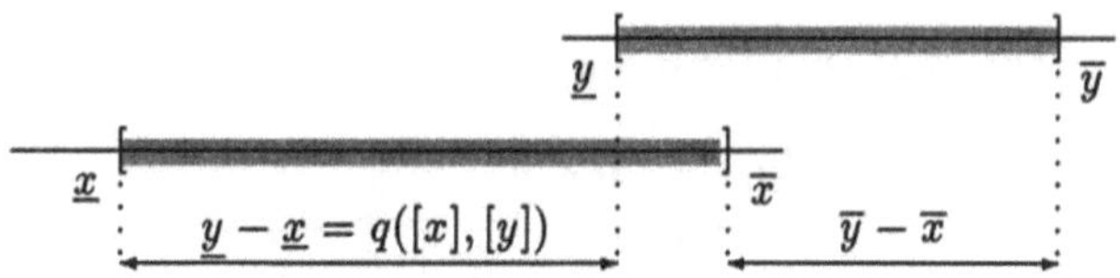

Figure 3.2: Distance between two intervals $[x]$ and $[y]$

Since $(I\mathbb{R}, q)$ is a metric space, the concepts of convergence and continuity may be introduced in the usual manner. A sequence $\{[x]^{(k)}\}$ of intervals converges to an interval $[x]$ if and only if the sequences of their bounds converge to the bounds of $[x]$. More precisely, we have

$$\begin{aligned}\lim_{k\to\infty}[x]^{(k)}=[x] \quad &\Leftrightarrow \quad \lim_{k\to\infty} q([x]^{(k)},[x])=0\\ &\Leftrightarrow \quad \left(\lim_{k\to\infty}\underline{x}^{(k)}=\underline{x} \;\wedge\; \lim_{k\to\infty}\overline{x}^{(k)}=\overline{x}\right).\end{aligned}$$

The elementary real operations $\circ \in \{+,-,\cdot,/\}$ are extended to interval arguments $[x],[y]$ by defining the result of an *elementary interval operation* to be the set of real numbers which results from combining any two numbers contained in $[x]$ and in $[y]$. That is,

$$[x]\circ[y] := \{x\circ y \mid x\in[x], y\in[y]\}. \tag{3.4}$$

Of course, the definition of $[x]/[y]$ is restricted to intervals $[y]$ with $0\notin[y]$. The operator $\cdot$ is used for clarity only and will be dropped, in general. The right-hand side of (3.4) is an interval, since the corresponding real operations are continuous. By using monotonicity properties, we get the more convenient formulae

$$\begin{aligned}
[x]+[y] &= [\underline{x}+\underline{y}, \overline{x}+\overline{y}],\\
[x]-[y] &= [\underline{x}-\overline{y}, \overline{x}-\underline{y}],\\
[x]\cdot[y] &= [\min\{\underline{x}\underline{y},\underline{x}\overline{y},\overline{x}\underline{y},\overline{x}\overline{y}\}, \max\{\underline{x}\underline{y},\underline{x}\overline{y},\overline{x}\underline{y},\overline{x}\overline{y}\}], \text{ and}\\
[x]\,/\,[y] &= [x]\cdot[1/\overline{y}, 1/\underline{y}], \quad 0\notin[y].
\end{aligned} \tag{3.5}$$

For each elementary operation, the bounds of the resulting interval may be expressed in terms of the bounds of its left and right operands. The rules given here for

Table 3.1: Multiplication of intervals $[x] \cdot [y]$

$[x] \cdot [y]$	$0 \leq \underline{y}$	$\underline{y} < 0 < \overline{y}$	$\overline{y} \leq 0$
$0 \leq \underline{x}$	$[\underline{x}\underline{y}, \overline{x}\overline{y}]$	$[\overline{x}\underline{y}, \overline{x}\overline{y}]$	$[\overline{x}\underline{y}, \underline{x}\overline{y}]$
$\underline{x} < 0 < \overline{x}$	$[\underline{x}\overline{y}, \overline{x}\overline{y}]$	$[\min\{\underline{x}\overline{y}, \overline{x}\underline{y}\}, \max\{\underline{x}\underline{y}, \overline{x}\overline{y}\}]$	$[\overline{x}\underline{y}, \underline{x}\underline{y}]$
$\overline{x} \leq 0$	$[\underline{x}\overline{y}, \overline{x}\underline{y}]$	$[\underline{x}\overline{y}, \underline{x}\underline{y}]$	$[\overline{x}\overline{y}, \underline{x}\underline{y}]$

Table 3.2: Division of intervals $[x]/[y]$ with $0 \notin [y]$

$[x]/[y]$	$0 < \underline{y}$	$\overline{y} < 0$
$0 \leq \underline{x}$	$[\underline{x}/\overline{y}, \overline{x}/\underline{y}]$	$[\overline{x}/\overline{y}, \underline{x}/\underline{y}]$
$\underline{x} < 0 < \overline{x}$	$[\underline{x}/\underline{y}, \overline{x}/\underline{y}]$	$[\overline{x}/\overline{y}, \underline{x}/\overline{y}]$
$\overline{x} \leq 0$	$[\underline{x}/\underline{y}, \overline{x}/\overline{y}]$	$[\overline{x}/\underline{y}, \underline{x}/\overline{y}]$

elementary interval operations assume exact arithmetic. We describe a floating-point interval arithmetic in Section 3.6. We give the more compact rules for multiplication and division in Tables 3.1 and 3.2.

Example 3.2 Those readers not yet familiar with interval arithmetic should check the following equalities by applying Equations (3.5) and Tables 3.1 and 3.2:

$$[-1,0] + [0,\pi] = [-1,\pi], \qquad -1 \cdot [2,5] = [-5,-2],$$

$$[1,4] - [1,4] = [-3,3], \qquad [-2,3][-2,3] = [-6,9],$$

$$[\tfrac{1}{2},1] - [0,\tfrac{1}{6}] = [\tfrac{1}{3},1], \qquad [1,\sqrt{2}][-1,1] = [-\sqrt{2},\sqrt{2}],$$

$$[2,4] - 3 = [-1,1], \qquad [1,2]/[-2,-1] = [-2,-\tfrac{1}{2}].$$

It is very important that the elementary operations are *inclusion isotonic*. That means, if $[x]$ is contained in another interval $[x']$, and $[y]$ is contained in $[y']$, then the combination of $[x]$ and $[y]$ is contained in the interval computed by combining the bigger intervals $[x']$ and $[y']$. More precisely,

$$[x] \subseteq [x'],\ [y] \subseteq [y'] \implies [x] \circ [y] \subseteq [x'] \circ [y'], \quad \circ \in \{+,-,\cdot,/\}.$$

By Definition (3.5), the operations of addition and multiplication satisfy the commutative and associative laws

$$\begin{aligned} [x] \circ [y] &= [y] \circ [x], \text{ and} \\ [x] \circ ([y] \circ [z]) &= ([x] \circ [y]) \circ [z], \quad \circ \in \{+,\cdot\}. \end{aligned}$$

The neutral elements of addition and multiplication are the thin intervals $[0] = 0$ and $[1] = 1$, respectively. Thus, we get $-[x] = 0 - [x] = [-\overline{x}, -\underline{x}]$. In general, there exists neither an additive nor a multiplicative inverse element. On the other hand, we have

$$\begin{aligned} 0 = [0] &\subseteq [x] - [x], \text{ and} \\ 1 = [1] &\subseteq [x] \;/\; [x], \end{aligned}$$

where equality holds if and only if $[x]$ is a thin interval. It is another important property of interval arithmetic that the distributive law is not generally satisfied. However, there is a weaker form called the *subdistributive law*

$$[x] \cdot ([y] + [z]) \subseteq [x] \cdot [y] + [x] \cdot [z]. \tag{3.6}$$

Equality holds only in special cases. For instance, equality holds in (3.6) if both intervals $[y]$ and $[z]$ have the same sign (see Ratschek [70] and Spaniol [81]).

Example 3.3 With $[x] = [1, 2]$, $[y] = [2, 3]$, and $[z] = [-4, -3]$, a short computation yields

$$[x] \cdot ([y] + [z]) = [-4, 0] \subset [-6, 3] = [x] \cdot [y] + [x] \cdot [z].$$

Let $\varphi : D \subset I\!R \to I\!R$ denote a real-valued *elementary function*, continuous on every closed interval in its domain D. We extend φ to interval arguments $[x] \in D$ by

$$\varphi([x]) := \{\varphi(x) \mid x \in [x]\}. \tag{3.7}$$

That is, $\varphi([x])$ denotes the range of the real-valued function φ over $[x]$. Since φ was assumed to be continuous, $\varphi([x])$ is an interval. By definition, an elementary interval function is inclusion isotonic, i.e. $[x] \subseteq [y] \Rightarrow \varphi([x]) \subseteq \varphi([y])$. We use the set of elementary functions listed in Table 2.4 supplemented by the power function $[x]^n$ defined in Example 3.5. It is possible to express the interval result of most elementary functions in terms of the argument bounds.

Example 3.4 By using monotonicity properties of the corresponding real-valued function, where $[x]$ is restricted to the domain of φ, we get

$$\begin{aligned} && \text{abs}([x]) &= [\langle [x] \rangle, |[x]|], & \\ && \varphi([x]) &= [\varphi(\underline{x}), \varphi(\overline{x})], & \varphi \in \{\text{arctan}, \text{arsinh}, \ln, \sinh\}, \\ && \varphi([x]) &= [\varphi(\overline{x}), \varphi(\underline{x})], & \varphi \in \{\text{arccot}, \text{arcoth}\}, \\ [x]^2 &= & \text{sqr}([x]) &= [\langle [x] \rangle^2, |[x]|^2], & \\ \sqrt{[x]} &= & \text{sqrt}([x]) &= [\sqrt{\underline{x}}, \sqrt{\overline{x}}], & \\ e^{[x]} &= & \exp([x]) &= [e^{\underline{x}}, e^{\overline{x}}]. & \end{aligned}$$

We remark that $[-1, 2]^2 = [0, 4] \neq [-1, 2] \cdot [-1, 2] = [-2, 4]$. We get only $[x]^2 \subset [x] \cdot [x]$ if $0 \in [x]$. Thus, it is highly recommended to use the square function instead of a multiplication to get a narrow enclosure of x^2 for $x \in [x]$.

Example 3.5 The real power function x^n with positive integer exponent $n \in \mathbb{N}$ is increasing for $x > 0$ and n even, or for all x with n odd, and decreasing otherwise. Thus, we get

$$[x]^n = \begin{cases} [\underline{x}^n, \overline{x}^n] & \text{if } 0 < \underline{x} \text{ or } n \text{ odd,} \\ [0, |[x]|^n] & \text{if } 0 \in [x] \text{ and } n \text{ even,} \\ [\overline{x}^n, \underline{x}^n] & \text{if } \overline{x} < 0 \text{ and } n \text{ even.} \end{cases}$$

For negative exponents with $0 \notin [x]$, a brief computation yields $[x]^n = 1/[x]^{-n}$. For a zero exponent, we set $[x]^0 = 1$.

Perhaps *the* fundamental problem of interval arithmetic is to compute an enclosure with only a slight overestimation of the range of a function $f : D \subset \mathbb{R} \to \mathbb{R}$ defined by an arithmetic expression $f(x)$ including arithmetic operations and elementary functions (see Alefeld and Herzberger [3], Ratschek and Rokne [71], Moore [62], or Mayer [59]). It is easy to get an enclosure of $f([x])$ by simply substituting $[x]$ for x in the defining expression of f, and then evaluating f using interval arithmetic. Assuming all interval arithmetical operations are well defined, this kind of evaluation is called an *interval evaluation* or an *interval extension* of f and is denoted by $f_{[]}([x])$. By definition we have

$$f([x]) \subseteq f_{[]}([x]),$$

where equality holds only in rare cases. Unfortunately, an interval extension usually overestimates the range. It is more difficult to get a *tight* enclosure.

An interval extension is inclusion isotonic, since elementary interval operations and functions are inclusion isotonic, i.e. $[x] \subseteq [y] \Rightarrow f_{[]}([x]) \subseteq f_{[]}([y])$. A real-valued function may have several interval extensions, since it may be defined by several equivalent arithmetic expressions. Mathematically equivalent expressions do not necessarily give rise to equivalent interval extensions, as Example 3.6 shows.

Example 3.6 Let us consider three different, but mathematically equivalent, expressions of a real-valued function f (see Figure 3.3)

$$f(x) := \underbrace{\frac{1}{2-x} + \frac{1}{2+x}}_{=:f^{(1)}(x)} = \underbrace{\frac{4}{4 - x \cdot x}}_{=:f^{(2)}(x)} = \underbrace{\frac{4}{4 - x^2}}_{=:f^{(3)}(x)} \quad \text{for } |x| < 2.$$

Since f is symmetric, decreasing for $x < 0$, and increasing otherwise, the range of f for $[x] \in (-2, 2)$ is given by $f([x]) = [f(\langle [x] \rangle), f(|[x]|)]$. Let $[x] = [-\frac{1}{2}, \frac{3}{2}]$. If we evaluate the interval extensions of $f^{(i)}$, $i = 1, 2, 3$, we get

$$f_{[]}^{(1)}([x]) = [\tfrac{24}{35}, \tfrac{8}{3}] \supset f_{[]}^{(2)}([x]) = [\tfrac{16}{19}, \tfrac{16}{7}] \supset f_{[]}^{(3)}([x]) = [1, \tfrac{16}{7}].$$

The true range $f([x]) = f_{[]}^{(3)}([x])$. In general, it is difficult to find the best possible interval extension. However, it is an empirical fact that the fewer occurrences of $[x]$ within an expression, the better is the result of the corresponding interval evaluation. If $[x]$ appears only once in the expression, and there are no interval-valued parameters (as in $f_{[]}^{(3)}([x])$), then Moore [62] showed that the naive interval extension yields a tight enclosure.

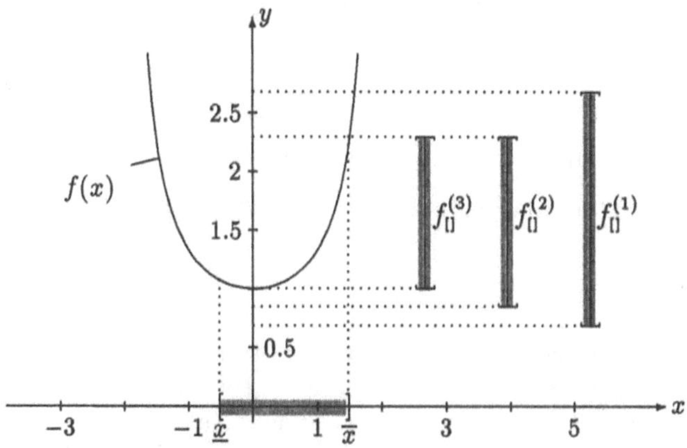

Figure 3.3: Different interval extensions of $f(x)$ as defined in Example 3.6

We have seen that, in general, the range of a function f over some interval $[x]$ is overestimated by evaluating its interval extension. There are different approaches to reduce this overestimation. *Centered forms* often are used to get a narrower enclosure for sufficiently small $[x]$ (see Alefeld and Herzberger [3], Neumaier [64] or Ratschek and Rokne [71]). A centered form is derived from the mean-value theorem. Suppose f is differentiable on its domain D. Then, $f(x) = f(c) + f'(\xi)(x - c)$ with some fixed $c \in D$ and ξ between x and c. Let $c, x \in [x]$, so $\xi \in [x]$. Therefore

$$\begin{aligned} f(x) &= f(c) + f'(\xi)(x - c) \in f(c) + f'([x])(x - c) \\ &\subseteq \underbrace{f(c) + f'([x])([x] - c)}_{=: f_{[],c}([x])}, \quad c, x \in [x]. \end{aligned} \tag{3.8}$$

Here, $f_{[],c}([x])$ is called a *standard centered form* of f on $[x]$ with *center c*. Since (3.8) does not depend on $x \in [x]$, the centered form is an enclosure for the range of f on $[x]$. If $c = m(x)$, we write $f_{[],m}([x])$ and call $f_{[],m}([x])$ the *mean-value form* of f. A centered form is not inclusion isotonic unless the midpoint is used as center. We only use the mean-value form in this book. A centered form does not depend on the expression of f, but it depends on the interval evaluation of f'. Thus, different expressions for f' lead to different values of the centered form.

Example 3.7 Let $f(x) = x + \sin(x)$, and $[x] = [2.7, 3]$. Since f is monotonically increasing, its range on $[x]$ is given by $f([x]) = [f(\underline{x}), f(\overline{x})]$. Hence, we have $f([x]) = [2.7 + \sin(2.7), 3 + \sin(3)] \subset [3.127, 3.142]$. Table 3.3 gives some enclosures computed for different interval arguments $[y]$ containing $[x]$ using both interval evaluation of f and its mean-value form with $f'(x) = 1 + \cos(x)$. The results are rounded to 4 decimal digits.

Example 3.7 shows that for rough interval arguments, the mean-value form may yields overestimated enclosures, too. For the arguments in the first row of Table 3.3,

Table 3.3: Interval evaluation compared to the mean-value form

$[y]$	$f_{[]}([y])$	$f_{[],m}([y])$
$[0,\pi] \subset [0.000, 3.142]$	$[0, \pi+1] \subset [0.000, 4.142]$	$1 + [-\frac{\pi}{2}, \frac{3}{2}\pi] \subset [-1.571, 4.713]$
$[\frac{\pi}{2}, \pi] \subset [1.570, 3.142]$	$[\frac{\pi}{2}, \pi+1] \subset [1.570, 4.142]$	$\frac{\sqrt{2}}{2} + [\frac{\pi}{2}, \pi] \subset [2.277, 3.849]$
$[\frac{3}{4}\pi, \pi] \subset [2.356, 3.142]$	$[\frac{3}{4}\pi, \pi + \frac{\sqrt{2}}{2}] \subset [2.356, 3.849]$	$f(\frac{7}{8}\pi) + \frac{\pi}{8}(1 + \cos(\frac{3}{4}\pi))[-1, 1]$ $\subset [3.016, 3.247]$

the standard interval evaluation delivers significantly tighter results. The effect changes as the diameter of the argument gets smaller. Thus, the mean-value form is recommended for narrow interval arguments. More precisely, the following inequalities relate the quality of an interval evaluation to the quality of a centered form:

$$q(f([x]), f_{[]}([x])) \leq \alpha\, d([x]), \text{ and} \tag{3.9}$$

$$q(f([x]), f_{[],m}([x])) \leq \beta\, d^2([x]), \tag{3.10}$$

where α and β are nonnegative constants independent from $[x]$. In this sense, an interval extension approximates the range of f linearly, whereas a centered form approximates it quadratically as the width of the argument tends to zero.

3.2 Complex Interval Arithmetic

In this section, we will introduce complex intervals, i.e. intervals in the complex plane. We only summarize those terms which are required for the understanding of the following chapters of this book. See Alefeld and Herzberger [3] for more details. Let $[x_{\text{re}}]$, $[x_{\text{im}}] \in I\mathbb{R}$. Then the set

$$[x] := [x_{\text{re}}] + i[x_{\text{im}}] := \{x = x_{\text{re}} + ix_{\text{im}} \mid x_{\text{re}} \in [x_{\text{re}}], x_{\text{im}} \in [x_{\text{im}}]\}$$

is called a *complex interval*, where i denotes the imaginary unit. The set of complex intervals is denoted by $I\mathbb{C}$. A complex interval $[x]$ is said to be *thin* or a point interval if both its *real part* $[x_{\text{re}}]$ and its *imaginary part* $[x_{\text{im}}]$ are thin. It is called *thick* otherwise. A complex interval may also be written as an ordered pair $([x_{\text{re}}], [x_{\text{im}}])$ of real intervals. We use rectangular intervals with sides parallel to the coordinate axes, but a complex interval could also be defined as a disk in the complex plane given by its midpoint and its radius. See Alefeld and Herzberger [2], [3] for more details concerning circular interval arithmetic.

Let $[x]$, $[y] \in I\mathbb{C}$. Relations like the equality relation or the inner inclusion relation are valid if and only if they are valid for both the real and the imaginary parts of their operands, i.e.

$$[x] \circ [y] :\Leftrightarrow ([x_{\text{re}}] \circ [y_{\text{re}}] \wedge [x_{\text{im}}] \circ [y_{\text{im}}]), \quad \circ \in \{=, \mathring{\subset}, \subseteq\}. \tag{3.11}$$

However, the proper subset relation is defined by $[x] \subset [y] :\Leftrightarrow ([x] \subseteq [y] \wedge [x] \neq [y])$. Figure 3.4 illustrates some subset relations.

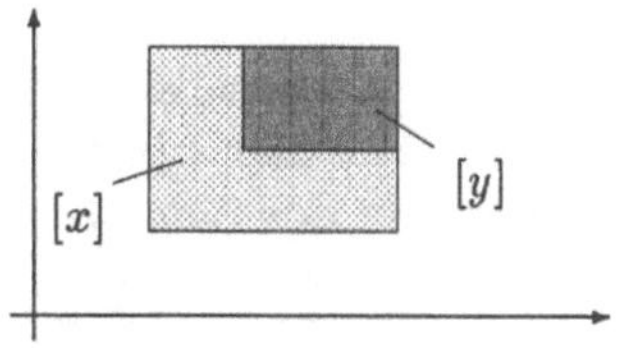

Figure 3.4: Some subset relations: $[y] \subseteq [x]$ and $[y] \subset [x]$, but $[y] \overset{\circ}{\not\subset} [x]$

The lattice operators for the intersection and the interval hull of two complex intervals may also be defined by reduction to the corresponding operators for the real and the imaginary parts, i.e.

$$[x] \circ [y] := ([x_{\text{re}}] \circ [y_{\text{re}}]) + i([x_{\text{im}}] \circ [y_{\text{im}}]), \quad \circ \in \{\cap, \underline{\cup}\}. \tag{3.12}$$

See Figure 3.5 for a graphical illustration of an intersection of complex intervals.

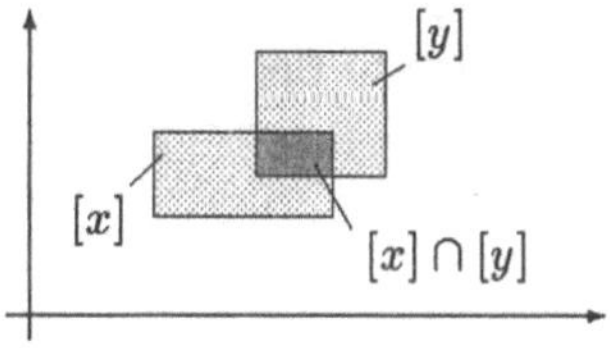

Figure 3.5: Intersection of complex intervals

As for real intervals, we may define a distance in $I\mathbb{C}$ (cf. Definition 3.3). It is easy to prove that $I\mathbb{C}$ provided with an appropriate distance is a complete metric space, so the concepts of convergence and continuity may be transferred to the space of complex intervals, too. For more details, we again refer to Alefeld and Herzberger [2], [3].

Let us now define the *elementary operations* $+$, $-$, $\cdot$, and $/$ for complex intervals. Let $[x], [y] \in I\mathbb{C}$. According to the standard rules of complex arithmetic, we get

$$\begin{aligned}
[x] + [y] &= [x_{\text{re}}] + [y_{\text{re}}] + i([x_{\text{im}}] + [y_{\text{im}}]), \\
[x] - [y] &= [x_{\text{re}}] - [y_{\text{re}}] + i([x_{\text{im}}] - [y_{\text{im}}]), \\
[x] \cdot [y] &= [x_{\text{re}}][y_{\text{re}}] - [x_{\text{im}}][y_{\text{im}}] + i([x_{\text{re}}][y_{\text{im}}] - [x_{\text{im}}][y_{\text{re}}]), \text{ and} \\
[x] \;/\; [y] &= \frac{[x_{\text{re}}][y_{\text{re}}] + [x_{\text{im}}][y_{\text{im}}]}{[y_{\text{re}}]^2 + [y_{\text{im}}]^2} + i\frac{[x_{\text{im}}][y_{\text{re}}] - [x_{\text{re}}][y_{\text{im}}]}{[y_{\text{re}}]^2 + [y_{\text{im}}]^2}.
\end{aligned} \tag{3.13}$$

Of course, the definition of $[x]/[y]$ is restricted to intervals with $0 \notin [y_{\text{re}}]^2 + [y_{\text{im}}]^2$. We point out that $[x]/[y]$ is evaluated using the elementary interval square function to guarantee $0 \notin [y_{\text{re}}]^2 + [y_{\text{im}}]^2$ for $[y]$ with $0 \notin [y]$.

Example 3.8 Let $[y] = [-2,1] + i[1,2]$. Then $0 = (0,0) \notin [y]$, and thus $0 \notin [y_{re}]^2 + [y_{im}]^2 = [1,8]$. Using multiplications instead of elementary square functions yields $0 \in [y_{re}][y_{re}] + [y_{im}][y_{im}] = [-1,8]$. Thus, the division would fail.

There is an important difference between the definitions of elementary operations for real and complex intervals. The continuous image of a complex interval is not necessarily another complex interval. In Equation (3.4), the result of an elementary real interval operation was defined as the set $\{x \circ y \mid x \in [x], y \in [y]\}$, with $[x]$, $[y] \in I\!R$. By definition, this set is a real interval. However, for $[x]$, $[y] \in I\!C$ the set $S := \{x \circ y \mid x \in [x], y \in [y]\}$ is not necessarily a complex interval. That is, S may not be a rectangle with sides parallel to the coordinate axes (see Example 3.9). To get a result in $I\!C$, we take the smallest rectangle enclosing S with sides parallel to the coordinate axes. This is established by Definition (3.13). In this way, we get easily implementable rules for complex interval arithmetic. Unfortunately, we have a loss of information by overestimating the true shape of S. This effect of overestimation is known as the *wrapping effect* (cf. Moore [62]).

Example 3.9 Let $[x] \in I\!C$ be a thick interval, and let $y = \cos\alpha + i\sin\alpha \in C$ be a thin complex interval. Multiplication of any $x \in [x]$ with y results in a rotation of x by the angle α. Thus, unless α is a multiple of $\frac{\pi}{2}$, the set $S = \{x \cdot y \mid x \in [x]\}$ is a rectangle with sides not parallel to the coordinate axes. The complex interval multiplication $[x] \cdot y$ wraps the set S in a rectangle with sides parallel to the axes as shown in Figure 3.6.

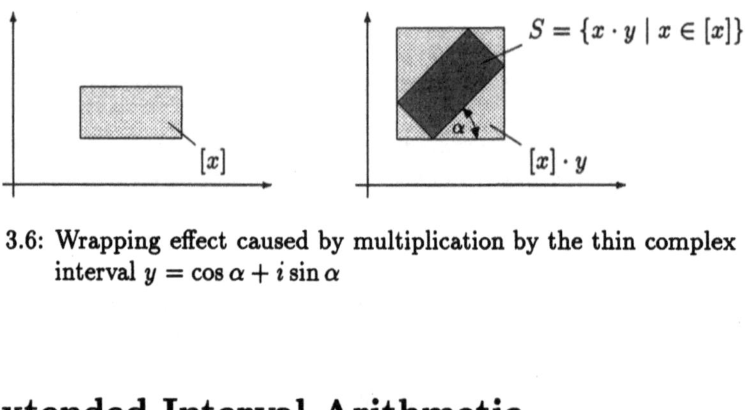

Figure 3.6: Wrapping effect caused by multiplication by the thin complex interval $y = \cos\alpha + i\sin\alpha$

3.3 Extended Interval Arithmetic

So far, we have introduced real and complex interval arithmetic and given easy-to-implement rules in Tables 3.1 and 3.2 and Equations (3.13). For these rules, we excluded a division by zero. We will now describe how to remove this restriction by defining a special kind of *extended interval arithmetic*. It is sufficient for our needs in this book to extend the definition of real intervals only. We extend the real number system by adjoining the two ideal points plus and minus infinity $I\!R^* := I\!R \cup \{-\infty\} \cup \{+\infty\}$. We now permit the infimum and the supremum of a real interval to be an ideal point. The set of extended real intervals is

$$I\!R^* := I\!R \cup \{[-\infty, r] \mid r \in I\!R\} \cup \{[l, +\infty] \mid l \in I\!R\} \cup \{[-\infty, +\infty]\}. \tag{3.14}$$

That is, $I\!R^*$ is the set of real intervals completed by those intervals whose lower and/or upper bound tends to infinity. For instance, $[-\infty, l]$ is another notation for the set $\{x \in I\!R \mid x \le l\}$, and $[-\infty, +\infty]$ denotes the entire real axis. We give an example to motivate the rules of extended interval arithmetic. See Ratschek and Rokne [72] for a detailed discussion of extended interval arithmetic.

Example 3.10 Let $[x]$, $[y] \in I\!R$ with $[x] = [4,5]$ and $[y] = [-1,2]$. By Definition (3.4), the quotient $[x]/[y]$ is the set $S := \{x/y \mid x \in [x], y \in [y]\}$. Let us now split the denominator at 0 to get $[y] = [y_1] \cup [y_2] = [-1,0] \cup [0,2]$. S is divided into two subsets $S = \{x/y \mid x \in [x], y \in [y_1]\} \cup \{x/y \mid x \in [x], y \in [y_2]\}$. Since $x \in [4,5]$, the quotient x/y tends to $-\infty$ for $y \to 0^-$, and it tends to $+\infty$ for $y \to 0^+$. Thus, the set S may be represented as the union of two extended real intervals, $S = [-\infty, -4] \cup [2, +\infty]$.

Example 3.10 is typical for the common case of an extended interval division. Its result may be represented by one or two extended intervals and may be illustrated by punching out a little gap around the origin of the real axis. For finite intervals $[x], [y] \in I\!R$ with $0 \in [y]$, the extended interval division is

$$[x]/[y] := \begin{cases} [-\infty, +\infty] & \text{if } \underline{x} < 0 < \overline{x} \text{ or } [x] = 0 \text{ or } [y] = 0 \\ [\overline{x}/\underline{y}, +\infty] & \text{if } \overline{x} \le 0 \text{ and } \underline{y} < \overline{y} = 0 \\ [-\infty, \overline{x}/\overline{y}] \cup [\overline{x}/\underline{y}, +\infty] & \text{if } \overline{x} \le 0 \text{ and } \underline{y} < 0 < \overline{y} \\ [-\infty, \overline{x}/\overline{y}] & \text{if } \overline{x} \le 0 \text{ and } 0 = \underline{y} < \overline{y} \\ [-\infty, \underline{x}/\underline{y}] & \text{if } 0 \le \underline{x} \text{ and } \underline{y} < \overline{y} = 0 \\ [-\infty, \underline{x}/\underline{y}] \cup [\underline{x}/\overline{y}, +\infty] & \text{if } 0 \le \underline{x} \text{ and } \underline{y} < 0 < \overline{y} \\ [\underline{x}/\overline{y}, +\infty] & \text{if } 0 \le \underline{x} \text{ and } 0 = \underline{y} < \overline{y}. \end{cases} \tag{3.15}$$

In the first case of Definition (3.15), we can not decide whether x/y with $x \in [x]$ and $y \in [y]$ tends to plus or minus infinity. Thus, we return the entire real axis $[-\infty, +\infty]$ as result. All the other cases are treated as in Example 3.10. In our algorithms, an extended arithmetic division often is followed by an intersection with some finite interval (see Chapter 6). The result after intersection is an empty set or one or two finite intervals. In this sense, extended arithmetic is a tool to generate and deal temporarily with infinite intervals.

Example 3.11 Consider Example 3.10. We look for $([x]/[y]) \cap [-5,4]$. The division yields $[-\infty, -4] \cup [2, +\infty]$. Intersection with $[-5,4]$ yields two finite intervals $[-5,-4] \cup [2,4]$. Similarly, $([x]/[y]) \cap [-2,4] = [2,4]$, and $([x]/[y]) \cap [-3,1] = \emptyset$.

In addition to the operation of an extended division, we only need one more extended operation in this book. If $x \in I\!R$, i.e. x may be interpreted as thin interval, and $[y] \in I\!R^*$ has at least one infinite endpoint, we define

$$x - [y] := \begin{cases} [-\infty, +\infty] & \text{if } [y] = [-\infty, +\infty] \\ [-\infty, x - \underline{y}] & \text{if } [y] = [\underline{y}, +\infty] \\ [x - \overline{y}, +\infty] & \text{if } [y] = [-\infty, \overline{y}]. \end{cases} \tag{3.16}$$

An extended arithmetic for complex intervals may be defined using the rules above for the real and imaginary parts.

3.4 Interval Vectors and Matrices

An *interval vector* is a vector whose elements are intervals. An *interval matrix* is a matrix whose elements are intervals. The sets of all n-dimensional real or complex interval vectors are denoted by $I\mathbb{R}^n$ or $I\mathbb{C}^n$, respectively. In the same manner, $I\mathbb{R}^{n\times m}$ and $I\mathbb{C}^{n\times m}$ denote the sets of all real and complex $n \times m$ interval matrices, respectively. We use the notations

$$[x] := ([x]_i)_{i=1,\ldots,n} := ([x]_1, \ldots, [x]_n)^T \quad \text{for } [x] \in I\mathbb{R}^n \text{ or } I\mathbb{C}^n$$

and

$$[A] := ([a]_{ij})_{\substack{i=1,\ldots,n \\ j=1,\ldots,m}} := \begin{pmatrix} [a]_{11} & \ldots & [a]_{1m} \\ \vdots & & \vdots \\ [a]_{n1} & \ldots & [a]_{nm} \end{pmatrix} \quad \text{for } [A] \in I\mathbb{R}^{n\times m} \text{ or } I\mathbb{C}^{n\times m}.$$

A real interval vector may be interpreted as the set of points in the n-dimensional space bounded by a parallelepiped with sides parallel to the coordinate axes (see Figure 3.7). For this reason, we often speak of a *box* as a synonym for an interval vector.

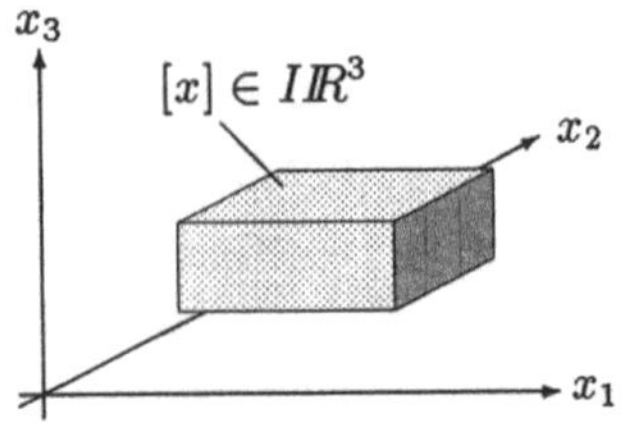

Figure 3.7: A three-dimensional real interval vector or box

The relations $=$, $\overset{\circ}{\subset}$, and $\subseteq$ are defined componentwise. For instance, the inner inclusion relation is defined by $[x] \overset{\circ}{\subset} [y] :\Leftrightarrow [x]_i \overset{\circ}{\subset} [y]_i$, $i = 1, \ldots, n$, for $[x], [y] \in I\mathbb{R}^n$. On the other hand, the proper subset relation is defined by $[x] \subset [y] :\Leftrightarrow ([x] \subseteq [y] \wedge [x] \neq [y])$. The *midpoint* and the *diameter* of an interval vector or matrix are also defined componentwise. For example, $m([x]) := (m([x]_i))$, and $d([A]) := (d([a]_{ij}))$, for $[x] \in I\mathbb{R}^n$, $[A] \in I\mathbb{R}^{n\times m}$. The *maximum norm* is extended to real interval vectors and matrices by

$$\|[x]\|_\infty := \max_{1\leq i\leq n} |[x]_i| \quad \text{for } [x] \in I\mathbb{R}^n, \text{ and}$$

$$\|[A]\|_\infty := \max_{1\leq i\leq n} \sum_{j=1}^{m} |[a]_{ij}| \quad \text{for } [A] \in I\mathbb{R}^{n\times m}.$$

Finally, we introduce notations for that component of a real interval vector with maximum absolute or relative diameter

$$\begin{aligned} d_\infty([x]) &:= \max_{1\le i\le n} d([x]_i), \text{ and} \\ d_{\mathrm{rel},\infty}([x]) &:= \max_{1\le i\le n} d_{\mathrm{rel}}([x]_i) \quad \text{for } [x] \in I\mathbb{R}^n. \end{aligned}$$

3.5 Floating-Point Arithmetic

Computers support only finite sets of numbers. In general, these numbers are represented in a semilogarithmic manner as *floating-point* numbers. A floating-point or *machine* number is of the form

$$x = \pm m \cdot b^e = \pm 0.m_1 m_2 \dots m_l \cdot b^e,$$

where m is a signed mantissa of fixed length l, b is the base, and e is the exponent. The digits of the mantissa are restricted to $1 \le m_1 \le b-1$, and $0 \le m_i \le b-1$, $i = 2,\dots,l$. Because $\frac{1}{b} \le m < 1$, x is called a normalized floating-point number. Its exponent is bounded by $e_{\min} \le e \le e_{\max}$. Floating-point numbers are usually represented in binary format, i.e. with base $b = 2$. The set of numbers is characterized by the above conditions, with $+0.0\dots0 \cdot b^{e_{\min}}$ as the unique representation of zero, and forms a *floating-point system* $R = R(b, l, e_{\min}, e_{\max})$. The elements of smallest and largest absolute value in R are $x_{\min} = 0.10\dots0 \cdot b^{e_{\min}}$ and $x_{\max} = 0.(b-1)(b-1)\dots(b-1) \cdot b^{e_{\max}}$. A floating-point system R is also called a *screen* for all real numbers lying in the interval $[-x_{\max}, +x_{\max}]$. Obviously, the elements of R are not uniformly distributed on the screen, but they are symmetrically ordered around zero. The IEEE standards 754 and 854 provide detailed descriptions of binary and radix-independent floating-point systems [4], [5].

Since it is our aim to carry out arithmetic calculations on a digital computer, we have to approximate real numbers by floating-point numbers. This is done by a special mapping, called a *rounding* $\bigcirc : \mathbb{R} \to R$, defined by the two conditions

$$\bigcirc x = x \qquad \text{for all } x \in R, \text{ and} \tag{3.17}$$

$$x \le y \quad \Rightarrow \quad \bigcirc x \le \bigcirc y \qquad \text{for all } x, y \in \mathbb{R}. \tag{3.18}$$

The first condition guarantees that elements of the screen are not changed by a rounding. The second condition means that a rounding is *monotone*, i.e. the order of elements is maintained if they are rounded. We distinguish the roundings

- $\square$: Rounding *to the nearest* element of the screen
- $\triangledown$: Rounding *toward* $-\infty$ or *downwardly* directed
- $\triangle$: Rounding *toward* $+\infty$ or *upwardly* directed

Figure 3.8 illustrates how the different roundings map an element of $\mathbb{R}$ to an element of the screen R.

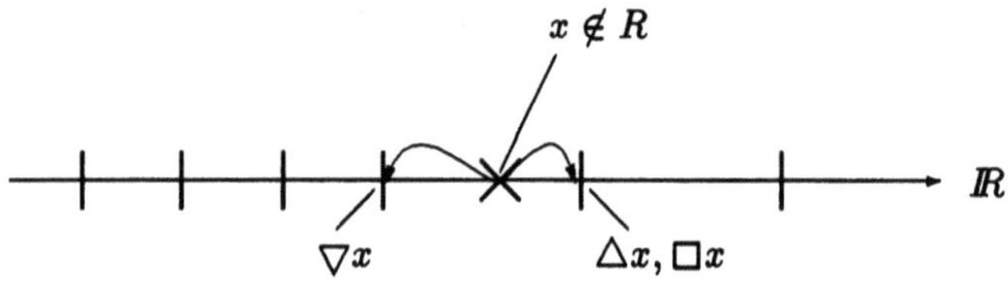

Figure 3.8: The different roundings: To the nearest number ($\square$), downwardly (∇), and upwardly (Δ) directed. The elements of the screen R are sketched in as vertical lines.

A rounding is said to be *antisymmetric* if it has the property

$$\bigcirc(-x) = -\bigcirc x \qquad \text{for all } x \in I\!R. \tag{3.19}$$

Thus, the rounding to the nearest number is antisymmetric, but the directed roundings ∇ and Δ are not antisymmetric. Instead, we have $\nabla(-x) = -\Delta x$, and $\Delta(-x) = -\nabla x$. The directed roundings satisfy the additional conditions $\nabla x \leq x$, and $x \leq \Delta x$ for all $x \in I\!R$.

One of the essential assumptions in verified numerical computing is that the elementary floating-point operations $\circledcirc$ with $\circ \in \{+, -, \cdot, /\}$ and $\bigcirc \in \{\nabla, \Delta, \square\}$ satisfy the condition

$$x \circledcirc y = \bigcirc(x \circ y) \qquad \text{for all } x, y \in R. \tag{3.20}$$

Formally, Equation (3.20) means that the result of a floating-point operation is defined to be the rounded result of the exactly computed real operation. In general, the exact result is not representable on the number screen of the computer. Thus, for an actual implementation, an auxiliary result $x \mathbin{\tilde{\circ}} y$ is computed which satisfies $x \circledcirc y = \bigcirc(x \circ y) = \bigcirc(x \mathbin{\tilde{\circ}} y)$. That is, we compute an approximate result that is rounded to the same floating-point number as the exact result.

A mapping satisfying Properties (3.17)–(3.20) is called a *semimorphism*. The principle of semimorphism may also be applied to define arithmetic operations for complex floating-point numbers as well as for real and complex floating-point vectors and matrices. For this purpose, the set of arithmetic operations $+$, $-$, $\cdot$, and $/$ is augmented by a fifth operation, the exact dot product, defined according to (3.20) by

$$x \odot y = \bigcirc(x \cdot y) \qquad \text{for all } x, y \in R^n. \tag{3.21}$$

A *complex floating-point number* is a complex number whose real and imaginary parts are elements of a floating-point system R. We denote the set of complex floating-point numbers by

$$C = \{x \in \mathbb{C} \mid x_{\text{re}}, x_{\text{im}} \in R\}.$$

Any operation defined according to the principle of semimorphism delivers a result of maximum accuracy. That is, there is no element of the screen lying between $x \circledcirc y$ and $x \circ y$. In case of operations for complex or higher dimensional operands, this is to be understood componentwise. For a detailed discussion of a semimorphic definition of computer arithmetics and their implementation on digital computers, we refer to Kulisch [52] and Kulisch and Miranker [54]. We stress that all arithmetic operations provided by PASCAL–XSC are implemented semimorphically. Moreover, PASCAL–XSC comes with a more sophisticated tool, the *accurate expressions* (see Chapter 2.7), which may be used to compute certain algebraic expressions with maximum accuracy. For instance, accurate expressions are used to realize a semimorphic implementation of the operators for vectors and matrices. In our algorithms, we use the notation $\Box(\ldots)$ to indicate that an expression is evaluated with maximum accuracy.

According to Property (3.20), the floating-point implementation $\varphi_\Box$ of an elementary real function $\varphi : D \subset I\!R \to I\!R$ should satisfy the condition

$$\varphi_\Box(x) = \Box(\varphi(x)) \qquad \text{for all } x \in D \cap R.$$

Thus, $\varphi_\Box$ delivers a result of maximum accuracy, too. An equivalent condition should hold for the floating-point implementations of complex elementary functions. Again, we emphasize that the elementary functions provided by PASCAL–XSC (see Table 2.4) satisfy this condition.

To point out that an arithmetic expression f composed of elementary operators and functions is to be evaluated using floating-point arithmetic, we provide it with a subscript $\Box$-symbol.

Example 3.12 Let $f(x) = \sin(x) + \sqrt{x} - x$ be a real-valued function. By specifying $f_\Box(x) = \sin(x) + \sqrt{x} - x$, we indicate that the argument of $f_\Box$ is a floating-point number, and that the operations on the right-hand side of the equality sign are floating-point operations. For clarity, we even might write $f_\Box(x) = \sin_\Box(x) \boxplus \mathrm{sqrt}_\Box(\mathrm{x}) \boxminus \mathrm{x}$.

3.6 Floating-Point Interval Arithmetic

Interval arithmetic as introduced in the earlier sections assumed exact arithmetic to compute the endpoints of the resulting intervals. Now, we must build it on an actual machine. A *floating-point* or *machine interval* is a real interval whose endpoints are floating-point numbers. Let R be a floating-point system. Then the set of floating-point intervals over R is denoted by

$$IR = \{[x] \in I\!I\!R \mid \underline{x}, \overline{x} \in R\}.$$

A machine interval $[x] \in IR$ denotes the continuum of numbers lying between its bounds. It is a very important fact that, though $\underline{x}$ and $\overline{x}$ are elements of the basic number screen R, $[x]$ contains not only every floating-point number between $\underline{x}$ and $\overline{x}$, but also every real number within that range (see Figure 3.9). That is, if we have proved a certain property to hold for a floating-point interval, this property holds for any real number lying in that interval.

To compute with a computer representation of intervals, we introduce a rounding $\Diamond : I\!I\!R \to IR$ which maps an interval to a machine interval. This *interval rounding*

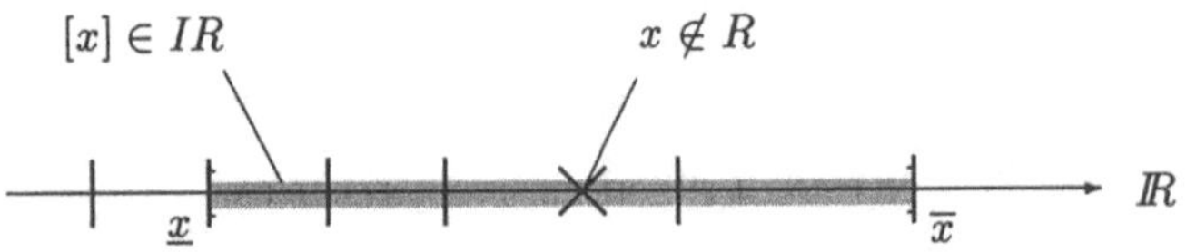

Figure 3.9: A floating-point interval $[x] \in IR$ contains not only elements of R (vertical lines), but also any real number x with $\underline{x} \leq x \leq \overline{x}$.

is assumed to satisfy the Conditions (3.17)–(3.19), where the real spaces $\mathbb{R}$ and R and the relation $\leq$ are replaced by $I\mathbb{R}$, IR, and the $\subseteq$ relation, respectively. Additionally, we assume

$$[x] \subseteq \Diamond([x]) \qquad \text{for all } [x] \in I\mathbb{R}. \tag{3.22}$$

This assumption is quite natural since the rounded image of an interval should always contain its original.

According to Property (3.20), an elementary floating-point interval operation is defined by

$$[x] \mathbin{\Diamond\!\!\!\circ} [y] = \Diamond([x] \circ [y]) \qquad \text{for all } [x], [y] \in IR. \tag{3.23}$$

The principle of semimorphism implies that any operation defined by (3.23) delivers a result of maximum accuracy in the sense that the resulting interval is the smallest machine interval which contains $[x] \circ [y]$. For an actual implementation of (3.23), we may use the rules given in (3.5), using directed-rounding operations to get proper lower and upper bounds.

Example 3.13 Let $[x], [y] \in IR$. Applying (3.5) for the difference of machine intervals with directed-rounding operations yields $[x] \mathbin{\Diamond\!\!\!-} [y] = \Diamond([x] - [y]) = [\underline{x} \bigtriangledown \overline{y}, \overline{x} \bigtriangleup \underline{y}]$.

A *complex floating-point interval* is an interval whose real and imaginary parts are floating-point intervals. We denote the set of complex floating-point intervals by

$$IC = \{[x] \in I\mathbb{C} \mid [x]_{\text{re}}, [x]_{\text{im}} \in IR\}.$$

See [52] and [54] for a detailed discussion of how to implement a semimorphic floating-point arithmetic for complex intervals and for real and complex interval vectors and matrices. All real and complex interval operations provided by PASCAL–XSC are implemented semimorphically. The notation $\Diamond(\ldots)$ used in our algorithms indicates that the specified interval expression is evaluated with maximum accuracy.

According to Definition (3.23), the floating-point implementation $\varphi_\Diamond$ of an elementary interval function $\varphi : D \subset I\mathbb{R} \rightarrow I\mathbb{R}$ should satisfy the condition

$$\varphi_\Diamond([x]) = \Diamond(\varphi([x])) \qquad \text{for all } [x] \in D \cap IR.$$

Thus, $\varphi_\Diamond$ also delivers a result of maximum accuracy. An equivalent condition should hold for the floating-point implementations of elementary complex interval functions. All elementary interval functions provided by PASCAL–XSC (see Table 2.4) satisfy this condition.

As already mentioned, we use $f_\square$ to indicate that an expression f is to be evaluated using floating-point arithmetic. We introduce an equivalent notation with a subscript $\Diamond$-symbol to indicate that the expression is to be evaluated using interval floating-point arithmetic.

Example 3.14 The notation $f_\Diamond([x]) = (e^{[x]} - 1) \cdot [x]$ indicates that the argument of $f_\Diamond$ is a floating-point interval and that floating-point interval operations are to be used for evaluation. For clarity, we might write $f_\Diamond([x]) = (\exp_\Diamond([x]) \diamondminus 1) \diamonddot [x]$.

In Section 3.8, we will see that sometimes it helps to accelerate certain verification steps by slightly enlarging a given interval. Let $\varepsilon \in R$ be some positive machine number. The *epsilon inflation* or *ε-inflation* of a real floating-point interval $[x] \in IR$ is defined by

$$[x] \bowtie \varepsilon := \begin{cases} [x] + [-\varepsilon, +\varepsilon] \cdot d([x]) & \text{if } d([x]) \neq 0 \\ [x] + [-x_{\min}, +x_{\min}] & \text{otherwise,} \end{cases} \tag{3.24}$$

where, $x_{\min}$ denotes the smallest positive element of the floating-point system R as defined in the preceding section. The ε-inflation is defined componentwise for complex intervals and interval vectors and matrices.

We conclude our discussion of machine interval arithmetics with the definition of a measure for the accuracy of a floating-point interval. A real floating-point interval with non-vanishing diameter is said to be accurate to n *ulp* if its interior contains $n-1$ or fewer elements of the basic number screen R. Thus, an accuracy of one ulp is equivalent to maximum accuracy, since the infimum and the supremum of the interval are successive elements of the number screen. An accuracy of n ulp indicates that the infimum and the supremum differ in at most n *units* in the *last place* of their mantissae. The ulp-accuracy of complex intervals and higher dimensional interval types is defined to be the maximum of the ulp-accuracies of their components.

Example 3.15 The shaded interval $[x]$ marked in Figure 3.9 has an accuracy of four ulp.

3.7 The Problem of Data Conversion

It is *critical* that you understand this section in order to use any of the toolbox routines in this book to achieve the validation they promise. The problem of data conversion is a common problem in numerical computing. It is based on the fact that numerical computations are executed in almost every case on non-decimal floating-point systems. A programmer and, most important, a user of numerical software should always be aware of the facts discussed below, yet even the most experienced interval expert gets trapped from time to time by forgetting data conversion questions. Be warned! We emphasize that the problem of data conversion is less a problem of verified numerical computing than of all digital computing (see Auzinger and Stetter [7]).

The floating-point format used internally by almost all modern digital computers is a binary format. For instance, the runtime system of PASCAL–XSC is based on the IEEE standard format for binary floating-point numbers [4] with a mantissa of 53 binary digits and an exponent range from 2^{-1021} to 2^{1024}, $R = R(2, 53, -1021, 1024)$.

Since people are accustomed to thinking in decimal notation, we must convert the decimal numbers of our thoughts to binary numbers the computer can use. Any real constant or input data specified in a user's program has to be converted at runtime. This is done by a rounding $\bigcirc \in \{\Box, \triangledown, \triangle\}$. In general, a decimal number has no binary representation of finite length. Thus, the conversion of constants and input data is generally afflicted with a small *conversion error*. It is a common misapprehension, as the following example demonstrates, that decimal numbers with a short mantissa may be converted without conversion error.

Example 3.16 The decimal number $x = 0.1$ has no exact representation in any binary floating-point system with a fixed-length mantissa. Since

$$\sum_{k=1}^{\infty} 2^{-4k} + 2^{-(4k+1)} = \frac{3}{2}\sum_{k=1}^{\infty}(2^{-4})^k = \frac{3}{2}\left(\frac{1}{1-2^{-4}} - 1\right) = \frac{1}{10} = x,$$

the unique, but infinite, binary representation of x is $0.000\overline{1100}_2$. Thus, conversion of the mantissa to any finite length results in a conversion error.

It is another common fallacy that integers may always be converted exactly to a binary floating-point format. Large integers also may be afflicted by a conversion error, because the mantissa of the objective format is of finite length.

Example 3.17 $R = R(2, 53, -1021, 1024)$ is the binary floating-point system used by PASCAL–XSC. Let

$$x = 9007199254740993 = 2^{53} + 2^0 = \underbrace{100000\ldots000001}_{54 \text{ digits}}{}_2.$$

Error-free conversion of x to the internal binary format requires at least 54 binary digits in the mantissa. Thus, the real value x cannot be mapped exactly to an element of R.

If $R(b, l, e_{\min}, e_{\max})$ is the objective floating-point system, then *any* integer within the range of $\pm b^l$ is converted without error. There are other integers outside that range which are also exactly representable in R, but most integers between b^l and $x_{\max}$ (or $-x_{\max}$ and $-b^l$) are subject to conversion errors. For the floating-point system used by PASCAL–XSC, we have $b^l = 9007199254740992$ and $x_{\max} \approx 1.8 \cdot 10^{308}$.

How can we check at runtime if a real input parameter was converted exactly? At first glance, it appears one could enter the real value, write it to a file, and compare these two values. However, this is not the correct way to solve our problem, since the method described above results in two successive conversions — one from decimal to binary format and the other from binary to decimal format. Unfortunately, in general, these two conversions cancel each other out.

To solve the problem correctly, PASCAL–XSC allows you to enter a real data as an interval. This guarantees that its exact value will be enclosed in a machine interval. If the diameter of that interval is zero, the parameter was exactly representable. The following sample program may be used to check if a real number was converted without error.

```
program Check_For_Conversion_Error;
use i_ari;
var x : interval;
begin
  write('Enter x: '); read(x);
  if (diam(x) = 0) then writeln('--> x is exactly representable')
                   else writeln('--> x is not exactly representable');
end.
```

This test program produces the results below:

```
Enter x: 0.1
--> x is not exactly representable

Enter x: 36452346
--> x is exactly representable

Enter x: 0.50390625
--> x is exactly representable

Enter x: 1E50
--> x is not exactly representable
```

As you use the modules and programs of the following chapters, you should always remember the problem of input conversion. If some of the input data are afflicted with conversion errors, you should be aware that the results of a verification procedure are proved for the converted input data only, but not for the original decimal problem. To avoid confusion about the results, you may wish to check whether the input parameters are exactly representable as demonstrated in the above program. However, this is no restriction in the common case, where the input parameters of a verification procedure result from previous computations.

Some of the problem-solving routines of the following chapters accept interval input parameters. They may be used to solve point problems by entering thin interval parameters. In this context, we want to point out some pitfalls in specifying interval expressions in PASCAL–XSC. Except at input (see the program above), real data are treated as thin intervals.

Example 3.18 Let $r \in R$ and $[x], [y], [z] \in IR$ be variables

```
var
  r       : real;
  x, y, z : interval;
```

Then the following statements

```
  r := 0.1;
  x := intval(r);
  y := r;
  z := 0.1;
```

result in identical thin (!) intervals for $[x]$, $[y]$, and $[z]$, none of which contains the real number 0.1! The literal constant 0.1 is rounded to the internally used binary format before being used for an assignment. Thus, actually, we get $[x] = [y] = [z] = [\square(0.1)] = [\square(0.1), \square(0.1)]$.

It is not our intention to cause confusion, but we have to remark that any of the intervals of the previous example, printed using the standard *write* procedure is printed as

```
[  1.000000000000000E-001,  1.000000000000001E-001 ].
```

Hence, it *appears* that 0.1 is contained in $[x]$, for instance. But remember, to display $[x]$ which actually is a thin interval stored in binary format, its lower and upper bounds are converted to a decimal format using the downwardly and upwardly directed roundings, respectively. Thus, the conversion error from decimal to binary format was canceled out by the subsequent rounding from binary to decimal format. This demonstrates that the effects of conversion are small, but they may not be neglected in a program claiming to produce validated results.

For the special case of literal constants, PASCAL–XSC also provides the notations (< ...) and (> ...) to force downwardly and upwardly directed roundings for literal constants to get a proper interval enclosure. So, specifying

```
x := intval( (<0.1), (>0.1) );
```

results in an interval $[x] = [\nabla(0.1), \Delta(0.1)]$. That is, the binary representation of $[x]$ is a proper enclosure of the decimal literal constant 0.1. In practice, the most natural notation to achieve an enclosure of 0.1 is probably

```
x := 1 / intval(10);
```

This notation emphasizes the role of type recognition for overloaded operators.

Our final remark on the problem of conversion concerns the evaluation of arithmetic expressions. Since PASCAL–XSC supports operator and function overloading, one should always bear in mind that the type of an operator or function is defined by the type of its operands. Thus, we must be careful in specifying expressions which include both real and interval operands if we want to compute enclosures for these expressions.

Example 3.19 Let $[x]$, $[y]$, and $[z]$ be defined as in Example 3.18. The statements

```
x := y + exp(1/3);
z := y + exp(intval(1)/3);
```

are not equivalent. Since the operands of the exponential's argument in the first expression are both of type *real*, it corresponds to $[x] = [y] \diamondplus [\exp_{\square}(1 \boxslash 3)]$. In particular, the second summand is not an enclosure of exp(1/3), but a thin interval. On the other hand, the second statement corresponds to $[z] = [y] \diamondplus \exp_{\diamond}(\diamond(1) \diamondslash 3)$. Here, the second summand is a proper enclosure of exp(1/3).

The example demonstrates that it is recommended to convert any integer or real operands to intervals before they are used within an expression of type interval.

3.8 Principles of Numerical Verification

The automatic verification of numerical results is based on two major prerequisites:

1. the theory of interval arithmetic,
2. appropriate algorithms.

The easiest technique for computing verified numerical results is to replace any real or complex operation by its interval equivalent and then to perform the computations using interval arithmetic. This procedure leads to reliable, verified results. However, the diameter of the computed enclosures may be so wide as to be practically useless. We need more sophisticated methods which combine the benefits of interval arithmetic with a mechanism for refining already computed, but rough, enclosures.

In developing algorithms for validated enclosures, we must be careful what we compute an enclosure *of*. For example, if we take a program for the approximate solution of initial value problems in ordinary differential equations by a Runge-Kutta method and replace all floating-point computations by computations with floating-point intervals, we get an enclosure, as suggested in the preceding paragraph. However, we get an enclosure not of the solution to the differential equation, but of the Runge-Kutta approximation. We have enclosed the roundoff errors but not the truncation errors. An algorithm must enclose all sources of error that come into the computations, including conversion errors discussed in Section 3.7, in order to achieve validated enclosures of the desired answer.

A simple mechanism for the verified solution of point problems, i.e. problems with non-interval input data, is the principle of *iterative refinement.* After computing a first approximation, its error is enclosed using machine interval arithmetic. If the diameter of the error interval is less than a desired accuracy, then a verified enclosure of the solution is given by the sum of the approximation and the enclosure of its error. Otherwise, the approximation may be refined by adding the midpoint of the error interval and repeating the process. See also Stetter [82]. We use iterative refinement methods in this book for the accurate evaluation of polynomials (Chapter 4) and arithmetic expressions (Chapter 8).

Many algorithms for numerical verification are based on the application of well-known fixed-point theorems with respect to interval sets. As an example, we cite the following theorem (cf. [64]).

Theorem 3.1 (Brouwer's Fixed-Point Theorem) *Let $f : I\!R^n \to I\!R^n$ be a continuous mapping and $X \subseteq I\!R^n$ be a closed, convex, and bounded set. If $f(X) \subseteq X$, then f has at least one fixed-point x^* in X.*

Let $X = [x] \in IR^n$ be a machine interval vector. As a box in the n-dimensional space, $[x]$ satisfies the conditions of Brouwer's Fixed-Point Theorem. Suppose we can find a box with $f([x]) \subseteq [x]$. Then $[x]$ is proved to be an enclosure of at least one fixed-point x^* of f. The assertion remains valid if we replace f by its floating-point interval evaluation $f_\diamond$ because $f_\diamond([x]) \subseteq [x]$ implies $f([x]) \subseteq [x]$, since $f_\diamond([x])$ is a superset of $f([x])$.

Viewing the Brouwer Fixed-Point Theorem in an interval context motivates a template for the design of algorithms that compute a verified solution of numerical problems. First, find a fixed-point form $x = f(x)$ equivalent to the original problem. As an example, we refer to Newton's method for problems of finding a zero. Replace the generating function of the right-hand side by its floating-point interval extension $f_\diamond$. Start the following iteration scheme with some approximate solution $[x]^{(0)}$

$$[x]^{(k+1)} = f_\diamond([x]^{(k)}) \qquad \text{for } k = 0, 1, 2, \ldots \tag{3.25}$$

Stop the iteration if $[x]^{(k+1)} \subseteq [x]^{(k)}$ for some $k \geq 0$. If the iteration succeeded, then we have proven in the mathematical sense that the original problem has at least one solution x^* contained in $[x]^{(k)}$.

We distinguish *a priori* and *a posteriori* methods for the starting approximation of (3.25). For an *a priori* method, the starting approximation already includes the fixed-point desired. For this case, the iteration scheme (3.25) may be modified by intersecting successive iterates, that is,

$$[x]^{(k+1)} = f_\diamond([x]^{(k)}) \cap [x]^{(k)} \qquad \text{for } k = 0, 1, 2, \ldots$$

The iteration is halted if there are two successive iterates of same value or if a maximum number of iterations is exceeded. Figure 3.10 gives an illustration of how an *a priori* interval iteration works. Examples for *a priori* methods are found in Chapters 6 and 7.

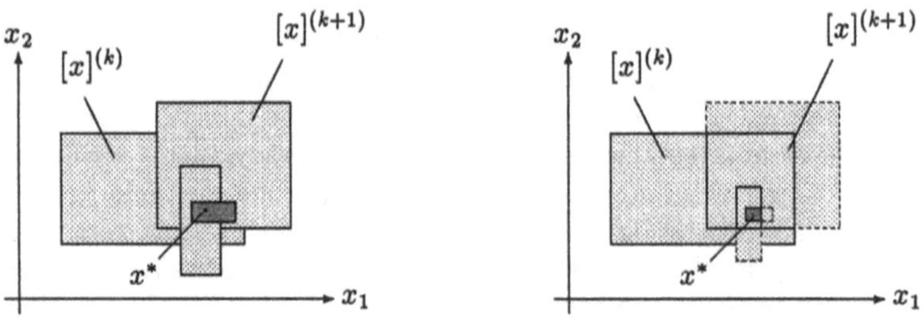

Figure 3.10: *A priori* method without (left picture) and with intersection

The starting approximation for an *a posteriori* method does not necessarily have to contain a desired fixed-point. Here, the hope is that successive iterates come closer and closer to a fixed-point and finally will enclose it. Of course, the better the starting approximation is, the faster the iteration will converge. However, in practice the iterates come closer and closer to the fixed-point, but they rarely catch it. A simple trick saves the method. Before starting a new iteration step, the actual iterate is slightly enlarged by means of the ε-inflation, as defined in Section 3.6. Thus, for an *a posteriori* method, the iteration scheme (3.25) is modified to

$$\left.\begin{array}{lcl} [x]^{(k)} & = & [x]^{(k)} \bowtie \varepsilon \\ [x]^{(k+1)} & = & f_\diamond([x]^{(k)}) \end{array}\right\} \quad \text{for } k = 0, 1, 2, \ldots$$

Figure 3.11 illustrates the effects of an ε-inflation on an *a posteriori* method. Examples of *a posteriori* methods appear in Chapters 10 and 11.

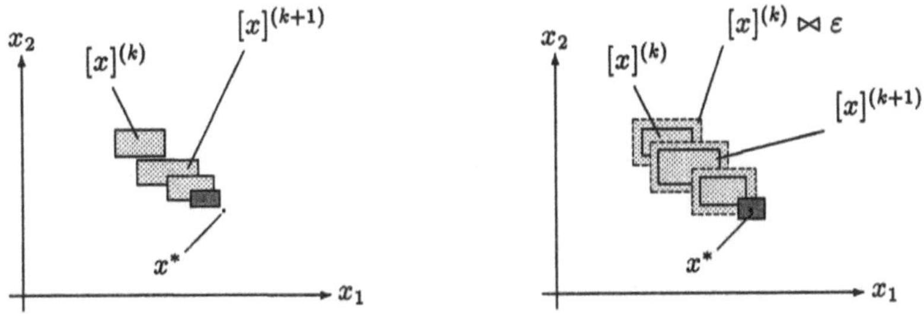

Figure 3.11: *A posteriori* method without (left picture) and with ε-inflation

Fixed-point methods may even be modified for some problems to prove the uniqueness of a fixed-point. For example, in Chapter 10 the uniqueness of the solution of a system of linear equations is proved by simply changing the stopping criterion of the fixed-point iteration to $[x]^{(k+1)} \overset{\circ}{\subset} [x]^{(k)}$.

So far, we have introduced the mathematical basics for verified numerical computing. We have defined the basic terms of real, complex, and extended interval arithmetics and have discussed aspects of an implementation on a computer. We also have discussed the problem of data conversion. Finally, we have introduced the principles of numerical verification by iterative refinement, *a priori*, and *a posteriori* methods. All these terms are used in the following chapters to describe algorithms that compute verified solutions of problems from numerical analysis.

Part II

One-Dimensional Problems

Chapter 4

Evaluation of Polynomials

In this chapter, we consider the evaluation of a polynomial function of a single variable. We usually compute the value of an arithmetic function by replacing each arithmetic operation by its corresponding floating-point machine operation (see Section 3.5). Roundoff errors and cancellations sometimes cause the calculated result to be drastically wrong. For similar reasons, a naive interval evaluation of a polynomial may lead to intervals so large as to be practically useless. Roundoff and cancellation errors are especially dangerous if we are evaluating a function close to a root, as we will see in Chapter 9 when we compute verified enclosures of zeros of polynomials.

We present an algorithm to evaluate a real polynomial $p : I\!R \to I\!R$ defined as

$$p(t) = \sum_{i=0}^{n} p_i t^i, \qquad p_i, t \in I\!R, \quad i = 0, .., n, \qquad p_n \neq 0. \tag{4.1}$$

We assume the coefficients p_i to be representable in the floating-point number system of the host computer. The algorithm achieves maximum accuracy, even in the neighborhood of a root where cancellation dooms an ordinary floating-point evaluation.

4.1 Theoretical Background

4.1.1 Description of the Problem

The basic idea of the algorithm is to transform the evaluation of a polynomial into a linear system of equations. The linear system can be solved efficiently and with maximal accuracy using vector and matrix operations based on the accurate scalar product.

We rewrite the polynomial $p(t)$ in Horner's nested multiplication form

$$p(t) = (\cdots(p_n t + p_{n-1})t + \cdots + p_1)t + p_0. \tag{4.2}$$

Labeling the intermediate results in Equation (4.2) as

$$\begin{aligned} x_n &= p_n \\ x_i &= x_{i+1}t + p_i, \quad i = n-1, \ldots, 0, \end{aligned} \tag{4.3}$$

we get the linear system

$$Ax = p \tag{4.4}$$

with $A \in I\!R^{(n+1)\times(n+1)}$, x and $p \in I\!R^{n+1}$ where

$$A = \begin{pmatrix} 1 & & & \\ -t & 1 & & \\ & \ddots & \ddots & \\ & & -t & 1 \end{pmatrix}, \quad x = \begin{pmatrix} x_n \\ \vdots \\ x_0 \end{pmatrix}, \quad \text{and} \quad p = \begin{pmatrix} p_n \\ \vdots \\ p_0 \end{pmatrix}.$$

The last unknown x_0 is the desired value of the polynomial $p(t)$. There are well known techniques and algorithms to solve a linear system of equations with maximum accuracy and verification of the result (see Chapter 10), but a special method for this very special bidiagonal Toeplitz form of A is much faster than the general methods.

4.1.2 Iterative Solution

First, we compute an approximation $\widetilde{x} \in I\!R^{n+1}$ for the solution of (4.4) by direct forward substitution in floating-point arithmetic

$$\begin{aligned} \widetilde{x}_n &= p_n \\ \widetilde{x}_i &= \widetilde{x}_{i+1} \cdot t + p_i, \quad i = n-1, \ldots, 0, \end{aligned}$$

such that $\widetilde{x}_0 \approx p(t)$. We will use a residual iteration to improve this approximation, to achieve the desired maximum accuracy, and to get verified bounds for the result. To determine the accuracy of the first approximation $\widetilde{x}$, we have to calculate an enclosure of the residual with sharp bounds (in floating-interval arithmetic)

$$[r]^{(1)} := \Diamond(p - A\widetilde{x}).$$

The dot product operation is essential to avoid both catastrophic cancellation and gross overestimation of the residual.

Let x^* denote the true solution to the system $Ax = p$. Let the error in the approximate solution be $y := x^* - \widetilde{x}$. Then $p = Ax^* = A(\widetilde{x} + y) = A\widetilde{x} + Ay$, and the residual $r := p - A\widetilde{x} = Ay$. Now the true residual $r \in [r]^{(1)} := \Diamond(p - A\widetilde{x})$ (evaluated in floating-interval arithmetic), so the true error y is contained in the solution (using floating-interval arithmetic) to the linear system

$$A[y]^{(1)} = [r]^{(1)} \tag{4.5}$$

Because of the special form of A, Equation (4.5) can be solved using the same forward sweep as used to determine $\widetilde{x}$, except that we use floating-interval arithmetic.

$$\begin{aligned} [y]_n^{(1)} &= [r]_n^{(1)} \\ [y]_i^{(1)} &= [y]_{i+1}^{(1)} \cdot t + [r]_i^{(1)}, \quad i = n-1, \ldots, 0. \end{aligned}$$

It is guaranteed that the exact value of $p(t)$ lies in the interval $[z] = \tilde{x}_0 + [y]_0^{(1)}$. If the diameter of this interval is not sufficiently small, the residual iteration is continued. Let $y^{(1)} = m([y]^{(1)})$ denote the midpoint of the vector $[y]^{(1)}$, and use $\tilde{x} + y^{(1)}$ as the new approximation for the solution of the linear system.

Let $y^{(0)} := \tilde{x}$. At the beginning of the $(k+1)^{st}$ iteration step, we have $k+1$ vectors $y^{(0)}, \ldots, y^{(k)}$ determining the approximate solution $\sum_{j=0}^{k} y^{(j)}$ for $Ax = p$. The residual of this approximation is enclosed in the interval vector

$$[r]^{(k+1)} = \Diamond\left(p - A \cdot \sum_{j=0}^{k} y^{(j)}\right) \in I\!R^{n+1}.$$

The solution of the linear system $Ay^{(k+1)} = r^{(k+1)}$ is enclosed by the solution of the linear system $A[y]^{(k+1)} = [r]^{(k+1)}$. Hence, the enclosure

$$p(t) = x_0^* \in [z] = \sum_{j=0}^{k} y_0^{(j)} + [y]_0^{(k+1)}$$

is always guaranteed. It is easy to prove that the sequence of interval vectors $[y]^{(k)}$ converges towards the zero vector for $k \to \infty$ as $\sum_{j=0}^{k} y^{(j)}$ approximates x^* (see [10]). With this we have

$$\lim_{k\to\infty}\left(\sum_{j=0}^{k} y_0^{(j)}\right) = x_0^* = p(t)$$

as an exact evaluation of polynomial (4.1) at the point t. The rate of convergence of the sequence $\left([y]_0^{(k)}\right)_{k=0}^{\infty}$ to zero, or of $\left(\sum_{j=0}^{k} y_0^{(j)}\right)_{k=0}^{\infty}$ to $p(t)$ is linear and is proportional to the condition number of matrix A. In fact, the number of residual iterations necessary to achieve maximum accuracy is a rough indicator for the condition number of A.

The complexity of the algorithm is only linear in the degree of the polynomial because of the special form of matrix A. However, execution time tends to grow faster than linearly because the condition number of A tends to grow with the degree, forcing more iterations to achieve maximum accuracy.

4.2 Algorithmic Description

We present the algorithm RPolyEval for the evaluation of a real polynomial $p(t) = \sum_{i=0}^{n} p_i t^i$ with maximum accuracy. Except for the special cases $n = 0$ and $n = 1$, which can be calculated directly, an iterative solution method is used. As described in the preceding section, we first compute a floating-point approximation of $p(t)$. We then carry out a residual iteration by solving a linear system of equations. Because of the shape of the matrix A (see (4.4)), this can be done by a direct forward sweep. The new solution interval determined in the next step is checked for being of maximum accuracy, i.e. for being exact to one unit in the last place of the mantissa (1 ulp) (see Section 3.6).

Algorithm 4.1: RPolyEval $(p, t, z, [z], k, Err)$ {Procedure}

1. {Initialization and treatment of special cases}
   $Err :=$ "No Error"; $k := 0$;
   if $(n = 0)$ **then** $z := p_0$; $[z] := p_0$;
   if $(n = 1)$ **then** $z := t \cdot p_1 + p_0$; $[z] := \Diamond(t \cdot p_1 + p_0)$;
2. {Usual case}
   if $(n > 1)$ **then**
   (a) {Computation of a first approximation of $A\tilde{x} = p$}
   $$\tilde{x}_n := p_n;$$
   $$\tilde{x}_i := \tilde{x}_{i+1} \cdot t + p_i; \quad (i = n-1, \ldots, 0)$$
   $$z := \tilde{x}_0;$$
   (b) {Iterative refinement of the solution of $A\tilde{x} = p$}
   $y^{(0)} := \tilde{x}$;
   repeat
   i. **if** $(k > 0)$ **then** $y^{(k)} := m([y]^{(k)})$;
   ii. {Computation of the residual $[r]^{(k+1)}$ and evaluation}
   {of the interval system $A \cdot [y]^{(k+1)} = [r]^{(k+1)}$ (see 4.5)}
   $$[y]_n^{(k+1)} := 0;$$
   $$\left.\begin{aligned} [r]_i^{(k+1)} &:= \Diamond\left(p_i - \sum_{j=0}^{k} y_i^{(j)} + \sum_{j=0}^{k} y_{i+1}^{(j)} \cdot t\right); \\ [y]_i^{(k+1)} &:= [y]_{i+1}^{(k+1)} \cdot t + [r]_i^{(k+1)}; \end{aligned}\right\} (i = n-1, \ldots, 0)$$
   iii. {Determination of a new enclosure $[z]$ of $p(t)$}
   $$[z] := \Diamond\left(\sum_{j=0}^{k} y_0^{(j)} + [y]_0^{(k+1)}\right);$$
   iv. $k := k + 1$;
   until (UlpAcc $([z])$ **or** $(k > k_{\max})$)
3. **if not** UlpAcc $([z])$ **then** $Err :=$ "Iteration failed";
4. **return** $z, [z], k, Err$;

Applicability of the Algorithm

The theoretical statement about the convergence of the sequence of iterates $[y]^{(k)}$ in Section 4.1.2 is also true when the calculations are done on a computer, if a precisely defined interval arithmetic as the one of PASCAL–XSC is used. Since all components of the residual interval vectors $[r]^{(k)}$ are results of scalar products, they can be calculated with maximum accuracy. That is, each component of $[r]^{(k)}$ is calculated with just one rounding. This guarantees that the upper and lower bounds of each interval component $[r]_i^{(k)}$ are identical or adjacent machine numbers. Even if the width of $[y]_0^{(k+1)}$ vanishes, the PASCAL–XSC runtime system returns an enclosure of the floating-point element of smallest absolute value (see Section 3.5). With this we always obtain a true enclosure of the defect.

If the condition number of the matrix A is extremely large, then the convergence of the residual iteration is slow. To avoid the possibility of an unbounded number of iterations at Step 2b, we halt after k_{max} iterations. Our implementation uses $k_{max} = 10$. This is large enough to achieve maximal accuracy unless the condition number of A is larger than about $10^{10} \cdot \|p(t)\|$.

4.3 Implementation and Examples

4.3.1 PASCAL–XSC Program Code

We list the PASCAL–XSC program code for the evaluation of a real polynomial with maximum accuracy. Interval data are named with double characters, e.g. rr[i] denotes the interval $[r]_i$.

4.3.1.1 Module rpoly

The module *rpoly* supplies a global type definition for the type *RPolynomial* representing a real polynomial $p(t) = \sum_{i=0}^{n} p_i t^i$. The routines *read* and *write* for the input and output of real polynomials are defined and exported. Since no operations on polynomials are requested by Algorithm 4.1, no operators have been implemented in this module.

```
{----------------------------------------------------------------------------}
{ Purpose: Declaration of data type for representation of a real polynomial }
{    by its coefficients, and of I/O procedures for this data type.          }
{ Global types and procedures:                                               }
{    type RPolynomial      : representation of real polynomials              }
{    procedure read(...)   : input of data type RPolynomial                  }
{    procedure write(...) : output of data type RPolynomial                  }
{ Remark: Variables of type 'RPolynomial' should be declared with lower      }
{    bound 0 (zero).                                                         }
{----------------------------------------------------------------------------}
module rpoly;

use
  iostd;  { Needed for abnormal termination with 'exit' }

global type
  RPolynomial  = global dynamic array[*] of real;

global procedure read ( var t : text; var p : RPolynomial );
var
  i    : integer;
begin
  if (lb(p) <> 0) then
  begin
    write('Error: Variable of type RPolynomial was declared with ');
    writeln('lower bound <> 0!');
    exit(-1); { Abnormal program termination }
  end;
  write('  x↑0 * '); read(t,p[0]);
```

```
    for i := 1 to ub(p) do
      begin  write('+ x↑',i:0,' * '); read(t,p[i])  end;
  end;

  global procedure write ( var t : text; p : RPolynomial );
  var
    i          : integer;
    PolyIsZero : boolean;   { Signals 'p' is a zero polynomial }
  begin
    PolyIsZero := true;
    for i := 0 to ub(p) do
      if (p[i] <> 0) then
      begin
        if PolyIsZero then write(t,'   ')  else  write(t,'+ ');
        writeln(t,p[i],' * x↑',i:1);
        PolyIsZero := false;
      end;
    if PolyIsZero then writeln(t,'   0 (= zero polynomial)');
  end;

  {----------------------------------------------------------------------------}
  { Module initialization part                                                 }
  {----------------------------------------------------------------------------}
  begin
    { Nothing to initialize }
  end.
```

4.3.1.2 Module rpeval

The module *rpeval* supplies the global routine *RPolyEval* to determine an enclosure of the value of the real polynomial $p(t) = \sum_{i=0}^{n} p_i t^i$ in a point $t \in R$ with maximum accuracy according to Algorithm 4.1. The global function *RPolyEvalErrMsg* is defined to get an error message for the error code returned by *RPolyEval*.

If no error occurred during the calculation, *RPolyEval* returns a floating-point approximation of $p(t)$ computed by the Horner's scheme and an enclosure with maximum accuracy of the exact value of $p(t)$.

```
{----------------------------------------------------------------------------}
{ Purpose: Evaluation of a real polynomial p with maximum accuracy.          }
{ Method: Transformation of Horner's scheme for evaluating p(t) to a linear  }
{    system of equations A(t)*x = p, where A(t) is a bidiagonal, Toeplitz    }
{    matrix  and x = (x_n,...,x_0). By iterative refinement the floating-    }
{    point approximation is improved and the exact solution x is enclosed in }
{    an interval vector [x].                                                 }
{ Global procedures and functions:                                           }
{    procedure RPolyEval(...)        : computes an enclosure for the exact   }
{                                      value p(t) with maximum accuracy      }
{    function RPolyEvalErrMsg(...) : delivers an error message text          }
{----------------------------------------------------------------------------}
module rpeval; { Real polynomial evaluation        }
               { -    -          ----              }

use
  rpoly,        { Real polynomials                  }
  i_ari,        { Interval arithmetic               }
  i_util,       { Utilities for type real           }
  mv_ari,       { Matrix/Vector arithmetic          }
  mvi_ari;      { Matrix/Vector interval arithmetic }

const
  kmax = 10;    { Maximum number of iteration steps }
```

```
{----------------------------------------------------------------------------}
{ Error codes used in this module. In the comments below p[0],..., p[n] are  }
{ the coefficients of a polynomial p.                                        }
{----------------------------------------------------------------------------}
const
  NoError   = 0;     { No error occurred                              }
  ItFailed  = 1;     { Maximum number of iterations exceeded          }
  IllProb   = 2;     { Illegal Problem, i.e. polynomials with lower   }
                     { bound <> 0 was passed to procedure RPolyEval.  }

{----------------------------------------------------------------------------}
{ Error messages depending on the error code.                                }
{----------------------------------------------------------------------------}
global function RPolyEvalErrMsg(Err : integer) : string;
var
  Msg : string;
begin
  case Err of
    NoError : Msg := '';
    ItFailed: Msg := 'Maximum number of iterations (=' + image(kmax,0) +
                     ') exceeded';
    IllProb : Msg := 'Illegal polynomial with lower bound <> 0 occurred';
    else    : Msg := 'Code not defined';
  end;
  if (Err <> NoError) then Msg := 'Error: ' + Msg + '!';
  RPolyEvalErrMsg := Msg;
end;

{----------------------------------------------------------------------------}
{ Purpose: Determination of p(t) (a polynomial p with argument t) with       }
{    maximum accuracy.                                                       }
{ Parameters:                                                                }
{    In    : 'p'   : represents a real polynomial by its coefficients        }
{            't'   : specifies the point of evaluation                       }
{    Out   : 'z'   : floating-point approximation computed by Horner's scheme }
{            'zz'  : enclosure of p(t)                                       }
{            'k'   : number of iterations needed                             }
{            'Err' : error flag                                              }
{ Description: The polynomial 'p' is defined by its coefficients and, 't'    }
{    denotes the evaluation point. Horner's scheme for evaluating p is equi- }
{    valent to computing the solution of a linear system of equations        }
{    A(t)*x = p, where A(t) is a bidiagonal, Toeplitz matrix and             }
{    x =(x_n,...,x_0). The component x_0 of x is equal to p(t). The solution }
{    x is enclosed in an interval vector [x] by iterative refinement. The    }
{    first element of [x] is an enclosure of p(t). It is returned in the     }
{    variable 'zz'. A floating-point approximation computed by Horner's      }
{    scheme is returned in 'z'. The number of iterations is returned in 'k'  }
{    and the state of success is returned in 'Err'.                          }
{ Remark: The polynomial's coefficients and the argument are assumed to be   }
{    exact floating-point numbers!                                           }
{----------------------------------------------------------------------------}
global procedure RPolyEval (     p      : RPolynomial;
                                 t      : real;
                             var z      : real;
                             var zz     : interval;
                             var k, Err : integer     );
var
  i, j, n : integer;                         { The j-th column of the matrix  }
  rr, yy  : ivector[0..ub(p)];               { 'y' is used to store the j-th  }
  x       : rvector[0..ub(p)];               { correction for the solution of }
  y       : rmatrix[0..ub(p),0..kmax];       { A(t)*y = p.                    }
begin                                        {--------------------------------}
  Err := NoError; n := ub(p); k := 0;                        { Initialization }
                                                             {----------------}
```

```
  if (lb(p) <> 0) then                              { Illegal polynomial declaration }
    Err := IllProb                                  {--------------------------------}
  else { lb(p) = 0 }
    begin
      if (n = 0) then
        begin  z := p[0];  zz := p[0];  end
      else if (n = 1) then
        begin
          z  := #*( t*p[1] + p[0] );
          zz := ##( t*p[1] + p[0] );
        end
      else { n > 1 }
        begin
          x := Null(x); y := Null(y);            { Initialization x := 0 and y := 0 }
                                                 {----------------------------------}
          x[n] := p[n];                      { Computation of a first approximation }
          for i := n-1 downto 0 do           { using Horner's scheme                }
            x[i] := x[i+1]*t+p[i];           {--------------------------------------}
          z := x[0];

          y[*,0] := x;                              { Iterative refinement for the }
          repeat                                    { solution of A*x = p          }
                                                    {------------------------------}
            if (k > 0) then           { If a residual was computed, its middle is }
              y[*,k] := mid(yy);      { stored as the next correction of 'y'.     }
                                      {-------------------------------------------}
            yy[n] := 0;                        { Computation of the residual [r] and }
            for i := n-1 downto 0 do           { evaluation of the interval system   }
            begin                              { A*[y] = [r]                         }
                                               {-------------------------------------}
              rr[i] := ##( p[i] - for j := 0 to k sum( y[i,j] )
                                + for j := 0 to k sum( t*y[i+1,j] ) );
              yy[i] := yy[i+1]*t + rr[i]
            end;

                                   { Determination of a new enclosure [z] of p(t) }
            zz := ##( for j := 0 to k sum( y[0,j] ) + yy[0] );
            k := k + 1;
          until UlpAcc(zz,1) or (k > kmax);
          if not UlpAcc(zz,1) then Err := ItFailed;
        end; { n > 1 }
    end; { lb(p) = 0 }
end; { procedure RPolyEval }

{----------------------------------------------------------------------------}
{ Module initialization part                                                 }
{----------------------------------------------------------------------------}
begin
  { Nothing to initialize }
end.
```

4.3.2 Examples

We consider the polynomials

$$p(t) = t^4 - 8t^3 + 24t^2 - 32t + 16$$

to be evaluated in the neighborhood of the real value $t = 2.0001$, and

$$q(t) = -t^3 + 3t^2 - 3t + 1$$

to be evaluated in the neighborhood of the real value $t = 1.000005$. To make sure that the arguments are representable on the computer, we use the machine numbers $t = \square(2.0001)$, and $t = \square(1.000005)$ respectively.

To illustrate the difficulties that may occur with the calculation of a polynomial value using floating-point arithmetic, these two polynomials have been evaluated in floating-point arithmetic for 100 values in the neighborhood of their roots. The corresponding plots are given in Figures 4.1 and 4.2.

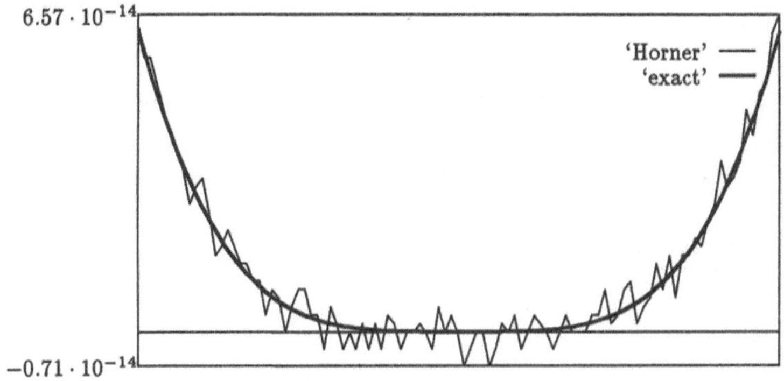

Figure 4.1: Polynomial $p(t)$ for $t \in [1.9995, 2.0005]$

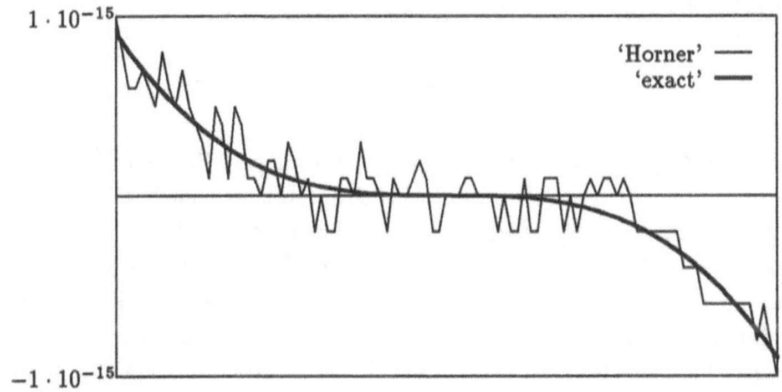

Figure 4.2: Polynomial $q(t)$ for $t \in [0.99999, 1.00001]$

These two examples are solved reliably using the following sample main program to call *RPolyEval*.

```
program rpeval_example(input, output);
use
  i_ari,                          { Interval arithmetic                          }
  rpoly,                          { Real polynomials                             }
  rpeval;                         { Polynomial evaluation with maximum accuracy }

procedure main(n : integer);
var
  ErrCode, No : integer;
  t, y        : real;
```

```
  yy           : interval;
  p            : RPolynomial[0..n];

begin
  writeln('Enter the coefficients of p in increasing order:'); read(p);
  writeln;
  write('Enter the argument t for evaluation: '); read(t); writeln;

  No := 0;
  RPolyEval(p, t, y, yy, No, ErrCode);

  if (ErrCode = 0) then
    begin
      writeln('Polynomial: '); writeln(p);
      writeln('Floating-point evaluation of p(t) using Horner''s scheme:');
      writeln('  ',y); writeln;
      writeln('Verified inclusion of p(t): '); writeln(yy); writeln;
      writeln('Number of iterations needed: ', No);
    end
  else
    writeln(RPolyEvalErrMsg(ErrCode));
end; { procedure main }

var
  n : integer;

begin
  n := -1;
  while (n < 0) do
  begin
    write('Enter the degree of the polynomial (>=0): ');
    read(n); writeln;
  end;
  main(n);
end.
```

Our implementation of Algorithm 4.1 produces the output listed below. In both cases the floating-point approximation, naively using Horner's nested multiplication form, does *not* lie within the verified enclosure of $p(\Box(t))$ and $q(\Box(t))$. An even worse side effect of rounding errors occurring during the evaluation using Horner's scheme is that the standard algorithm returns not only wrong digits but also an incorrect sign for the values of $p(\Box(2.0001))$ and $q(\Box(1.000005))$. To avoid an incorrect interpretation of the resulting interval, we stress that this interval is numerically proved to be an enclosure of the exact value of the given polynomials for the machine numbers $t = \Box(2.0001)$ and $t = \Box(1.000005)$.

Output for polynomial $p(t) = t^4 - 8t^3 + 24t^2 - 32t + 16$:

```
Enter the degree of the polynomial (>=0): 4

Enter the coefficients of p in increasing order:
  x^0 * 16
+ x^1 * -32
+ x^2 * 24
+ x^3 * -8
+ x^4 * 1

Enter the argument t for evaluation: 2.0001
```

```
Polynomial:
   1.600000000000000E+001 * x^0
+ -3.200000000000000E+001 * x^1
+  2.400000000000000E+001 * x^2
+ -8.000000000000000E+000 * x^3
+  1.000000000000000E+000 * x^4

Floating-point evaluation of p(t) using Horner's scheme:
  -3.552713678800501E-015

Verified inclusion of p(t):
[  1.000000000008441E-016,  1.000000000008442E-016 ]

Number of iterations needed: 2
```

Output for polynomial $q(t) = -t^3 + 3t^2 - 3t + 1$:

```
Enter the degree of the polynomial (>=0): 3

Enter the coefficients of p in increasing order:
  x^0 * 1
+ x^1 * -3
+ x^2 * 3
+ x^3 * -1

Enter the argument t for evaluation: 1.000005

Polynomial:
   1.000000000000000E+000 * x^0
+ -3.000000000000000E+000 * x^1
+  3.000000000000000E+000 * x^2
+ -1.000000000000000E+000 * x^3

Floating-point evaluation of p(t) using Horner's scheme:
   1.110223024625157E-016

Verified inclusion of p(t):
[ -1.250000000024568E-016, -1.250000000024566E-016 ]

Number of iterations needed: 2
```

4.3.3 Restrictions and Hints

Ours is a method for evaluating *real* polynomials, so we must be careful that the coefficients of p used for evaluation are exactly representable on the computer's number screen. To get an exact representation of real coefficients p_i such as 0.01 the entire polynomial might be scaled by a factor of 100. For more details about conversion of input data, see the remarks in Section 3.7.

4.4 Exercises

Exercise 4.1 Let $p(t) = 543339720t^3 - 768398401t^2 - 1086679440t + 1536796802$ and $q(t) = 67872320568t^3 - 95985956275t^2 - 135744641136t + 191971912515$ be real polynomials. Evaluate them at the machine number nearest to $t = \sqrt{2}$, i.e. $t =$

$\square(\sqrt{2})$. Compare the results of floating-point evaluation, high accuracy evaluation by residual correction, and the interval enclosure of the exact value.

Exercise 4.2 Let $p_n(t) = \sum_{i=0}^{n} t^i/i!$ be the truncated Taylor series for e^t. Use the example program to evaluate $p_n(-100)$ for various n. Graph the execution time as a function of the degree n. Can you ever elicit the error message "Maximum number of iterations exceeded"? Note that the coefficients $t^i/i!$ are not exactly representable. Therefore, you may either evaluate an approximation to $p_n(t)$, or else follow the suggestion of Section 4.3.3 and evaluate $n! \cdot p_n(t)$. Is the enclosure for $p_n(-100)$ also an enclosure for e^{-100}?

4.5 References and Further Reading

A first algorithm for the evaluation of polynomials with maximum accuracy was given by Böhm in [10] and [57] and later by Krämer [49]. This algorithm is easily modified for rational expressions. A more general approach which also includes the evaluation of certain arithmetic expressions is described in Chapter 8. Finally, we remark that there are similar commercial implementations of Algorithm 4.1 in the subroutine libraries ACRITH from IBM [32] and ARITHMOS from SIEMENS [80].

Chapter 5

Automatic Differentiation

In many applications of numerical and scientific computing, it is necessary to compute derivatives of functions. Simple examples are methods for finding the zeros, maxima, or minima of nonlinear functions. There are three different methods to get the values of the derivatives: *numerical differentiation*, *symbolic differentiation*, and *automatic differentiation*.

Numerical differentiation uses difference approximations to compute approximations of the derivative values. Symbolic differentiation computes explicit formulas for the derivative functions by applying differentiation rules. Those formulas must be evaluated afterwards to get numerical values. Automatic differentiation also uses the well-known differentiation rules, but it propagates numerical values for the derivatives. This combines the advantages of symbolic and numerical differentiation. Numbers instead of symbolic formulas must be handled, and the computation of the derivative values is done automatically together with the computation of the function values. The main advantage of this process is that only the algorithm or formula for the function is required. No explicit formulas for the derivatives are required.

In this chapter, we deal with automatic differentiation based on interval operations to get guaranteed enclosures for the function value and the derivative values. Automatic differentiation is a fundamental enabling technology for validated computation since so many interval algorithms require an enclosure for high-order derivative terms to capture truncation errors in the algorithms.

5.1 Theoretical Background

Automatic differentiation evaluates functions specified by algorithms or formulas where all operations are executed according to the rules of a *differentiation arithmetic* given below (see also [66], [68]).

First Derivatives

In the one-dimensional case, first order differentiation arithmetic is an arithmetic for ordered pairs of the form

$$U = (u, u') \text{ with } u, u' \in I\!R.$$

The first component of U contains the value $u(x)$ of the function $u : I\!R \to I\!R$ at the point $x \in I\!R$. The second component contains the value of the derivative $u'(x)$. The rules for the arithmetic are

$$\begin{array}{ccccccccl} U & + & V & = & (u,u') & + & (v,v') & = & (u+v, u'+v') \\ U & - & V & = & (u,u') & - & (v,v') & = & (u-v, u'-v') \\ U & \cdot & V & = & (u,u') & \cdot & (v,v') & = & (u \cdot v, u \cdot v' + u' \cdot v) \\ U & / & V & = & (u,u') & / & (v,v') & = & (u/v, (u' - u/v \cdot v')/v), \quad v \neq 0, \end{array} \tag{5.1}$$

The familiar rules of calculus are used in the second component. The operations inside the parentheses in these definitions are operations on real numbers. An independent variable x and the arbitrary constant c correspond to the ordered pairs

$$X = (x, 1) \quad \text{and} \quad C = (c, 0),$$

because $\frac{dx}{dx} = 1$, and $\frac{dc}{dx} = 0$.

Let the independent variable x of a formula for a function $f : I\!R \to I\!R$ be replaced by $X = (x, 1)$, and let all constants c be replaced by their $(c, 0)$ representation. Then evaluation of f using the rules of differentiation arithmetic gives the ordered pair

$$f(X) = f((x,1)) = (f(x), f'(x)).$$

Example 5.1 We want to compute the function value and the derivative value of the function $f(x) = x \cdot (4 + x)/(3 - x)$ at the point $x = 1$. Using the differentiation arithmetic defined by (5.1) we get

$$\begin{aligned} f(X) = (f, f') &= (x,1) \cdot ((4,0) + (x,1))/((3,0) - (x,1)) \\ &= (1,1) \cdot ((4,0) + (1,1))/((3,0) - (1,1)) \\ &= (1,1) \cdot (5,1)/(2,-1) \\ &= (5,6)/(2,-1) \\ &= (2.5, 4.25). \end{aligned}$$

That is, $f(1) = 2.5$, and $f'(1) = 4.25$.

For an elementary function $s : I\!R \to I\!R$ (see Section 3.1 for the set of elementary function used within this scope), the rules of differentiation arithmetic must be extended using the chain rule

$$s(U) = s((u,u')) = (s(u), u' \cdot s'(u)). \tag{5.2}$$

For example, the sine function is defined by

$$\sin U = \sin(u, u') = (\sin u, u' \cdot \cos u).$$

Second Derivatives

For a second order differentiation arithmetic, we use triples

$$U = (u, u', u''), \quad \text{with } u, u', u'' \in I\!R,$$

for the description of the arithmetic rules. Here u, u', and u'' denote the function value, the value of the first derivative, and the value of the second derivative, respectively, each evaluated at a fixed point $x \in I\!R$. For the constant function $u(x) = c$, we set $U = (u, u', u'') = (c, 0, 0)$. For the function $u(x) = x$, we define $U = (u, u', u'') = (x, 1, 0)$.

We define the arithmetic operations $W = U \circ V$ with $\circ \in \{+, -, \cdot, /\}$ for two triples $U = (u, u', u'')$ and $V = (v, v', v'')$ using the rules of first order differentiation arithmetic in the first and second components. The third components are defined by

$$\begin{array}{lllllll} W &=& U &+& V &\Rightarrow& w'' = u'' + v'', \\ W &=& U &-& V &\Rightarrow& w'' = u'' - v'', \\ W &=& U &\cdot& V &\Rightarrow& w'' = u \cdot v'' + 2 \cdot u' \cdot v' + v \cdot u'', \\ W &=& U &/& V &\Rightarrow& w'' = (u'' - 2 \cdot w' \cdot v' - w \cdot v'')/v, \quad v \neq 0. \end{array} \tag{5.3}$$

For an elementary function $s : I\!R \to I\!R$ and $U = (u, u', u'')$, we define

$$s(U) = (s(u),\ s'(u) \cdot u',\ s'(u) \cdot u'' + s''(u) \cdot (u')^2). \tag{5.4}$$

Here $s' : I\!R \to I\!R$ and $s'' : I\!R \to I\!R$ are the first and second derivatives of s, assuming they exist.

We compute *enclosures* for the true values of the function and its derivatives using a differentiation arithmetic based on interval arithmetic. That is, the components u, u', and u'' are replaced by the corresponding interval values, and all arithmetic operations and function evaluations performed to compute the components are replaced by the corresponding interval operations and function evaluations. The evaluation of a function $f : I\!R \to I\!R$ for an argument $[x] \in II\!R$ using interval differentiation arithmetic delivers

$$f(X) = f(([x], 1, 0)) = ([f], [f'], [f''])$$

satisfying

$$f([x]) \subseteq [f], \quad f'([x]) \subseteq [f'], \quad \text{and} \quad f''([x]) \subseteq [f''].$$

5.2 Algorithmic Description

We now give the description of the algorithms for the elementary operators $+$, $-$, $\cdot$, and $/$, and for an elementary function $s \in$ {sqr, sqrt, power, exp, ln, sin, cos, tan, cot, arcsin, arccos, arctan, arccot, sinh, cosh, tanh, coth, arsinh, arcosh, artanh, arcoth}. For the operands and arguments U, we use the triple representation $U = ([u_f], [u_{df}], [u_{ddf}])$ with $[u_f], [u_{df}], [u_{ddf}] \in II\!R$ to be close to the notation of our implementation.

Algorithm 5.1: $+\ (U,V)$ {Operator}

1. $[w_f] := [u_f] + [v_f]$; {Function value}
2. $[w_{df}] := [u_{df}] + [v_{df}]$; {First derivative}
3. $[w_{ddf}] := [u_{ddf}] + [v_{ddf}]$; {Second derivative}
4. **return** $W = ([w_f], [w_{df}], [w_{ddf}])$;

Algorithm 5.2: $-\ (U,V)$ {Operator}

1. $[w_f] := [u_f] - [v_f]$; {Function value}
2. $[w_{df}] := [u_{df}] - [v_{df}]$; {First derivative}
3. $[w_{ddf}] := [u_{ddf}] - [v_{ddf}]$; {Second derivative}
4. **return** $W = ([w_f], [w_{df}], [w_{ddf}])$;

Algorithm 5.3: $\cdot\ (U,V)$ {Operator}

1. $[w_f] := [u_f] \cdot [v_f]$; {Function value}
2. $[w_{df}] := [u_f] \cdot [v_{df}] + [u_{df}] \cdot [v_f]$; {First derivative}
3. $[w_{ddf}] := [u_f] \cdot [v_{ddf}] + 2 \cdot [u_{df}] \cdot [v_{df}] + [v_f] \cdot [u_{ddf}]$; {Second derivative}
4. **return** $W = ([w_f], [w_{df}], [w_{ddf}])$;

In Algorithm 5.4, we do not take care of the case $0 \in [v_f]$ because it does not make sense to go any further in computations when this case occurs. In an implementation, the standard error handling (runtime error) should be invoked if a division by zero occurs while computing the function value. We chose a special form of the rule for the differentiation of a quotient to be close to our implementation, where this form can save some computation time.

Algorithm 5.4: $/\,(U,V)$ {Operator}

1. $[w_f] := [u_f]/[v_f]$; {Function value}
2. $[w_{df}] := ([u_{df}] - [w_f] \cdot [v_{df}])/[v_f]$; {First derivative}
3. $[w_{ddf}] := ([u_{ddf}] - 2 \cdot [w_{df}] \cdot [v_{df}] - [w_f] \cdot [v_{ddf}])/[v_f]$; {Second derivative}
4. **return** $W = ([w_f], [w_{df}], [w_{ddf}])$;

In Algorithm 5.5, the derivative rules for the elementary functions are applied to compute the temporary values. For a better understanding of our implementation, these functions are listed in Table 5.1. There we use $\tan^2 + 1$ as the derivative of $\tan x$ to avoid a separate evaluation of $\cos x$.

Except for $\sqrt{0}$, we do not have to consider non-differentiability because all of these elementary functions are differentiable whenever they are defined. That is,

Table 5.1: Elementary functions and their derivatives

$s(x)$	$s'(x)$	$s''(x)$	$s(x)$	$s'(x)$	$s''(x)$
x^k	kx^{k-1}	$k(k-1)x^{k-2}$	e^x	e^x	e^x
$\sqrt{x}$	$1/(2\sqrt{x})$	$-1/(4\sqrt{x}^3)$	$\ln x$	$1/x$	$-1/x^2$
$\sin x$	$\cos x$	$-\sin x$	$\sinh x$	$\cosh x$	$\sinh x$
$\cos x$	$-\sin x$	$-\cos x$	$\cosh x$	$\sinh x$	$\cosh x$
$\tan x$	$\tan^2 x + 1$	$2(\tan^3 x + \tan x)$	$\tanh x$	$1 - \tanh^2 x$	$-2(\tanh^3 x - \tanh x)$
$\cot x$	$-\cot^2 x - 1$	$2(\cot^3 x + \cot x)$	$\coth x$	$1 - \coth^2 x$	$-2(\coth^3 x - \coth x)$
$\arcsin x$	$1/\sqrt{1-x^2}$	$x/(1-x^2)^{\frac{3}{2}}$	$\operatorname{arsinh} x$	$1/\sqrt{x^2+1}$	$-x/(x^2+1)^{\frac{3}{2}}$
$\arccos x$	$-1/\sqrt{1-x^2}$	$-x/(1-x^2)^{\frac{3}{2}}$	$\operatorname{arcosh} x$	$1/\sqrt{x^2-1}$	$-x/(x^2-1)^{\frac{3}{2}}$
$\arctan x$	$1/(1+x^2)$	$-2x/(1+x^2)^2$	$\operatorname{artanh} x$	$1/(1-x^2)$	$2x/(x^2-1)^2$
$\operatorname{arccot} x$	$-1/(1+x^2)$	$2x/(1+x^2)^2$	$\operatorname{arcoth} x$	$1/(1-x^2)$	$2x/(x^2-1)^2$

except for sqrt, the domains for s, s', and s'' are the same. However, note that many derivative values may overflow, even when the function values are evaluated safely. In an implementation of Algorithm 5.5, the standard error handling of interval arithmetic should be invoked for domain violations in step 1 (or step 2 for sqrt). For details on the domains of the interval functions, see [65].

Algorithm 5.5: s (U) {Function}

1. $[w_f] := s([u_f])$; {Function value}
2. $[h_1] := s'([u_f])$; $[h_2] := s''([u_f])$; {Temporary values}
3. $[w_{df}] := [h_1] \cdot [u_{df}]$; {First derivative}
4. $[w_{ddf}] := [h_1] \cdot [u_{ddf}] + [h_2] \cdot ([u_{df}])^2$; {Second derivative}
5. **return** s $:= ([w_f], [w_{df}], [w_{ddf}])$;

The computational complexity of this forward mode of differentiation arithmetic depends somewhat on the mix of arithmetic operations and elementary functions. The cost of computing first and second derivatives is at most a small multiple of the cost of evaluating f itself.

5.3 Implementation and Examples

5.3.1 PASCAL–XSC Program Code

5.3.1.1 Module ddf_ari

The module *ddf_ari* supplies type definition, operators and elementary functions for an interval differentiation arithmetic for derivatives up to second order. The local variable *DerivOrder* is used to select the highest order of derivative which is computed up to second order. This enables the user to save computation time computing only the function value or the first derivative. The default value of *DerivOrder* is 2, so normally the first and the second derivatives are computed. The procedures *fEval*, *dfEval*, and *ddfEval* simplify the mechanism of function evaluating and automate the setting and resetting of the *DerivOrder* variable. For a function of type *DerivType*, *fEval* sets *DerivOrder* to 0 before the evaluation is done, computes, and delivers only the function value. If the first derivative also is desired, *dfEval* sets *DerivOrder* to 1 before the evaluation is done. The procedure *ddfEval* uses the default value of *DerivOrder*, computes, and returns the values of $f(x)$, $f'(x)$, and $f''(x)$.

Module *ddf_ari* can easily be modified to get a *real* version of this second order differentiation arithmetic by replacing data type *interval* by data type *real* everywhere it occurs.

```
{----------------------------------------------------------------------------}
{ Purpose: Definition of an interval differentiation arithmetic which allows }
{    function evaluation with automatic differentiation up to second order.  }
{ Method: Overloading of operators and elementary functions for operations   }
{    of data type 'DerivType'.                                               }
{ Global types, operators, functions, and procedures:                        }
{    type       DerivType          : data type for differentiation arithmetic }
{    operators  +, -, *, /         : operators of differentiation arithmetic  }
{    functions  DerivConst,                                                  }
{               DerivVar           : to define derivative constants/variables }
{    functions  fValue,                                                      }
{               dfValue,                                                     }
{               ddfValue           : to get function and derivative values   }
{    functions  sqr, sqrt, power,                                            }
{               exp, sin, cos,...  : elementary functions of diff. arithmetic }
{    procedure  fEval(...)         : to compute function value only          }
{    procedure  dfEval(...)        : to compute function and first derivative }
{                                    value                                   }
{    procedure  ddfEval(...)       : to compute function, first, and second  }
{                                    derivative value                        }
{----------------------------------------------------------------------------}
module ddf_ari;

use i_ari, i_util;        { interval arithmetic, interval utility functions }

{----------------------------------------------------------------------------}
{ Global type definition and local variable                                  }
{----------------------------------------------------------------------------}
global type
  DerivType     = record f, df, ddf : interval; end;

var                     { The local variable 'DerivOrder' is used to select the }
  DerivOrder : 0..2;    { highest order of derivative which is computed. Its    }
                        { default value is 2, and normally the first and the    }
                        { second derivatives are computed.                      }
{----------------------------------------------------------------------------}
{ Transfer functions for constants and variables                             }
{----------------------------------------------------------------------------}
global function DerivConst (c: real) : DerivType;          { Generate constant }
begin                                                      {-------------------}
  DerivConst.f := c;  DerivConst.df := 0;  DerivConst.ddf := 0;
end;

global function DerivConst (c: interval) : DerivType;      { Generate constant }
begin                                                      {-------------------}
  DerivConst.f := c;  DerivConst.df := 0;  DerivConst.ddf := 0;
```

```
end;

global function DerivVar (v: real) : DerivType;          { Generate variable }
begin                                                    {-------------------}
  DerivVar.f := v;  DerivVar.df := 1;  DerivVar.ddf := 0;
end;

global function DerivVar (v: interval) : DerivType;      { Generate variable }
begin                                                    {-------------------}
  DerivVar.f := v;  DerivVar.df := 1;  DerivVar.ddf := 0;
end;

{----------------------------------------------------------------------------}
{ Access functions for function and derivative values                        }
{----------------------------------------------------------------------------}
global function fValue (u: DerivType) : interval;       { Get function value }
begin                                                   {--------------------}
  fValue:= u.f;
end;

global function dfValue (u: DerivType) : interval; { Get 1. derivative value }
begin                                              {-------------------------}
  dfValue:= u.df;
end;

global function ddfValue (u: DerivType) : interval;{ Get 2. derivative value }
begin                                              {-------------------------}
  ddfValue:= u.ddf;
end;

{----------------------------------------------------------------------------}
{ Monadic operators + and - for DerivType operands                           }
{----------------------------------------------------------------------------}
global operator + (u: DerivType) res: DerivType;
begin
  res:= u;
end;

global operator - (u: DerivType) res: DerivType;
begin
  res.f := -u.f;
  if (DerivOrder > 0) then
  begin
    res.df := -u.df;
    if (DerivOrder > 1) then res.ddf:= -u.ddf;
  end;
end;

{----------------------------------------------------------------------------}
{ Operators +, -, *, and / for two DerivType operands                        }
{----------------------------------------------------------------------------}
global operator + (u,v: DerivType) res: DerivType;
begin
  res.f := u.f + v.f;
  if (DerivOrder > 0) then
  begin
    res.df := u.df + v.df;
    if (DerivOrder > 1) then res.ddf:= u.ddf + v.ddf;
  end;
end;

global operator - (u,v: DerivType) res: DerivType;
begin
  res.f := u.f   - v.f;
```

```
  if (DerivOrder > 0) then
  begin
    res.df := u.df - v.df;
    if (DerivOrder > 1) then res.ddf:= u.ddf - v.ddf;
  end;
end;

global operator * (u,v: DerivType) res: DerivType;
begin
  res.f := u.f*v.f;
  if (DerivOrder > 0) then
  begin
    res.df := u.df*v.f + u.f*v.df;
    if (DerivOrder > 1) then res.ddf:= u.ddf*v.f + 2*u.df*v.df + u.f*v.ddf;
  end;
end;

global operator / (u,v: DerivType) res: DerivType;
var h1, h2: interval;
begin
  h1 := u.f/v.f;  { Can propagate 'division by zero' error }
  res.f := h1;
  if (DerivOrder > 0) then
  begin
    h2 := (u.df - h1*v.df)/v.f;  res.df := h2;
    if (DerivOrder > 1) then res.ddf:= (u.ddf - h1*v.ddf - 2*h2*v.df)/v.f;
  end;
end;

{-----------------------------------------------------------------------------}
{ Operators +, -, *, and / for one interval and one DerivType operand        }
{-----------------------------------------------------------------------------}
global operator + (u: interval; v: DerivType) res: DerivType;
begin
  res.f := u + v.f;
  if (DerivOrder > 0) then
  begin
    res.df := v.df;
    if (DerivOrder > 1) then res.ddf:= v.ddf;
  end;
end;

global operator - (u: interval; v: DerivType) res: DerivType;
begin
  res.f := u - v.f;
  if (DerivOrder > 0) then
  begin
    res.df := - v.df;
    if (DerivOrder > 1) then res.ddf:= - v.ddf;
  end;
end;

global operator * (u: interval; v: DerivType) res: DerivType;
begin
  res.f := u*v.f;
  if (DerivOrder > 0) then
  begin
    res.df := u*v.df;
    if (DerivOrder > 1) then res.ddf:= u*v.ddf;
  end;
end;

global operator / (u: interval; v: DerivType) res: DerivType;
var h1, h2: interval;
```

```
begin
  h1 := u/v.f;  { Can propagate 'division by zero' error }
  res.f := h1;
  if (DerivOrder > 0) then
  begin
    h2 := -h1*v.df/v.f;  res.df := h2;
    if (DerivOrder > 1) then res.ddf:= (-h1*v.ddf - 2*h2*v.df)/v.f;
  end;
end;

global operator + (u: DerivType; v: interval) res: DerivType;
begin
  res.f := u.f + v;
  if (DerivOrder > 0) then
  begin
    res.df := u.df;
    if (DerivOrder > 1) then res.ddf:= u.ddf;
  end;
end;

global operator - (u: DerivType; v: interval) res: DerivType;
begin
  res.f := u.f - v;
  if (DerivOrder > 0) then
  begin
    res.df := u.df;
    if (DerivOrder > 1) then res.ddf:= u.ddf;
  end;
end;

global operator * (u: DerivType; v: interval) res: DerivType;
begin
  res.f := u.f * v;
  if (DerivOrder > 0) then
  begin
    res.df := u.df * v;
    if (DerivOrder > 1) then res.ddf:= u.ddf * v;
  end;
end;

global operator / (u: DerivType; v: interval) res: DerivType;
begin
  res.f := u.f / v;  { Can propagate 'division by zero' error }
  if (DerivOrder > 0) then
  begin
    res.df := u.df / v;
    if (DerivOrder > 1) then res.ddf:= u.ddf / v;
  end;
end;

{----------------------------------------------------------------------------}
{ Operators +, -, *, and / for one real and one DerivType operand            }
{----------------------------------------------------------------------------}
global operator + (u: real; v: DerivType) res: DerivType;
begin
  res := intval(u) + v;
end;

global operator - (u: real; v: DerivType) res: DerivType;
begin
  res := intval(u) - v;
end;

global operator * (u: real; v: DerivType) res: DerivType;
```

```
begin
  res := intval(u) * v;
end;

global operator / (u: real; v: DerivType) res: DerivType;
begin
  res := intval(u) / v;  { Can propagate 'division by zero' error }
end;

global operator + (u: DerivType; v: real) res: DerivType;
begin
  res := u + intval(v);
end;

global operator - (u: DerivType; v: real) res: DerivType;
begin
  res := u - intval(v);
end;

global operator * (u: DerivType; v: real) res: DerivType;
begin
  res := u * intval(v);
end;

global operator / (u: DerivType; v: real) res: DerivType;
begin
  res := u / intval(v);  { Can propagate 'division by zero' error }
end;

{----------------------------------------------------------------------------}
{ Elementary functions for DerivType arguments                               }
{----------------------------------------------------------------------------}
global function sqr (u: DerivType) : DerivType;
begin
  sqr.f  := sqr(u.f);
  if (DerivOrder > 0) then
  begin
    sqr.df := 2*u.f*u.df;
    if (DerivOrder > 1) then sqr.ddf:= 2 * (sqr(u.df) + u.f*u.ddf);
  end;
end;

global function power (u: DerivType; k: integer) : DerivType;
var
  h1 : interval;
begin
  if (k = 0) then
    power:= DerivConst(1)
  else if (k = 1) then
    power:= u
  else if (k = 2) then
    power:= sqr(u)
  else
    begin
      power.f:= power(u.f, k);
      if (DerivOrder > 0) then
      begin
        h1 := k * power(u.f, k-1);  power.df:= h1 * u.df;
        if (DerivOrder > 1) then
          power.ddf:= h1 * u.ddf + k*(k-1)*power(u.f, k-2)*sqr(u.df);
      end;
    end;
end;
```

```
global function sqrt (u: DerivType) : DerivType;
var h1, h2: interval;
begin
  h1 := sqrt(u.f);  { Can propagate domain error }
  sqrt.f := h1;
  if (DerivOrder > 0) then
  begin
    h1 := 0.5/h1;  h2 := u.df*h1;  sqrt.df := h2;
    if (DerivOrder > 1) then sqrt.ddf:= u.ddf*h1 - 0.5*u.df/u.f*h2;
  end;
end;

global function exp (u: DerivType) : DerivType;
var h1: interval;
begin
  h1 := exp(u.f);  exp.f := h1;
  if (DerivOrder > 0) then
  begin
    exp.df := h1*u.df;
    if (DerivOrder > 1) then exp.ddf:= (u.ddf + sqr(u.df))*h1;
  end;
end;

global function ln (u: DerivType) : DerivType;
var h: interval;
begin
  ln.f := ln(u.f);  { Can propagate domain error }
  if (DerivOrder > 0) then
  begin
    h := u.df/u.f;   ln.df := h;
    if (DerivOrder > 1) then ln.ddf:= (u.ddf - h*u.df) / u.f;
  end;
end;

global function sin (u: DerivType) : DerivType;
var h0, h1: interval;
begin
  h0 := sin(u.f);  sin.f := h0;
  if (DerivOrder > 0) then
  begin
    h1 := cos(u.f);  sin.df := h1*u.df;
    if (DerivOrder > 1) then sin.ddf:= h1*u.ddf - h0*sqr(u.df);
  end;
end;

global function cos (u: DerivType) : DerivType;
var h0, h1: interval;
begin
  h0 := cos(u.f);  cos.f := h0;
  if (DerivOrder > 0) then
  begin
    h1 := -sin(u.f);  cos.df := h1*u.df;
    if (DerivOrder > 1) then cos.ddf:= h1*u.ddf - h0*sqr(u.df);
  end;
end;

global function tan (u: DerivType) : DerivType;
var h0, h1, h2: interval;
begin
  h0 := tan(u.f);  { Can propagate domain error }
  tan.f := h0;
  if (DerivOrder > 0) then
  begin                                { The subdistributive law implies  }
    h1 := sqr(h0)+1;  h2 := 2*h0*h1;   {   h0 * (h0↑2 + 1) <= h0↑3 + h0   }
```

```
    tan.df := h1*u.df;                   { So, we use the first form.       }
    if (DerivOrder > 1) then tan.ddf:= h1*u.ddf + h2*sqr(u.df);
  end;
end;

global function cot (u: DerivType) : DerivType;
var h0, h1, h2: interval;
begin
  h0 := cot(u.f);  { Can propagate domain error }
  cot.f := h0;
  if (DerivOrder > 0) then
  begin                                { The subdistributive law implies  }
    h1 := -(sqr(h0)+1);  h2 := -2*h0*h1;  {   h0 * (h0↑2 + 1) <= h0↑3 + h0    }
    cot.df := h1*u.df;                   { So, we use the first form.       }
    if (DerivOrder > 1) then cot.ddf:= h1*u.ddf + h2*sqr(u.df);
  end;
end;

global function arcsin (u: DerivType) : DerivType;
var h, h1, h2: interval;
begin
  arcsin.f := arcsin(u.f);  { Can propagate domain error }
  if (DerivOrder > 0) then
  begin
    h:= 1 - sqr(u.f);  h1:= 1/sqrt(h);  arcsin.df:= h1*u.df;  h2:= u.f*h1/h;
    if (DerivOrder > 1) then arcsin.ddf:= h1*u.ddf + h2*sqr(u.df);
  end;
end;

global function arccos (u: DerivType) : DerivType;
var h, h1, h2: interval;
begin
  arccos.f := arccos(u.f);  { Can propagate domain error }
  if (DerivOrder > 0) then
  begin
    h:= 1 - sqr(u.f);  h1:= -1/sqrt(h);  arccos.df:= h1*u.df;  h2:= u.f*h1/h;
    if (DerivOrder > 1) then arccos.ddf:= h1*u.ddf + h2*sqr(u.df);
  end;
end;

global function arctan (u: DerivType) : DerivType;
var h1,h2: interval;
begin
  arctan.f := arctan(u.f);  { Can propagate domain error }
  if (DerivOrder > 0) then
  begin
    h1 := 1 / (1 + sqr(u.f));  arctan.df := h1*u.df;  h2 := -2*u.f*sqr(h1);
    if (DerivOrder > 1) then arctan.ddf:= h1*u.ddf + h2*sqr(u.df);
  end;
end;

global function arccot (u: DerivType) : DerivType;
var h1, h2: interval;
begin
  arccot.f := arccot(u.f);  { Can propagate domain error }
  if (DerivOrder > 0) then
  begin
    h1 := -1 / (1 + sqr(u.f));  arccot.df := h1*u.df;  h2 := 2*u.f*sqr(h1);
    if (DerivOrder > 1) then arccot.ddf:= h1*u.ddf + h2*sqr(u.df);
  end;
end;

global function sinh (u: DerivType) : DerivType;
var h0, h1: interval;
```

```
begin
  h0 := sinh(u.f);  sinh.f := h0;
  if (DerivOrder > 0) then
  begin
    h1 := cosh(u.f);  sinh.df := h1*u.df;
    if (DerivOrder > 1) then sinh.ddf:= h1*u.ddf + h0*sqr(u.df);
  end;
end;

global function cosh (u: DerivType) : DerivType;
var h0, h1: interval;
begin
  h0 := cosh(u.f);  cosh.f := h0;
  if (DerivOrder > 0) then
  begin
    h1 := sinh(u.f);  cosh.df := h1*u.df;
    if (DerivOrder > 1) then cosh.ddf:= h1*u.ddf + h0*sqr(u.df);
  end;
end;

global function tanh (u: DerivType) : DerivType;
var h0, h1, h2: interval;
begin
  h0 := tanh(u.f);  tanh.f := h0;
  if (DerivOrder > 0) then
  begin                                   { The subdistributive law implies  }
    h1 := 1 - sqr(h0);  h2 := -2*h0*h1;   {   h0 * (h0↑2 - 1) <= h0↑3 - h0   }
    tanh.df := h1*u.df;                   { So, we use the first form.       }
    if (DerivOrder > 1) then tanh.ddf:= h1*u.ddf + h2*sqr(u.df);
  end;
end;

global function coth (u: DerivType) : DerivType;
var h0, h1, h2: interval;
begin
  h0 := coth(u.f);  { Can propagate domain error }
  coth.f := h0;
  if (DerivOrder > 0) then
  begin                                   { The subdistributive law implies  }
    h1 := 1 - sqr(h0);  h2 := -2*h0*h1;   {   h0 * (h0↑2 - 1) <= h0↑3 - h0   }
    coth.df := h1*u.df;                   { So, we use the first form.       }
    if (DerivOrder > 1) then coth.ddf:= h1*u.ddf + h2*sqr(u.df);
  end;
end;

global function arsinh (u: DerivType) : DerivType;
var h, h1, h2: interval;
begin
  arsinh.f := arsinh(u.f);  { Can propagate domain error }
  if (DerivOrder > 0) then
  begin
    h:= 1 + sqr(u.f);  h1:= 1/sqrt(h);  arsinh.df:= h1*u.df;  h2:= -u.f*h1/h;
    if (DerivOrder > 1) then arsinh.ddf:= h1*u.ddf + h2*sqr(u.df);
  end;
end;

global function arcosh (u: DerivType) : DerivType;
var h, h1, h2: interval;
begin
  arcosh.f := arcosh(u.f);  { Can propagate domain error }
  if (DerivOrder > 0) then
  begin
    h:= sqr(u.f) - 1;  h1:= 1/sqrt(h);  arcosh.df:= h1*u.df;  h2:= -u.f*h1/h;
    if (DerivOrder > 1) then arcosh.ddf:= h1*u.ddf + h2*sqr(u.df);
```

```
  end;
end;

global function artanh (u: DerivType) : DerivType;
var h1,h2: interval;
begin
  artanh.f := artanh(u.f);  { Can propagate domain error }
  if (DerivOrder > 0) then
  begin
    h1 := 1 / (1 - sqr(u.f));  artanh.df := h1*u.df;  h2 := 2*u.f*sqr(h1);
    if (DerivOrder > 1) then artanh.ddf:= h1*u.ddf + h2*sqr(u.df);
  end;
end;

global function arcoth (u: DerivType) : DerivType;
var h1, h2: interval;
begin
  arcoth.f := arcoth(u.f);  { Can propagate domain error }
  if (DerivOrder > 0) then
  begin
    h1 := 1 / (1 - sqr(u.f));  arcoth.df := h1*u.df;  h2 := 2*u.f*sqr(h1);
    if (DerivOrder > 1) then arcoth.ddf:= h1*u.ddf + h2*sqr(u.df);
  end;
end;

{----------------------------------------------------------------------------}
{ Predefined routines for evaluation of DerivType-functions                  }
{----------------------------------------------------------------------------}
{ Purpose: Evaluation of function 'f' for argument 'x' in differentiation    }
{    arithmetic computing only the function value.                           }
{ Parameters:                                                                }
{    In     : 'f'      : function of 'DerivType'.                            }
{             'x'      : argument for evaluation of 'f'.                     }
{    Out    : 'fx'     : returns the function value 'f(x)'.                  }
{ Description: This procedure sets 'DerivOrder' to 0, evaluates 'f(x)' in    }
{    differentiation arithmetic, and returns the function value only.        }
{----------------------------------------------------------------------------}
global procedure fEval (function f(x:DerivType) : DerivType;
                                               x : interval;
                                var           fx : interval);
var
  fxD : DerivType;
begin
  DerivOrder := 0;  fxD := f(DerivVar(x));  fx := fxD.f;  DerivOrder := 2;
end;

{----------------------------------------------------------------------------}
{ Purpose: Evaluation of function 'f' for argument 'x' in differentiation    }
{    arithmetic computing the function value and the value of the first      }
{    derivative.                                                             }
{ Parameters:                                                                }
{    In     : 'f'      : function of 'HessType'.                             }
{             'x'      : argument for evaluation of 'f'.                     }
{    Out    : 'fx'     : returns the function value 'f(x)'.                  }
{             'dfx'    : returns the first derivative value 'f'(x)'.         }
{ Description: This procedure sets 'DerivOrder' to 1, evaluates 'f(x)' in    }
{    differentiation arithmetic, and returns the function value and the      }
{    value of the first derivative.                                          }
{----------------------------------------------------------------------------}
global procedure dfEval (function f(x:DerivType) : DerivType;
                                                x : interval;
                                 var      fx, dfx : interval);
var
  fxD : DerivType;
```

```
begin
  DerivOrder:= 1;  fxD:= f(DerivVar(x));  fx:= fxD.f;  dfx:= fxD.df;
  DerivOrder:= 2;
end;

{----------------------------------------------------------------------------}
{ Purpose: Evaluation of function 'f' for argument 'x' in differentiation    }
{    arithmetic computing the function value, the value of the first, and    }
{    the value of the second derivative.                                     }
{ Parameters:                                                                }
{    In     : 'f'     : function of 'HessType'.                              }
{             'x'     : argument for evaluation of 'f'.                      }
{    Out    : 'fx'    : returns the function value 'f(x)'.                   }
{             'dfx'   : returns the value of the first derivative 'f'(x)'.   }
{             'ddfx'  : returns the value of the second derivative 'f''(x)'. }
{ Description: This procedure keeps 'DerivOrder' = 2, evaluates 'f(x)' in    }
{    differentiation arithmetic, and returns the function value, the value   }
{    of the first, and the value of the second derivative.                   }
{----------------------------------------------------------------------------}
global procedure ddfEval (function f(x:DerivType) : DerivType;
                                                x : interval;
                                var  fx, dfx, ddfx : interval);
var
  fxD : DerivType;
begin
  fxD:= f(DerivVar(x));  fx:= fxD.f;  dfx:= fxD.df;  ddfx:= fxD.ddf;
end;

{----------------------------------------------------------------------------}
{ Module initialization part                                                 }
{----------------------------------------------------------------------------}
begin
  DerivOrder := 2;
end.
```

5.3.2 Examples

We first illustrate how to write a program using our differentiation arithmetic in Example 5.1:

Example 5.2 We must define a PASCAL–XSC function for the function $f(x) = x \cdot (4+x)/(3-x)$:

```
function f (x: DerivType) : DerivType;
begin
  f := x * (4 + x) / (3 - x);
end;
```

With the declarations

```
var x, fx             : DerivType;
    y, fy, dfy, ddfy : interval;
```

we are able to compute function and derivative values

```
      y  := 123;
      x  := DerivVar(y);
     fx  := f(x);
     fy  := fValue(fx);        { Function value f(y)       }
    dfy  := dfValue(fx);       { First derivative f'(y)    }
   ddfy  := ddfValue(fx);      { Second derivative f''(y)  }
```

Even easier is

```
y := 123;
ddfEval(f,y,fy,dfy,ddfy);
```

Example 5.3 Newton's method for approximating a zero of a nonlinear function $f : I\!R \rightarrow I\!R$ requires f'. Other methods use derivatives of second or higher order. Halley's method uses the first and the second derivatives of f. Starting from an approximation $x^{(0)} \in I\!R$, Halley's method iterates according to

$$\left.\begin{array}{rcl} a^{(k)} & := & -\dfrac{f(x^{(k)})}{f'(x^{(k)})} \\[2ex] b^{(k)} & := & a^{(k)} \cdot \dfrac{f''(x^{(k)})}{f'(x^{(k)})} \\[2ex] x^{(k+1)} & := & x^{(k)} + \dfrac{a^{(k)}}{1 + \frac{b^{(k)}}{2}} \end{array}\right\} \quad k = 0, 1, 2, \ldots$$

In our sample program *ddf_ex* for Halley's method, all real operations are replaced by interval operations, because we make use of the interval differentiation arithmetic. The influence of rounding errors is demonstrated directly by the increasing interval diameters. We terminate the iteration if an interval evaluation of f over the current interval iterate $[x]^{(k)}$ contains zero, or if the number of iterations is greater than or equal to 100. This approximate method can *not* guarantee to find and enclose a zero of the function f because we have enclosed only roundoff errors in Halley's method, not truncation errors. We present in Chapter 6 a more sophisticated interval method for finding *all* zeros of a nonlinear function with guarantees and high accuracy.

```
{----------------------------------------------------------------------------}
{ This is an implementation of Halley's method for approximating a zero of   }
{ a twice continuously differentiable function.                              }
{                                                                            }
{    given:       the function f(x) and the starting value x[0]              }
{    iteration:   x[k+1] := x[k] + a[k]/(1+b[k]/2)          \                }
{    with:        a[k]   := - f(x[k])/f'(x[k])               ) k=0,1,2,...   }
{                 b[k]   := a[k]*f''(x[k])/f'(x[k])         /                }
{                                                                            }
{ All real operations are replaced by interval operations. All function and  }
{ derivative evaluations are calculated by differentiation arithmetic.       }
{ This approximate method can NOT guarantee to find and enclose a zero of    }
{ the function 'f' because we have enclosed only roundoff errors, not        }
{ truncation errors.                                                         }
{----------------------------------------------------------------------------}
program ddf_ex_halley_method;
use
  i_ari,     { Interval arithmetic        }
  ddf_ari;   { Differentiation arithmetic }
const
  MaxIter = 100;
var
  x, a, b, fx, dfx, ddfx : interval;
  n                      : integer;
  start                  : real;
```

```
function f (xD: DerivType) : DerivType;
begin
  f  := exp(xD) * sin(4*xD)
end;

begin
  writeln('Halley''s method for the function  f(x) = exp(x) * sin(4*x)');
  writeln;
  write('Starting value = ');  start:= 1.25;  writeln(start:10);
  writeln;  writeln('Iteration:');
  x:= start;  ddfEval(f,x,fx,dfx,ddfx);  n:= 0;
  repeat
    n:= n + 1;
    writeln('x                    : ', x);
    writeln('f(x)                 : ', fx);
    a:= - fx / dfx;  b:= ddfx / dfx * a;
    x:= x + a / (1 + 0.5*b);
    ddfEval(f,x,fx,dfx,ddfx);
  until (0 in fx) or (n >= MaxIter);
  writeln;
  writeln('Approximate zero : ', x);
  writeln('Function value   : ', fx);
  writeln;
  writeln('Expected zero    : 1.57079632679489661923...');
end.
```

If we run our sample program for Halley's method, we get the following runtime output.

```
Halley's method for the function  f(x) = exp(x) * sin(4*x)

Starting value =  1.25E+000

Iteration:
x                : [  1.250000000000000E+000,  1.250000000000000E+000 ]
f(x)             : [ -3.346974588809691E+000, -3.346974588809689E+000 ]
x                : [  1.271022734475337E+000,  1.271022734475338E+000 ]
f(x)             : [  -3.32107871440321E+000,  -3.32107871440319E+000 ]
x                : [   1.33069880673815E+000,   1.33069880673817E+000 ]
f(x)             : [   -3.1004103481201E+000,   -3.1004103481199E+000 ]
x                : [    1.4635835150601E+000,    1.4635835150603E+000 ]
f(x)             : [    -1.796959342436E+000,    -1.796959342435E+000 ]
x                : [     1.565154617724E+000,     1.565154617725E+000 ]
f(x)             : [      -1.0793736948E-001,      -1.0793736947E-001 ]
x                : [     1.570795749328E+000,     1.570795749329E+000 ]
f(x)             : [          -1.111156E-005,          -1.111155E-005 ]

Approximate zero : [     1.570796326794E+000,     1.570796326796E+000 ]
Function value   : [               -4.8E-012,                4.8E-012 ]

Expected zero    : 1.57079632679489661923...
```

Mathematically, Halley's method is converging to the correct root. The floating-point algorithm converges, but grows wider. This shows the importance of using true *interval* algorithms, as opposed to *point* algorithms in interval arithmetic. The algorithm presented in Chapter 6 uses a contractive map to generate a sequence of intervals that both

- is guaranteed to enclose the root, and
- can achieve one ulp accuracy.

5.3.3 Restrictions and Hints

The implementation in module *ddf_ari* uses the standard error handling of PASCAL–XSC if the interval argument of an elementary function does not lie in the domain specified (see [65]). The same holds for an interval containing zero as second operand of the division operator. These cases can also occur during a function evaluation using differentiation arithmetic because of the known overestimation effects of interval arithmetic (see Section 3.1).

The rules for getting true enclosures in connection with conversion errors discussed in Section 3.7 also apply to interval differentiation arithmetic. If you want to compute enclosures for the values of the derivatives of the function f in Example 5.2 at the point $x = 0.1$, you must insure that the machine interval argument $[x]$ used as argument for the interval function evaluation satisfies $x \in [x]$.

5.4 Exercises

Exercise 5.1 Apply program *ddf_ex_halley_method* to find positive roots of the function $f(x) = x - \tan x$.

Exercise 5.2 Implement a *real* differentiation arithmetic by replacing in module *ddf_ari* the data type *interval* by the data type *real*. Then use the new module for Halley's method.

5.5 References and Further Reading

The method used in the implementation above is called the *forward method* of automatic differentiation, because all differentiation arithmetic operations can be executed in the same order as the corresponding floating-point operations. Automatic differentiation for gradients and Hessian matrices (see Chapter 12) can optimize the time complexity using the *backward method*, also called fast automatic differentiation (see [16]).

Automatic differentiation methods can also be used to compute interval slopes (see [64]).

Further applications and differentiation arithmetics such as Taylor arithmetics, power series arithmetics, or parallel implementation of differentiation arithmetics can be found in [20], [21], and [69]. A large bibliography on automatic differentiation is given in [12].

Chapter 6

Nonlinear Equations in One Variable

One of the most important tasks in scientific computing is the problem of finding zeros (or roots) of nonlinear functions. In classical numerical analysis, root-finding methods for nonlinear functions begin with an approximation and apply an iterative method (such as Newton's or Halley's methods), which hopefully improves the approximation. It is a myth that no numerical algorithm is able to compute *all* zeros of a nonlinear equation with guaranteed error bounds, or even more, that no method is able to give concrete information about the existence and uniqueness of solutions of such a problem.

In this chapter, we consider a method for finding *all* zeros of a nonlinear continuously differentiable function $f : \mathbb{R} \to \mathbb{R}$ in a given interval and consider the domain of applicability of that method. It computes tight bounds on the roots, and it delivers information about existence and uniqueness of the computed solutions.

The method we present is an extension of *interval Newton's method* which makes use of the extended interval operations defined in Section 3.3. Interval Newton's method was derived by Moore [61]. Modifications, extensions, and applications are given by Alefeld [3], Hansen [28], and many others.

6.1 Theoretical Background

We address the problem of finding all solutions of the equation

$$f(x) = 0 \tag{6.1}$$

for a continuously differentiable function $f : \mathbb{R} \to \mathbb{R}$ and $x \in [x]$. Interval Newton's method for solving (6.1) can easily be derived from the mean value form

$$f(m([x])) - f(x^*) = f'(\xi) \cdot (m([x]) - x^*),$$

where $x^*, \xi \in [x]$. If we assume x^* to be a zero of f, we get

$$x^* = m([x]) - \frac{f(m([x]))}{f'(\xi)} \in m([x]) - \frac{f(m([x]))}{f'([x])} =: N([x]). \tag{6.2}$$

Hence, every zero of f in $[x]$ also lies in $N([x])$, and therefore in $N([x]) \cap [x]$. Using standard interval arithmetic, interval Newton's method starts with an interval $[x]^{(0)}$ satisfying $0 \notin f'([x]^{(0)})$ and iterates according to

$$[x]^{(k+1)} := [x]^{(k)} \cap N([x]^{(k)}), \quad k = 0, 1, 2, \ldots \tag{6.3}$$

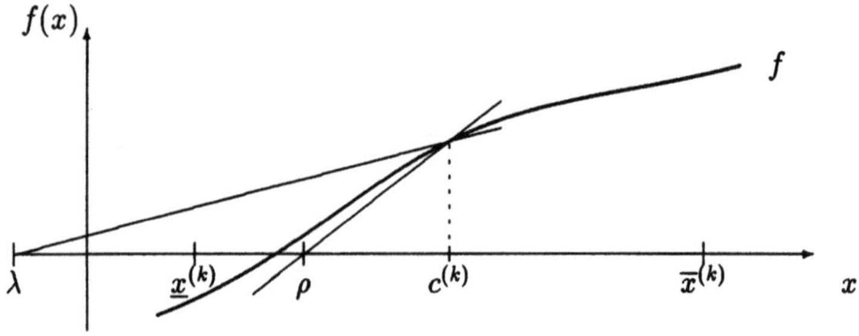

Figure 6.1: Interval Newton step with $0 \notin f'([x]^{(k)})$

The method cannot diverge due to the intersection of $N([x]^{(k)})$ with $[x]^{(k)}$. Like classical Newton's method, interval Newton's method can be geometrically interpreted as drawing two lines from the point $(c^{(k)}, f(c^{(k)}))$ with $c^{(k)} = m([x]^{(k)})$, and intersecting them with the x-axis. The lines have the slope $\underline{g}$ (the smallest slope of f in $[x]$) and $\overline{g}$ (the largest slope of f in $[x]$), respectively, where $[\underline{g}, \overline{g}] = f'([x]^{(k)})$. The points of intersection with the x-axis, let us call them λ and ρ, form the new interval $N([x]^{(k)}) = \lambda \,\underline{\cup}\, \rho$. Figure 6.1 demonstrates a single interval Newton step with resulting interval

$$[x]^{(k+1)} = [\lambda, \rho] \cap [x]^{(k)} = [\underline{x}^{(k)}, \rho].$$

If the intersection is empty, we know that there is no root of f in $[x]^{(k)}$.

Using extended interval arithmetic as defined in Section 3.3, we are able to treat the case $0 \in f'([x]^{(0)})$ that occurs if there are several zeros in the starting interval $[x]^{(0)}$. In this case, $N([x]^{(k)})$ is given by one or two extended intervals resulting from the interval division. Even though $N([x]^{(k)})$ is not finite, the intersection $[x]^{(k+1)} = N([x]^{(k)}) \cap [x]^{(k)}$ is finite and may be a single interval, the union of two intervals, or the empty set. Then, the next step of interval Newton's iteration must be applied to each of the resulting intervals. In this way it is possible to enclose *all* zeros of f in the starting interval $[x]^{(0)}$.

In Figure 6.2, we illustrate one extended interval Newton step geometrically. Again we draw lines through the point $(c^{(k)}, f(c^{(k)}))$ with $c^{(k)} = m([x]^{(k)})$. The first line with the smallest (negative) slope of f in $[x]^{(k)}$ intersects the x-axis in point ρ. The line with the largest slope intersects the x-axis in point λ. Therefore, we get

$$N([x]^{(k)}) = [-\infty, \rho] \cup [\lambda, \infty]$$

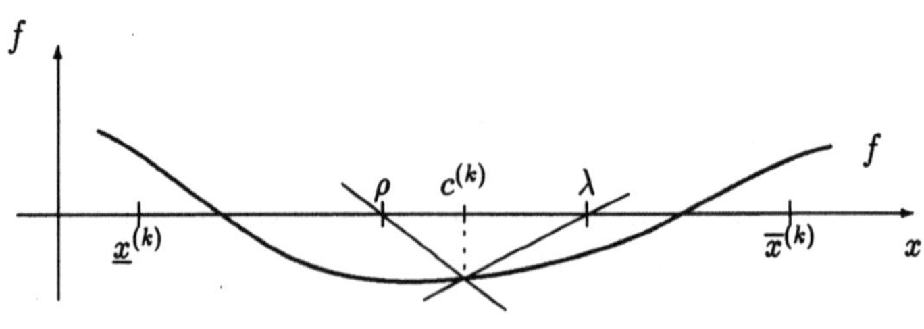

Figure 6.2: Extended interval Newton step with $0 \in f'([x]^{(k)})$

and

$$[x]^{(k+1)} = [\underline{x}^{(k)}, \rho] \cup [\lambda, \overline{x}^{(k)}].$$

The following theorem summarizes the most important properties of interval Newton's method.

Theorem 6.1 *Let $f : D \subseteq I\!R \to I\!R$ be a continuously differentiable function, and let $[x] \in II\!R$ be an interval with $[x] \subseteq D$. Then*

$$N([x]) := m([x]) - \frac{f(m([x]))}{f'([x])}$$

has the following properties:

1. *Every zero $x^* \in [x]$ of f satisfies $x^* \in N([x])$.*
2. *If $N([x]) \cap [x] = \emptyset$, then there exists no zero of f in $[x]$.*
3. *If $N([x]) \overset{\circ}{\subset} [x]$, then there exists a unique zero of f in $[x]$ and hence in $N([x])$.*

The proofs appear in [28], [61], or [64].

Remark: The conditions of Theorem 6.1 can be checked on a computer. For example, if $N_\diamond([x])$ denotes the evaluation of $N([x])$ in floating-interval arithmetic and if condition 3 is satisfied for $N_\diamond$, then we have

$$N([x]) \subseteq N_\diamond([x]) \overset{\circ}{\subset} [x].$$

Thus, the condition is fulfilled for $N([x])$, too. On the other hand, if we cannot fulfill conditions 2 and 3 of the Theorem, then $[x]$ *may* contain one or more zeros.

6.2 Algorithmic Description

Algorithm 6.3 AllZeros consists of two parts. The first part is the extended interval Newton iteration itself, including intermediate checks for the uniqueness of a zero in a computed enclosure. The second part is an additional verification step which tries to verify the uniqueness for enclosures that have not already been marked as enclosing a unique zero.

Algorithm 6.1 is a recursive procedure for the execution of the extended interval Newton method for the function f. The input parameter $[y]$ specifies the actual interval. The input parameter ε corresponds to the desired relative accuracy or tolerance for the resulting interval enclosures of the zeros. The calling procedure AllZeros guarantees that ε is not chosen less than the relative machine accuracy (1 ulp). The input parameter *yUnique* signals whether we have already verified that the incoming interval $[y]$ contains a unique zero.

If $0 \notin f([y])$, then we know that no zero can lie within the interval $[y]$. Hence, a single interval function evaluation of f can guarantee that a complete range of real values cannot contain a zero of f. If $f([y])$ contains zero, then the extended interval

Newton step given by (6.3) is applied to $[y]$ resulting in at most two intervals $[y_p]_i$ (Step 3 and 4). Interval arithmetic must be used to evaluate $f(c)$ in an implementation in order to bound all rounding errors. Therefore, we use the notation $f_\diamond(c)$ in Step 3 of the algorithm.

If no improvement for $[y]$ is achieved, $[y]$ gets bisected (Step 5). If the Newton step results in only one single interval and if the uniqueness of a zero in $[y]$ is not already proven, then the algorithm checks the Condition 3 of Theorem 6.1 for the resulting interval $[y_p]_1$, and sets the flag *yUnique* (Step 6).

In Step 7(b) the actual interval $[y_p]_i$ gets stored in the interval vector [*Zero*] if the tolerance criterion is satisfied. Otherwise, the procedure **XINewton** is recursively called (else-branch of Step 7(b)). The corresponding information on the uniqueness of the zero is stored in the flag vector *Info*. N represents the number of enclosures of zeros stored in [*Zero*].

Procedure **XINewton** terminates when no more recursive calls are necessary, that is if $0 \notin f([y])$ or if $(d_{\text{rel}}([y_p]_i) < \varepsilon)$ for $i = 1, 2$. The bisection step (Step 5) guarantees that this second condition is fulfilled at some stage of the recursion.

Algorithm 6.1: XINewton $(f, [y], \varepsilon, yUnique, [Zero], Info, N)$ {Proced.}

1. **if** $0 \notin f([y])$ **then return**;
2. $c := m([y])$;
3. $[z] := c - f_\diamond(c)/f'([y])$; {Extended interval Newton step}
4. $[y_p] := [y] \cap [z]$; {Intersection $[y] \cap [z] = [y_p]_1 \cup [y_p]_2$}
5. **if** $[y_p]_1 = [y]$ **then** $[y_p]_1 := [\underline{y}, c]$; $[y_p]_2 := [c, \overline{y}]$; {Bisection}
6. **if** $[y_p]_1 \neq \emptyset$ **and** $[y_p]_2 = \emptyset$ **then**

   $yUnique := yUnique$ **or** $([y_p]_1 \overset{\circ}{\subset} [y])$; {Inner inclusion ⇒ uniqueness}

   else

   $yUnique := false$;
7. **for** $i := 1$ **to** 2 **do**

   (a) **if** $[y_p]_i = \emptyset$ **then next**$_i$;

   (b) **if** $(d_{\text{rel}}([y_p]_i) < \varepsilon)$ **then** {Store enclosure of zero and uniqueness info}

   if $(0 \in f([y_p]_i))$ **then**

   $N := N + 1$; $[Zero]_N := [y_p]_i$; $Info_N := yUnique$;

   else {Recursive call of XINewton for $[y_p]_i$}

   XINewton $(f, [y_p]_i, \varepsilon, yUnique, [Zero], Info, N)$;
8. **return** $[Zero], Info, N$;

Algorithm 6.2 describes an additional verification step which checks the uniqueness of the zero enclosed in the interval $[y]$. It can be used for intervals which have not yet been guaranteed to enclose a unique zero. This is done according to condition

3 of Theorem 6.1 by applying interval Newton steps including an epsilon inflation of the iterates $[y]$, which can help to verify zeros lying on the edge of $[y]$. We use $k_{\max} = 10$ as the maximum number of iterations, $\varepsilon = 0.25$ as the starting value for the epsilon inflation, and a factor of 8 to increase ε within the iterations. It turned out that these are good values for minimizing the effort if no verification is possible (see also [46]).

Algorithm 6.2: VerificationStep $(f, [y], yUnique)$ {Procedure}

1. $k_{\max} := 10$; $k := 0$; $[y_{\text{in}}] := [y]$; $\varepsilon := 0.25$; $yUnique := false$ {Initializations}
2. **while** (**not** $yUnique$) **and** $(k < k_{\max})$ **do** {Do $k_{\max}$ loops to achieve inclusion}
   (a) $[y_{\text{old}}] := [y] \bowtie \varepsilon$; {Epsilon inflation of $[y]$}
   (b) **if** $0 \in f'([y_{\text{old}}])$ **then exit**$_{\text{while-loop}}$; {No verification possible}
   (c) $k := k + 1$; $c := m([y_{\text{old}}])$;
   (d) $[y] := c - f_\diamond(c)/f'([y_{\text{old}}])$; {Interval Newton step}
   (e) **if** $[y] \cap [y_{\text{old}}] = \emptyset$ **then exit**$_{\text{while-loop}}$; {No root}
   (f) $yUnique := ([y] \overset{\circ}{\subset} [y_{\text{old}}])$; {Inner inclusion $\Rightarrow$ uniqueness}
   (g) $[y] := [y] \cap [y_{\text{old}}]$; {Intersection with old value}
   (h) **if** $[y] = [y_{\text{old}}]$ **then** $\varepsilon := \varepsilon \cdot 8$; {Increase ε}
3. **if not** $yUnique$ **then** $[y] := [y_{\text{in}}]$; {Reset $[y]$ to starting value}
4. **return** $[y], yUnique$;

Algorithm 6.3 now combines Algorithm 6.1 and Algorithm 6.2 to compute enclosures for all zeros of the function f within the input interval $[x]$ and tries to prove the local uniqueness of a zero in each enclosure computed. The desired accuracy (relative diameter) of the interval enclosures is specified by the input parameter ε. 1 ulp accuracy is chosen if the specified value of ε is too small (for example 0). The enclosures for the zeros of f are returned in the interval vector $[Zero]$. The corresponding information on the local uniqueness of the zeros are returned in the Boolean vector *Info*. The number of enclosures computed is returned in the integer variable N.

We use a function called **CheckParameters** as an abbreviation for the error checks for the parameters of **AllZeros** which are necessary in an implementation. If no error occurs, **AllZeros** delivers the N enclosures $[Zero]_i$, $i = 1, 2, \ldots, N$, and for each $i = 1, 2, \ldots, N$,

if $Info_i = true$, then $[Zero]_i$ encloses a locally unique zero of f,
if $Info_i = false$, then $[Zero]_i$ may enclose a zero of f.

If $N = 0$, then it is guaranteed that there is no zero of f in the starting interval $[x]$.

Algorithm 6.3: AllZeros (f, $[x]$, ε, $[Zero]$, $Info$, N, Err) {Procedure}

1. Err := CheckParameters;
2. **if** $Err \neq$ "No Error" **then return** Err;
3. $N := 0$;
4. Set ε to "1 ulp accuracy" if ε is too small;
5. XINewton (f, $[x]$, ε, *false*, $[Zero]$, $Info$, N);
6. **for** $i := 1$ **to** N **do**
   if $Info_i \neq$ *true* **then** VerificationStep (f, $[Zero]_i$, $Info_i$);
7. **return** $[Zero]$, $Info$, N, Err;

Applicability of the Algorithm

To keep our algorithm and implementation as easy to use as possible, Algorithm 6.1 uses only the very simple stopping criteria

$$d_{\mathrm{rel}}([y_p]_i) < \varepsilon.$$

Hence, the method reduces to a bisection method if the interval Newton steps do not improve the actual interval iterate. For more sophisticated stopping criteria, see [28] for example. On the other hand, if the actual interval iterate $[y]$ satisfies the condition

$$0 \notin f'([y]),$$

then the asymptotic rate of convergence to a zero of f in $[y]$ is quadratic.

The algorithm cannot verify the existence and the uniqueness of a multiple zero x^* of f in $[y]$. Nevertheless, the zero is bounded to the desired accuracy specified by ε. In this case, the corresponding component of the *Info*-vector is set to the value *false*. When the zero lies exactly at a splitting point, that zero is enclosed in two different intervals as a consequence of the bisection of the intervals.

The algorithm is well suited for parallel computers due to its natural parallelism given by the splitting of the intervals by the extended interval Newton step or by the bisection.

6.3 Implementation and Examples

6.3.1 PASCAL–XSC Program Code

First, we list our implementation of the operations of the extended interval arithmetic needed in the extended interval Newton step (see Section 3.3). Newton's method is the only application we make of extended interval arithmetic. Therefore, we provide only those operations on extended intervals we require for our implementation of Newton's method. An implementation of iteration (6.3) requires only operators for

interval	/	interval	$\rightarrow$	extended interval,
real	$-$	extended interval	$\rightarrow$	extended interval, and
interval	$\cap$	extended interval	$\rightarrow$	0, 1, or 2 intervals.

Second, we list the implementation of Algorithm 6.3 and its subalgorithms. We use module *ddf_ari* for differentiation arithmetic given in Chapter 5.

6.3.1.1 Module xi_ari

This module supplies the type *xinterval* representing a normal interval, a single extended interval, or a pair of extended intervals. The record component *kind* specifies these cases by

kind	interval form $(a, b \in I\!R)$
Finite	$[a,b], \quad a \leq b,$
PlusInfty	$[a,\infty],$
MinusInfty	$[-\infty,a],$
Double	$[-\infty,a] \cup [b,\infty], \quad a \leq b.$

The case $[-\infty,\infty]$ is represented by $[-\infty,a] \cup [a,\infty]$ for simplicity.

The global function *EmptyIntval* delivers an irregular value of type *interval* (i.e. an interval with infimum greater than supremum) representing the empty set and serving as result of the intersection of two disjoint (extended) intervals.

Overloaded operators in PASCAL-XSC are distinguished by their operands. Hence, the operator / could not be used for the extended interval division, and we used the operator **div**.

The operator ** for the intersection of an interval with an extended interval pair of type *xinterval* delivers a result of type *ivector*[1..2], where one or both of the interval components may be equal to *EmptyIntval*.

```
{----------------------------------------------------------------------------}
{ Purpose: Definition of an extended interval arithmetic which allows the    }
{    division by an interval containing zero.                                }
{ Method: Overloading of operators for arithmetic and lattice operations     }
{    of data type 'xinterval'.                                               }
{ Global types, operators, and functions:                                    }
{    types      KindType         : component type of extended intervals      }
{               xinterval        : data type for extended intervals          }
{    operators  div, -, **       : operators of extended interval arithmetic }
{    function   EmptyIntval      : delivers empty set as irregular interval  }
{----------------------------------------------------------------------------}
module xi_ari;

use i_ari;  { Interval arithmetic }

{----------------------------------------------------------------------------}
{ Global type definitions                                                    }
{----------------------------------------------------------------------------}
{ An extended interval 'x', represented by the type 'xinterval', is defined  }
```

```
{ according to the following rules (a <= b):                                  }
{                                                                              }
{ x = [a, b]               :  x.kind = Finite,      x.inf = a, x.sup = b        }
{ x = [a, +oo]             :  x.kind = PlusInfty,   x.inf = a, x.sup undefined  }
{ x = [-oo, a]             :  x.kind = MinusInfty,  x.sup = a, x.inf undefined  }
{ x = [-oo, a] v [b, +oo]  :  x.kind = Double,      x.inf = b, x.sup = a        }
{ x = [-oo, +oo]           :  x.kind = Double,      x.inf = a, x.sup = a        }
{                                                                              }
{ In this definition, 'v' stands for the set union and 'oo' for infinity.      }
{------------------------------------------------------------------------------}
global type
  KindType     = (Finite, PlusInfty, MinusInfty, Double);
  xinterval    = global record                        { Extended intervals  }
                   kind      : KindType;               { according to the    }
                   inf, sup  : real;                   { definition above    }
                 end;                                  {---------------------}

{------------------------------------------------------------------------------}
{ Function 'EmptyIntval' delivers an empty interval (empty set)                }
{------------------------------------------------------------------------------}
global function EmptyIntval : interval;
begin
  EmptyIntval.inf :=  999999999;     { Definition of an irregular interval  }
  EmptyIntval.sup := -999999999;     { EmptyIntval = [999999999,-999999999] }
end;                                 {--------------------------------------}

{------------------------------------------------------------------------------}
{ Operators -, 'div', and ** for extended intervals                            }
{------------------------------------------------------------------------------}
{ Subtraction of an extended interval 'B' from a real value 'a'. }
{----------------------------------------------------------------}
global operator - (a : real; B : xinterval) difference : xinterval;
  var
    D : xinterval;
  begin
    case B.kind of
      Finite     : begin                                  { D = [D.inf, D.sup] }
                     D.kind:= Finite;                     {--------------------}
                     D.inf := a -< B.sup;
                     D.sup := a -> B.inf;
                   end;
      PlusInfty  : begin                                      { D = [inf, +oo] }
                     D.kind:= MinusInfty;                     {----------------}
                     D.sup := a -> B.inf;
                   end;
      MinusInfty : begin                                      { D = [-oo, sup] }
                     D.kind:= PlusInfty;                      {----------------}
                     D.inf := a -< B.sup;
                   end;
      Double     : begin                  { D = [-oo, D.sup] v [D.inf, +oo] }
                     D.kind:= Double;     {---------------------------------}
                     D.inf := a -< B.sup;
                     D.sup := a -> B.inf;
                     if (D.inf < D.sup) then
                       D.inf := D.sup;
                   end;
    end;
    difference := D;
  end;

{----------------------------------------------------------------}
{ Extended interval division 'A / B', where 0 in 'B' is allowed. }
{----------------------------------------------------------------}
global operator div (A, B : interval) quotient : xinterval;
```

```
  var
    C : interval;
    Q : xinterval;
  begin
    if (0 in B) then
      begin
        if ((inf(A) < 0) and (0 < sup(A))) or (A = 0) or (B = 0)  then
          begin                       { Q = [-oo, +oo] = [-oo, 0] v [0, +oo] }
            Q.kind:= Double;          {--------------------------------------}
            Q.sup := 0;
            Q.inf := 0;
          end
        else if ((sup(A) <= 0) and (sup(B) = 0)) then
          begin                                           { Q = [Q.inf, +oo] }
            Q.kind:= PlusInfty;                           {------------------}
            Q.inf := sup(A) /< inf(B);
          end
        else if ((sup(A) <= 0) and (inf(B) < 0) and (sup(B) > 0)) then
          begin                            { Q = [-oo, Q.sup] v [Q.inf, +oo] }
            Q.kind:= Double;               {---------------------------------}
            Q.sup := sup(A) /> sup(B);
            Q.inf := sup(A) /< inf(B);
          end
        else if ((sup(A) <= 0) and (inf(B) = 0)) then
          begin                                           { Q = [-oo, Q.sup] }
            Q.kind:= MinusInfty;                          {------------------}
            Q.sup := sup(a) /> sup(b);
          end
        else if ((inf(A) >= 0) and (sup(B) = 0)) then
          begin                                           { Q = [-oo, Q.sup] }
            Q.kind:= MinusInfty;                          {------------------}
            Q.sup := inf(A) /> inf(B);
          end
        else if ((inf(A) >= 0) and (inf(B) < 0) and (sup(B) > 0)) then
          begin                            { Q = [-oo, Q.sup] v [Q.inf, +oo] }
            Q.kind:= Double;               {---------------------------------}
            Q.sup := inf(A) /> inf(B);
            Q.inf := inf(A) /< sup(B);
          end
        else if ((inf(A) >= 0) and (inf(B) = 0)) then
          begin                                           { Q = [Q.inf, +oo] }
            Q.kind:= PlusInfty;                           {------------------}
            Q.inf := inf(A) /< sup(B);
          end
        else ;
      end { 0 in B }
    else  { not (0 in B) }
      begin                                             { Q = [C.inf, C.sup] }
        C     := A / B;                                 {--------------------}
        Q.kind:= Finite;
        Q.inf := C.inf;
        Q.sup := C.sup;
      end;
    quotient := Q;
  end;

{----------------------------------------------------------------------------}
{ Intersection of an interval 'X' and an extended interval 'Y'. The result   }
{ is given as a pair (vector) of intervals, where one or both of them can    }
{ be empty intervals ('EmptyIntval').                                        }
{----------------------------------------------------------------------------}
global operator ** (X: interval; Y: xinterval) Intersect : ivector[1..2];
  var
    H : interval;
```

```
begin
  Intersect[1]:= EmptyIntval;
  Intersect[2]:= EmptyIntval;

  case Y.kind of
    Finite     : begin                       { [X.inf,X.sup]**[Y.inf,Y.sup] }
                   H := intval(Y.inf,Y.sup); {------------------------------}
                   if not (X >< H) then
                     Intersect[1]:= X ** H;
                 end;
    PlusInfty  : if (X.sup >= Y.inf) then    { [X.inf,X.sup] ** [Y.inf,+oo] }
                 begin                       {------------------------------}
                   if (X.inf > Y.inf) then
                     Intersect[1]:= X
                   else
                     Intersect[1]:= intval(Y.inf, X.sup);
                 end;
    MinusInfty: if (Y.sup >= X.inf) then     { [X.inf,X.sup] ** [-oo,Y.sup] }
                 begin                       {------------------------------}
                   if (X.sup < Y.sup) then
                     Intersect[1]:= X
                   else
                     Intersect[1]:= intval(X.inf, Y.sup)
                 end;
    Double     : if ((X.inf <= Y.sup) and (Y.inf <= X.sup)) then
                   begin
                     Intersect[1]:= intval(X.inf, Y.sup);  { X**[-oo,Y.sup] }
                     Intersect[2]:= intval(Y.inf, X.sup);  { X**[Y.inf,+oo] }
                   end                                     {----------------}
                 else if (Y.inf <= X.sup) then
                   begin                       { [X.inf,X.sup]**[Y.inf,+oo] }
                     if (X.inf >= Y.inf) then  {----------------------------}
                       Intersect[1]:= X
                     else
                       Intersect[1]:= intval(Y.inf, X.sup);
                   end
                 else if (X.inf <= Y.sup) then
                   begin                       { [X.inf,X.sup]**[-oo,Y.sup] }
                     if (X.sup <= Y.sup) then  {----------------------------}
                       Intersect[1]:=  X
                     else
                       Intersect[1]:=  intval(X.inf, Y.sup);
                   end;
  end; { CASE kind OF ... }
end; { OPERATOR ** ... }

{----------------------------------------------------------------------------}
{ Module initialization part                                                 }
{----------------------------------------------------------------------------}
begin
  { Nothing to initialize }
end.
```

6.3.1.2 Module nlfzero

The following module supplies the global routines *AllZeros* (the implementation of Algorithm 6.3) and the function *AllZerosErrMsg* to get an error message for the error code returned by *AllZeros*. The procedures *XINewton* and *VerificationStep* are defined locally. All derivative evaluations are done with the help of the differentiation arithmetic *ddf_ari* (see Chapter 5).

The procedure *AllZeros* uses the *DerivType* function f and the starting interval *Start* as input parameters and stores all computed enclosures in the interval vector *ZeroVector* which is also passed to and from the procedure *XINewton*. Before storing a further result interval in *ZeroVector*, we check whether the vector has free components. If this is not the case, the corresponding error code is returned together with the complete *ZeroVector* containing all solutions already computed. The user must then increase the upper index bound of *ZeroVector* to compute *all* zeros.

The same applies to the information about the uniqueness of the roots located, stored in the Boolean vector *InfoVector*. Hence, the user must declare both vectors with lower index bound equal to 1 and with upper index bounds which are equal. These conditions, as well as the condition *Epsilon* $\geq$ *MinEpsilon* are checked at the beginning of procedure *AllZeros*. *Epsilon* in our program corresponds to the parameter ε in the algorithms, and *MinEpsilon* corresponds to 1 ulp accuracy.

```
{----------------------------------------------------------------------------}
{ Purpose: Computing enclosures for all zeros of a continuously              }
{    differentiable one-dimensional, real-valued function.                   }
{ Method: Extended interval Newton's method.                                 }
{ Global procedures and functions:                                           }
{    procedure AllZeros(...)       : computes enclosures for all zeros       }
{    function  AllZerosErrMsg(..) : delivers an error message text           }
{----------------------------------------------------------------------------}
module nlfzero;

use
  b_util,     { Boolean utilities              }
  i_ari,      { Interval arithmetic            }
  i_util,     { Interval utilities             }
  xi_ari,     { Extended interval arithmetic }
  ddf_ari;    { Differentiation arithmetic     }

{----------------------------------------------------------------------------}
{ Error codes used in this module.                                           }
{----------------------------------------------------------------------------}
const
  NoError       = 0;  { No error occurred.                                   }
  lbZeroVecNot1 = 1;  { Lower bound of variable ZeroVector is not equal to 1.}
  lbInfoVecNot1 = 2;  { Lower bound of variable InfoVector is not equal to 1.}
  VecsDiffer    = 3;  { Bounds of ZeroVector and InfoVector do not match.    }
  VecTooSmall   = 4;  { ZeroVector too small. Not all zeros can be stored.   }

{----------------------------------------------------------------------------}
{ Error messages depending on the error code.                                }
{----------------------------------------------------------------------------}
global function AllZerosErrMsg ( Err : integer ) : string;
var
  Msg : string;
begin
  case Err of
    NoError       : Msg := '';
    lbZeroVecNot1 : Msg := 'Lower bound of ZeroVector is not equal to 1';
    lbInfoVecNot1 : Msg := 'Lower bound of InfoVector is not equal to 1';
    VecsDiffer    : Msg := 'Bounds of ZeroVector and InfoVector do not match';
    VecTooSmall   : Msg := 'Not all zeros found. ZeroVector is too small';
    else          : Msg := 'Code not defined';
  end;
  if (Err <> NoError) then Msg := 'Error: ' + Msg + '!';
  AllZerosErrMsg := Msg;
end;
```

```
{----------------------------------------------------------------------------}
{ Purpose: Recursive procedure for the execution of the extended interval    }
{    Newton's method for the function 'f'.                                   }
{ Parameters:                                                                }
{    In     : 'f'          : must be declared for the 'DerivType' to enable  }
{                            the internal use of the differentiation         }
{                            arithmetic 'ddf_ari'.                           }
{             'y'          : specifies the starting interval.                }
{             'Epsilon'    : specifies the desired relative accuracy         }
{                            (interval diameter) of the result intervals.    }
{             'yUnique'    : signals whether it is already verified that the }
{                            actual interval 'y' contains a unique zero.     }
{    Out    : 'ZeroVector' : stores the enclosures for the zeros of 'f'.     }
{             'InfoVector' : stores the corresponding information on the     }
{                            uniqueness of the zero in each enclosure.       }
{    In/Out : 'ZeroNo'     : represents the number of the zero computed last }
{                            (in) and the total number of enclosures         }
{                            computed at the end of the recursion (out).     }
{ Description:                                                               }
{    The procedure 'XINewton' is recursively called whenever the extended    }
{    interval division results in a bisection of the actual interval 'y' in  }
{    two intervals 'y1' and 'y2', and the tolerance condition is not ful-    }
{    filled yet. Otherwise the enclosures for the zeros of 'f' are stored in }
{    the interval vector 'ZeroVector', the corresponding information on the  }
{    uniqueness of the zero in each enclosure is stored in the Boolean       }
{    vector 'InfoVector'.                                                    }
{----------------------------------------------------------------------------}
procedure XINewton (function f (x: DerivType) : DerivType;
                                            y : interval;
                                      Epsilon : real;
                                      yUnique : boolean;
                        var        ZeroVector : ivector;
                        var        InfoVector : bvector;
                        var            ZeroNo : integer);
var
  z             : xinterval;
  fc, fy, dfy   : interval;
  yp            : ivector[1..2];
  c             : real;
  i             : integer;
begin
  dfEval(f, y, fy, dfy);                     { Compute f(y) and f'(y)            }
  if 0 in fy then                            { Start if 0 in f(y), else do nothing }
  begin                                      {-----------------------------------}
    c := mid(y);
    fEval(f, intval(c), fc);  { Compute f(c) and f'(c).                         }
    z := c - fc div dfy;      { Extended interval Newton step.                  }
    yp:= y ** z;              { Intersect interval y and extended interval z    }
                              { resulting in two intervals yp[1] and yp[2].     }
                              {-------------------------------------------------}
    if yp[1] = y then                                      { Stagnation, so 'y' }
    begin                                                  { must be bisected.  }
      yp[1]:= intval(y.inf,c);  yp[2]:= intval(c,y.sup);{--------------------}
    end;

    if (yp[1] <> EmptyIntval) and (yp[2] = EmptyIntval) then
      yUnique := yUnique or (yp[1] in y)    { Inner inclusion ===> uniqueness }
    else                                    {---------------------------------}
      yUnique := false;

    for i:=1 to 2 do
      if yp[i] <> EmptyIntval then
      begin
```

```
        if (RelDiam(yp[i]) < Epsilon) then
          begin { No more Newton steps }
            fEval(f, yp[i], fy);                  { Compute f(yp[i])          }
            if (0 in fy) then                     { Store enclosure and info  }
            begin                                 {---------------------------}
              ZeroNo := ZeroNo + 1;
              if (ZeroNo <= ub(ZeroVector)) then
              begin
                ZeroVector[ZeroNo] := yp[i];        { Store enclosure of zero }
                InfoVector[ZeroNo] := yUnique;      { Store uniqueness info   }
              end;                                  {-------------------------}
            end;
          end
        else { Recursive call of 'XINewton' for interval 'yp[i]' }
         XINewton(f,yp[i],Epsilon,yUnique,ZeroVector,InfoVector,ZeroNo);
      end;
  end;
end;

{----------------------------------------------------------------------------}
{ Purpose: Execution of a verification step including the use of an epsilon  }
{    inflation.                                                               }
{ Parameters:                                                                 }
{    In     : 'f'       : function of 'DerivType'.                            }
{    Out    : 'yUnique' : returns 'true' if the verification is successful.   }
{    In/Out : 'y'       : interval enclosure to be verified.                  }
{ Description: This procedure checks the uniqueness of the zero enclosed in   }
{    the variable 'y' by an additional verification step including the use    }
{    of an epsilon inflation of the iterates.                                 }
{----------------------------------------------------------------------------}
procedure VerificationStep (function f (x: DerivType) : DerivType;
                                var                 y : interval;
                                var           yUnique : boolean);
const
  kmax = 10;
var
  yIn, yOld, fc, fy, dfy : interval;
  c, eps                 : real;
  k                      : integer;
begin
  k := 0;  yIn := y;  eps:= 0.25;         { Initializations                     }
  yUnique := false;                       {-------------------------------------}
  while (not yUnique) and (k < kmax) do { Do kmax loops to achieve inclusion }
  begin                                   {-------------------------------------}
    yOld := blow(y, eps);                 { Epsilon inflation of 'y'            }
    dfEval(f, yOld, fy, dfy);             {-------------------------------------}

    if 0 in dfy then
      k := kmax                                     { No verification possible }
    else                                            {--------------------------}
      begin                                     { Perform interval Newton step }
        k := k+1;  c := mid(yOld);              {------------------------------}
        fEval(f,intval(c),fc);
        y := c - fc / dfy;

        if y >< yOld then                           { No verification possible }
          k := kmax                                 {--------------------------}
        else
          begin
            yUnique := y in yOld;              { Inner inclusion ===> uniqueness }
            y       := y ** yOld;              { Intersection with old value     }
            if y = yOld then                   {---------------------------------}
              eps := eps * 8                   { Increase the value of 'eps'.    }
          end;                                 {---------------------------------}
```

```
    end;
  end;
  if not yUnique then y := yIn;
end;

{----------------------------------------------------------------------------}
{ Purpose: Computation of enclosures for all zeros of a continuously         }
{    differentiable one-dimensional, real-valued function.                   }
{ Parameters:                                                                }
{    In     : 'f'            : objective function, must be declared for the }
{                              'DerivType' to enable the internal use of     }
{                              the differentiation arithmetic 'ddf_ari'.     }
{             'Start',       : specifies the starting interval.              }
{             'Epsilon'.     : specifies the desired relative accuracy       }
{                              (interval diameter) of the result intervals. }
{    Out    : 'ZeroVector'   : stores and returns the enclosures for the     }
{                              zeros of 'f'.                                 }
{             'InfoVector'   : stores the corresponding information on the  }
{                              uniqueness of the zeros in these enclosures. }
{             'NumberOfZeros' : returns the number of enclosures computed.   }
{             'Err'          : returns an error code.                        }
{ Description:                                                               }
{    The enclosures for the zeros of 'f' are computed by calling procedure   }
{    'XINewton'. Then an additional verification step is applied to those    }
{    enclosures which have not been verified.                                }
{    If an error occurs the value of 'Err' is different from 0.              }
{----------------------------------------------------------------------------}
global procedure AllZeros (function f (x: DerivType) : DerivType;
                                                  Start : interval;
                                                Epsilon : real;
                                  var        ZeroVector : ivector;
                                  var        InfoVector : bvector;
                                  var     NumberOfZeros : integer;
                                  var               Err : integer);
var
  i          : integer;
  MinEpsilon : real;
begin
  NumberOfZeros:= 0;
  if (lb(ZeroVector) <> 1) then        { Check index bounds of result vectors }
    Err := lbZeroVecNot1               {--------------------------------------}
  else if (lb(InfoVector) <> 1) then
    Err := lbInfoVecNot1
  else if (ub(InfoVector) <> ub(ZeroVector)) then
    Err := VecsDiffer
  else
    begin                             { Start extended interval Newton method }
      Err := NoError;                 {---------------------------------------}
      MinEpsilon:= succ(1.0) - 1.0;    { Relative machine accuracy (1 ulp)    }
                                       {--------------------------------------}
      if (Epsilon < MinEpsilon) then Epsilon := MinEpsilon;  { Set 'Epsilon' }
                                                             { to 1 ulp acc. }
      XINewton(f, Start, Epsilon, false,                     {---------------}
               ZeroVector, InfoVector, NumberOfZeros);

                                           { Check if there are more zeros }
      if ub(ZeroVector) < NumberOfZeros then { than storage space           }
      begin                                {-------------------------------}
        Err:= VecTooSmall;  NumberOfZeros:= ub(ZeroVector);
      end;
                                                        { Verification step  }
      for i:=1 to NumberOfZeros do                      { for the enclosures }
        if InfoVector[i] <> true then                   {--------------------}
          VerificationStep(f,ZeroVector[i],InfoVector[i]);
```

```
    end;
end;

{----------------------------------------------------------------------------}
{ Module initialization part                                                 }
{----------------------------------------------------------------------------}
begin
  { Nothing to initialize }
end.
```

6.3.2 Example

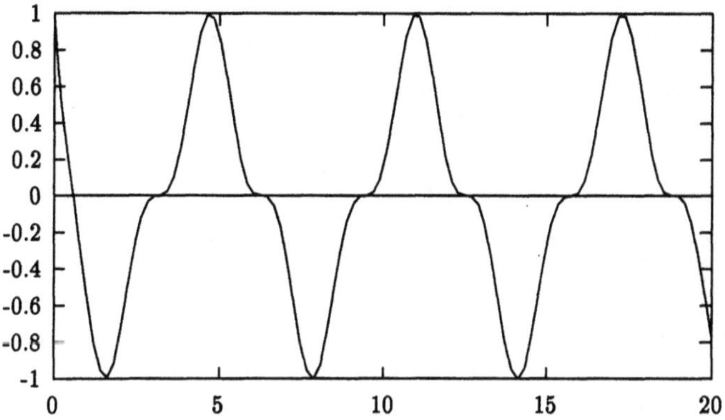

Figure 6.3: Function $f(x) = \mathrm{e}^{-3x} - \sin^3 x$

Our sample program uses *nlfzero* to compute all zeros of the function

$$f(x) = \mathrm{e}^{-3x} - \sin^3 x$$

in a specified starting interval (see Figure 6.3). The first positive root near $x = \frac{\pi}{2}$ is easy to locate numerically. As $x \longrightarrow \infty$, roots of f approach the triple roots of $\sin^3 x$ at $\frac{\pi}{2} + n\pi$. We first have to define the function f for the type *DerivType*. Then we use a procedure *compute* with the function f and a string (a description of the function) as parameters. This procedure prompts you to enter the necessary data for the call of procedure *AllZeros*. If we add further function definitions in our program, we can apply this procedure in the same way using another function as parameter.

```
{----------------------------------------------------------------------------}
{ This program uses module 'nlfzero' to compute the zeros of the function    }
{                                                                            }
{              exp(-3*x) - power(sin(x), 3)                                  }
{                                                                            }
{ A starting interval and a tolerance must be entered.                       }
{----------------------------------------------------------------------------}
program nlfz_ex;
use
  i_ari,     { Interval arithmetic       }
```

```
  ddf_ari,  { Differentiation arithmetic }
  b_util,   { Boolean utilities          }
  nlfzero;  { Nonlinear function zeros   }
const
  n = 20;   { Maximum number of zeros to be computed }

function f (x: DerivType) : DerivType;
begin
  f:= exp(-3*x) - power(sin(x),3);
end;

{----------------------------------------------------------------------------}
{ Procedure for prompting and reading information to call the procedure      }
{ 'AllZeros'. This procedure must be called with the function 'f' and a      }
{ string 'Name' containing a textual description of that function.           }
{----------------------------------------------------------------------------}
procedure compute (function f(x:DerivType): DerivType; Name: string);
var
  SearchInterval              : interval;
  Tolerance                   : real;
  Zero                        : ivector[1..n];
  Unique                      : bvector[1..n];
  NumberOfZeros, i, ErrCode   : integer;
begin
  writeln('Computing all zeros of the function  ', Name);
  write('Search interval      : ');  read(SearchInterval);
  write('Tolerance (relative) : ');  read(Tolerance);
  writeln;
  AllZeros(f,SearchInterval,Tolerance,Zero,Unique,NumberOfZeros,ErrCode);
  for i:=1 to NumberOfZeros do
  begin
    write(Zero[i]);
    if unique[i] then
      writeln(' encloses a locally unique zero!')
    else
      begin
        writeln(' may contain a zero');
        write(' ':52);  writeln(' (not verified unique)!')
      end;
  end;
  if ErrCode <> 0 then writeln(AllZerosErrMsg(ErrCode));
  writeln;  writeln(NumberOfZeros:1, ' interval enclosure(s)');  writeln;
end;

begin
  compute(f, 'EXP(-3x)-POWER(SIN(x),3)');
end.
```

If we execute this program, we get the following runtime output:

```
Computing all zeros of the function  EXP(-3x)-POWER(SIN(x),3)
Search interval      : [0,20]
Tolerance (relative) : 1e-10

[      5.88532743979E-001,      5.88532743985E-001 ] encloses a unique zero!
[    3.0963639324106E+000,    3.0963639324107E+000 ] encloses a unique zero!
[  6.285049273382585E+000,  6.285049273382588E+000 ] encloses a unique zero!
[      9.42469725468E+000,      9.42469725478E+000 ] encloses a unique zero!
[     1.256637410168E+001,     1.256637410170E+001 ] encloses a unique zero!
[     1.570796311724E+001,     1.570796311725E+001 ] encloses a unique zero!
[       1.8849555927E+001,       1.8849555929E+001 ] encloses a unique zero!

7 interval enclosure(s)
```

The seven interval enclosures are disjoint, so we can conclude that there are seven roots of our sample function within the specified starting interval.

6.3.3 Restrictions and Hints

The function f whose root we seek must be expressible as a finite sequence of arithmetic operations and elementary functions supported by the differentiation arithmetic module *ddf_ari* described in Chapter 5.

As already mentioned, the procedure *AllZeros* stores all enclosures in the interval vector *ZeroVector*, which must be of sufficient length. If the first run of *AllZeros* is not able to compute all zeros because there are not enough locations in *ZeroVector*, then *AllZeros* must be called again. If you wish, you may then use a smaller starting interval, namely the original starting interval *without* the area already fully treated which is given by the interval hull of the computed solutions. This works, because the recursive implementation of our method treats the starting interval from the left to the right.

The method is not very fast if a very small value of ε (*Epsilon*) is used and the interval Newton step does not improve the actual iterates because of rounding and overestimation effects of the machine interval arithmetic. In this case the method is equivalent with a bisection method.

In *XINewton*, the evaluation of the function with differentiation arithmetic can cause a runtime error if the interval argument of an elementary function does not lie in the domain specified for this interval function (see [65]) or if a division by an interval containing zero occurs, even though the domain is mathematically valid. For example, $\sqrt{1 + x \cdot x}$ for $x \in [-2, 2]$ attempts to take the square root of a negative number due to the overestimation effects of interval arithmetic (see Section 3.1). To get rid of these errors, you may rewrite f in a different (but mathematically equivalent) form ($\sqrt{1 + x^2}$ in this example), or you may try to split the starting interval in several parts and call *AllZeros* for these parts.

The rules for getting true enclosures in connection with conversion errors (see Section 3.7) also apply here. If you want to compute enclosures for the zeros of the function $f(x) = (x - \frac{1}{3}) \cdot (x - 0.1)$, for example, be careful that the values $\frac{1}{3}$ and 0.1 are enclosed in machine intervals when computing interval evaluations of f. A corresponding implementation of f could be

```
function f (x: DerivType) : DerivType;
begin
  f:= (x - intval(1) / 3) * (x - intval(1) / 10);
end;
```

6.4 Exercises

Exercise 6.1 Use the procedure *AllZeros* to compute the zero of the function $f(x) = (x - 1)^3$ with starting intervals $[-3, 4]$ and $[-3, 5]$, respectively, to demonstrate the behavior of our implementation when computing multiple zeros or zeros lying on the splitting point of the bisection step.

Exercise 6.2 In [14, Section 2.1], Dennis and Schnabel write:

> *"It is unlikely, that there will ever be a wonderful general purpose computer routine that would tell us for* $f_1(x) = x^4 - 12x^3 + 47x^2 - 60x$, $f_2(x) = f_1(x) + 24$, *and* $f_3(x) = f_1(x) + 24.1$ *the roots of* f_1 *are* $x = 0, 3, 4,$ *and 5, the real roots of* f_2 *are* $x = 1$ *and* $x = 0.888305\ldots$, *and* f_3 *has no real roots. In general, the question of existence and uniqueness are beyond the capabilities one can expect of algorithms that solve nonlinear problems."*

Use our *AllZeros* routine to compute the zeros of f_1, f_2, and f_3 in the starting interval $[-3, 8]$ and to demonstrate that the wonderful routine exists!

Exercise 6.3 Use the procedure *AllZeros* to verify an approximation of a zero of the function $f(x) = \sin\frac{1}{x}$. Compute the approximation using an arbitrary method of your choice (e.g. Newton's method). Then apply *AllZeros* to a neighborhood interval of this approximation.

6.5 References and Further Reading

The method we discussed in this chapter is an *a priori* method because the iteration starts with a (possibly large) interval enclosing all the solutions which have to be found. Here, the iterates of the method are subintervals of the previous iterates. There are also methods for finding (and bounding) one single solution of a nonlinear equation called *a posteriori* methods. These methods start with an approximation of a zero and apply a test procedure for a neighborhood interval of the approximation to verify a zero within that interval. Our method presented in this chapter can also be applied to verify such an approximation, if we start the process with a small interval containing the approximation (Exercise 6.3).

Many authors have dealt with the problem of computing enclosures of zeros of nonlinear functions. For further references in the field of *a priori* methods, see [3], [8], [28], [62], [63], [64], [71], or [72], for example. For *a posteriori* methods, see [11], [37], [38], [50], [53], [55], [60], [64], [77], [78], or [79], for example.

Chapter 7

Global Optimization

We want to find the global minimum in an interval $[x]$ of a function f that may have many local minima. We want to compute the minimum value of f and the point(s) at which the minimum value is attained. This is a very difficult problem for classical methods because narrow, deep valleys may escape detection. In contrast, the interval method presented here evaluates f on a continuum of points, including those points that are not finitely representable, so valleys, no matter how narrow, are recognized with certainty. Further, interval techniques often can reject large regions in which the optimum can be guaranteed not to lie, so they can be faster overall than classical methods for many problems.

In classical numerical analysis, global optimization methods proceed by iteration starting from some approximate trial points. Classical optimization methods sample the objective function at only a finite number of points and cannot guarantee that the function does not have some unexpectedly small values in between these trial points.

In this chapter, we consider an algorithm based on the method of Hansen [25] for computing all solutions of the global unconstrained optimization problem. The algorithm computes enclosures for all global minimizers in a given interval and for the global minimum value of a twice continuously differentiable function.

7.1 Theoretical Background

We address the problem of finding all solutions x^* of

$$\min_{x \in [x]} f(x) \tag{7.1}$$

for a twice continuously differentiable function $f : I\!R \to I\!R$ and $[x] \in II\!R$. We wish to find the set of all global minimizers x^* and the global minimum value $f^* = f(x^*)$. The existence of a global minimizer is assured since f is a continuous function on a compact set, but need not be unique. In fact, the minimizers may include continua of points.

Our algorithm for problem (7.1) is based on the method of Hansen (see [25] or [28]) and the modifications of Ratz (see [73]). Starting from the initial interval $[x]$, our algorithm subdivides $[x]$ and stores the subintervals $[y] \subset [x]$ in a list L. Subintervals which are guaranteed not to contain a global minimizer of f are discarded from that list, while the other subintervals get subdivided again until the desired accuracy (relative diameter) of the intervals in the list is achieved. The key to the speed of this method is its use of several tests to discard subintervals in which no minimizer may occur:

- midpoint test,
- monotonicity test,
- concavity test, and
- interval Newton step.

In the following sections, we consider each of these methods in detail. We use the notation $\underline{f_y}$ as abbreviation for the lower interval bound of the interval function evaluation $[f_y] := f_{\square}([y])$ for $[y] \in I\!R$.

7.1.1 Midpoint Test

If we are able to determine an upper bound $\widetilde{f}$ for the global minimum value f^*, then we can delete all subintervals $[y]$ for which

$$\underline{f_y} > \widetilde{f} \geq f^*. \tag{7.2}$$

The midpoint test first determines or improves such an upper bound for f^*. Initially, let $\widetilde{f} = +\infty$. We choose an interval $[y]$ out of the list L which satisfies $\underline{f_y} \leq \underline{f_z}$ for all intervals $[z]$ in the list L. That is, $[y]$ has the smallest lower bound for the range of f. Hence, it is a likely candidate to contain a minimizer. Let $c = m([y])$ (or any other point in $[y]$), and compute $\widetilde{f} = \min\{f(c), \widetilde{f}\}$. Such an upper bound can also be computed on a computer when rounding errors occur: we compute $f_{\diamond}(c)$ and use the upper bound of the resulting interval as the possibly new value $\widetilde{f}$.

Now, with a possibly improved (decreased) value of $\widetilde{f}$, we can discard all intervals $[z]$ from the list L for which $\widetilde{f} < \underline{f_z}$. The midpoint test is relatively inexpensive, and it often allows us to discard from consideration large portions of the original interval $[x]$. Figure 7.1 illustrates this procedure, which deletes the intervals $[y]_2$,

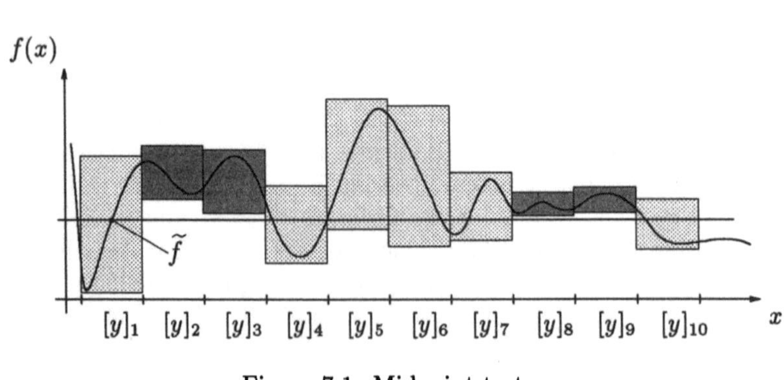

Figure 7.1: Midpoint test

$[y]_3$, $[y]_8$, and $[y]_9$ in this special case. The test remains valid if an arbitrary $c' \in [y]$ is used instead of $c = m([y])$. Our algorithm could be extended by using a local approximate search to find a $c' \in [y]$ that is likely to give a smaller upper bound for f than c gives. In Figure 7.1, we evaluate f at $c = m([y]_1)$ to get $\widetilde{f}$. A value of c near the left end of $[y]_1$ would have yielded a more effective $\widetilde{f}$. We pursue this modification in the Exercises.

The value $\widetilde{f}$ is also used when entering newly subdivided intervals $[w]$ in our list L. If we know that $[w]$ satisfies $\underline{f_w} > \widetilde{f}$, then $[w]$ cannot contain a global minimizer. Thus, we must only enter intervals $[w]$ that satisfy $\underline{f_w} \leq \widetilde{f}$ in the list L.

7.1.2 Monotonicity Test

The monotonicity test determines whether the function f is *strictly monotone* in an entire subinterval $[y] \subset [x]$. If f is strictly monotone in $[y]$, then $[y]$ cannot contain a global minimizer in its interior. Further, a global minimizer can only lie on a boundary point of $[y]$ if this point is also a boundary point of $[x]$. Therefore, if f satisfies

$$0 \notin f'([y]), \tag{7.3}$$

then the subinterval $[y]$ can be deleted (with the exception of boundary points of $[x]$).

Figure 7.2 demonstrates the monotonicity test for four subintervals of $[x]$. In this special case, $[y]_1$ can be reduced to the boundary point $\underline{x}$, $[y]_2$ remains unchanged, and $[y]_3$ can be deleted. Since f is monotonically increasing in $[y]_4$, this entire interval can also be deleted because we are looking for a minimum. In this example, the monotonicity test has reduced the list L from four elements $[y]_1$, $[y]_2$, $[y]_3$, and $[y]_4$ to two elements $[\underline{x}]$ and $[y]_2$. It appears that applying the midpoint test with $[\underline{x}]$ might also discard $[y]_2$, leaving the unique solution $x^* = \underline{x}$ at a cost of only four interval evaluations of f' and two interval evaluations of f.

The monotonicity test costs only one interval evaluation of $f'([y])$. As illustrated in Figure 7.2, it often allows us to discard large portions of the original interval $[x]$ from further consideration.

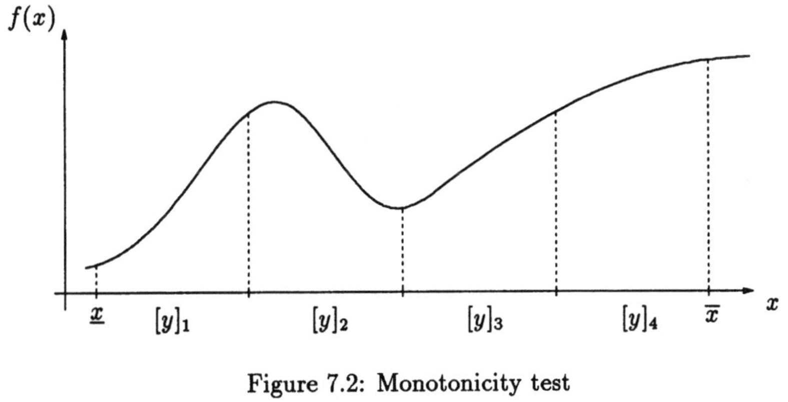

Figure 7.2: Monotonicity test

7.1.3 Concavity Test

The concavity test (non-convexity test) examines the concavity of f. If f is *not convex* in a subinterval $[y] \subset [x]$, then $[y]$ cannot contain a global minimizer in its interior. Further, a global minimizer can only lie on a boundary point if this point is also a boundary point of $[x]$.

A function f is convex in $[y]$ if the second derivative is positive or zero. Therefore, if f satisfies

$$\overline{f''([y])} < 0, \tag{7.4}$$

then f is concave over $[y]$, and the subinterval $[y]$ can be deleted (with the exception of boundary points of $[x]$). One of the reasons to include second-order derivatives in our automatic differentiation module in Chapter 5 was to provide the computational tool to apply the concavity test to non-trivial functions f.

Figure 7.3 demonstrates the concavity test for four subintervals of $[x]$. In this special case, $[y]_1$ can be reduced to the boundary point $\underline{x}$, $[y]_2$ remains unchanged, $[y]_3$ can be deleted, and $[y]_4$ can be reduced to the boundary point $\overline{x}$. In this example, the concavity test has reduced the list L from four elements $[y]_1$, $[y]_2$, $[y]_3$, and $[y]_4$ to a much smaller subset of $[x]$ ($[\underline{x}]$, $[y]_2$, $[\overline{x}]$) at a cost of only four interval evaluations of f''.

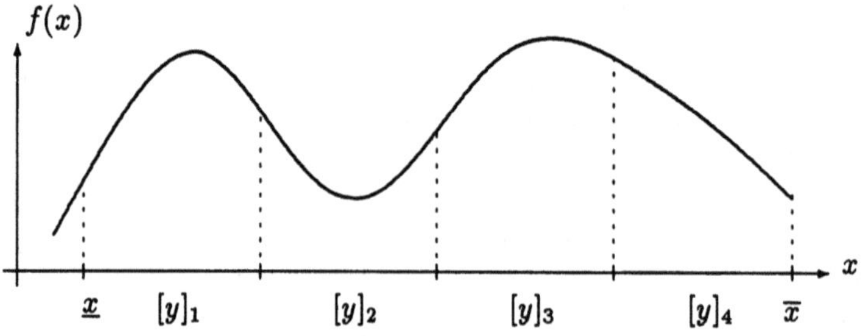

Figure 7.3: Concavity test

7.1.4 Interval Newton Step

In our global optimization method, we apply one step of the extended interval Newton's method (see Chapter 6) to the problem

$$f'(y) = 0, \quad y \in [y]. \tag{7.5}$$

Let

$$N'([y]) := m([y]) - \frac{f'(m([y]))}{f''([y])}, \tag{7.6}$$

and compute

$$[y_N] := [y] \cap N'([y]), \tag{7.7}$$

where $[y_N]$ is a single interval, the union of two intervals, or the empty set. Thus, the interval Newton step may eliminate parts of $[y]$ which cannot contain a stationary point of f. We do not apply several iterations of interval Newton because it would be inefficient to compute all stationary points which not necessarily are minimizers.

An interval Newton step is computationally more expensive than the midpoint, monotonicity, or concavity tests. However, it is often very effective in reducing the widths of subintervals to the point where overestimation by interval arithmetic is small enough that one of the less expensive tests can succeed.

7.1.5 Verification

We have to check two conditions to verify the existence and uniqueness of a local minimizer within an interval of the final list our method produces. The first condition is

$$N'([y]) \overset{\circ}{\subset} [y], \tag{7.8}$$

which guarantees the existence and uniqueness of a stationary point of f, i.e. a zero of f' in $[y]$ (cf. Theorem 6.1). The second condition is

$$\underline{f''([y])} > 0, \tag{7.9}$$

which guarantees that f is convex over $[y]$ (recall that we are assuming in this chapter that f is twice continuously differentiable). We know of no way to verify the uniqueness of a global minimizer in general. The global minimizer can even be one or more continua of points. Hence, we settle for attempting to verify that intervals we compute as candidates for containing a global minimizer contain unique local minimizers. Failure to verify the uniqueness of a local minimizer in a subinterval is not grounds for discarding that subinterval from the list of candidates.

In fact, our method produces a final list containing enclosures for *locally unique candidates for global minimizers*. If we have in the final list exactly one subinterval $[y]$ in which we can validate a local minimizer, then we have validated a unique global minimizer in the starting interval $[x]$. If we have two or more subintervals validated to contain unique local minimizers, then the best we can say is that each contains a candidate for a global minimizer, which need not be unique. In this case, the global minimizer could not be a continuum of points.

7.2 Algorithmic Description

The algorithm AllGOp1 (Algorithm 7.4) for determining all global optimizers of a 1-dimensional twice continuously differentiable function f in an interval $[x]$ consists of two parts. The first part is the subdivision method including the tests and the extended interval Newton steps described in Section 7.1. The second part is a verification step, which tries to verify the uniqueness of a local minimizer within the remaining intervals of the work list.

The algorithm for the execution of the extended interval Newton step (Algorithm 7.1), yields a pair of intervals $[v]$ and the number p, specifying the number of non-empty interval components of $[v]$.

Algorithm 7.1: NewtonStep $(f, [y], [v], p)$ {Procedure}

1. $c := m([y]);\quad p := 0;$
2. $[z] := c - f'_\diamond(c)/f''([y]);$ {Extended interval Newton step}
3. $[v] := [y] \cap [z];$ {Intersection $[y] \cap [z] = [v]_1 \cup [v]_2$ }
4. **if** $[v]_1 \neq \emptyset$ **then** $p := p + 1;$
5. **if** $[v]_2 \neq \emptyset$ **then** $p := p + 1;$
6. **return** $[v], p;$

Algorithm 7.2 manages the list L of pending subintervals that may contain global minimizers x^*. Subintervals are removed from the list and placed in an accepted list when they satisfy relative or absolute error acceptance criteria. Subintervals are also removed from the list by the midpoint, monotonicity, concavity tests, or by the interval Newton step. Subintervals are added to the pending list when an element from the list is bisected or when the extended interval Newton step yields two candidate intervals.

Our algorithm stores the subdivided intervals $[y]$ and the lower bound of the interval function evaluations $[f_y] := f_\Diamond([y])$ as pairs $([y], \underline{f_y})$. Pairs are stored in the list sorted in increasing order of lower bounds $\underline{f_y}$. Therefore, a newly computed pair is stored in the list L according to the ordering rule (cf. [73]):

$$\left.\begin{array}{ll} \bullet \text{ either} & \underline{f_w} \le \underline{f_y} < \underline{f_z} \text{ holds,} \\ \bullet \text{ or} & \underline{f_y} < \underline{f_z} \text{ holds, and } ([y], \underline{f_y}) \text{ is the first element of the list,} \\ \bullet \text{ or} & \underline{f_w} \le \underline{f_y} \text{ holds, and } ([y], \underline{f_y}) \text{ is the last element of the list,} \\ \bullet \text{ or} & ([y], \underline{f_y}) \text{ is the only element of the list,} \end{array}\right\} \quad (7.10)$$

where $([w], \underline{f_w})$ is the predecessor and $([z], \underline{f_z})$ is the successor of $([y], \underline{f_y})$ in L.

That is, the second components of the list elements may not decrease, and a new pair is entered behind all other pairs with the same second component. Thus, the first element of the list has the smallest second component, and we can directly use the corresponding interval to compute $f(c)$ for the improvement of $\widetilde{f}$ in performing the midpoint test. Because of the ordering of the list, we can also save some work when deleting elements in the midpoint test. When we have found the first element to be deleted, we can delete the rest of the list. At each step of our algorithm, $\widetilde{f}$ is the best known upper bound for the global minimum f^*, and $\underline{f_y}$ from the first element of the list is the best known lower bound for f^*. Hence in some applications, we might halt the algorithm when $\widetilde{f} - \underline{f_y} < \varepsilon_f$ (absolute tolerance) or when $\widetilde{f} - \underline{f_y} < |f_\Diamond([y])|\varepsilon_f$ (relative tolerance).

Given the list L and a list element E, we use the following notations in our algorithms:

Notation	Meaning
$L := \{\}$	Initialization by an empty list
$L := E$	Initialization by a single element
$L := L + E$	Enter element E in L in sorted order according to (7.10)
$L := L - E$	Discard element E from L
$E :=$ **Head** (L)	Set E to the first element of L
MultiDelete $(L, \tilde{f})$	Discard all elements from L satisfying condition (7.2)
Length (L)	Delivers the number of elements in L

In **GlobalOptimize**, we first compute an upper bound for the global minimum value and we do some initializations. Then in Step 3, we separately treat the two boundary points of the starting interval $[x]$. We enter them into the work list if they are candidates for minimizers. Step 4 is the main iteration. Here, we first do a bisection of the actual interval $[y]$. Then in Step 4(b), we apply the monotonicity test, a function value check using the centered form, the concavity test, and the extended interval Newton step to the bisected intervals $[u]_1$ and $[u]_2$. The interval Newton step may result in at most two intervals. We have to handle them both in step 4(b)vii, where we again apply a monotonicity test and a function value check with centered forms. If the actual interval $[v]_j$ is still a candidate for a minimizer, we store it in L.

In Step 4(d), we remove the first element from the list L, i.e. the element of L with the smallest lower bound of the interval function evaluation, and we perform the midpoint test. Then, we check the tolerance criterion for the new actual interval. If the desired accuracy is achieved, we store this interval in the result list L_{res}. Otherwise, we go to the bisection step.

When the iteration stops because the list L is empty, we compute a final enclosure $[f^*]$ for the global minimum value in Step 5, and we return L_{res} and $[f^*]$. Procedure **GlobalOptimize** terminates because the elements of L move to L_{res} if $(d_{rel}([y]) < \varepsilon)$ or if $(d_{rel}([f^*]) < \varepsilon)$. The bisection step (Step 4(a)) guarantees that the first condition is fulfilled at some stage of the iteration.

Algorithm 7.2: **GlobalOptimize** $(f, [x], \varepsilon, L_{res}, [f^*])$ {Procedure}

1. $c := m([x])$; $[f_c] := f_\diamond(c)$; $\tilde{f} := \overline{f_c}$; {Compute upper bound for f^*}
2. $[y] := [x]$; $L := \{\}$; $L_{res} := \{\}$;
3. {Treat boundary points separately}

   $[a] := f_\diamond(\underline{y})$; $[b] := f_\diamond(\overline{y})$; $\tilde{f} := \min\{\tilde{f}, \overline{a}, \overline{b}\}$

   if $\tilde{f} \geq \underline{a}$ **then** $L := L + ([\underline{y}, \underline{y}], \underline{a})$;

   if $\tilde{f} \geq \underline{b}$ **then** $L := L + ([\overline{y}, \overline{y}], \underline{b})$;
4. **repeat** {Start iteration}

   (a) $[u]_1 := [\underline{y}, c]$; $[u]_2 := [c, \overline{y}]$; {Bisect $[y]$}

for $i := 1$ **to** 2 **do**
  i. **if** $0 \notin f'([u]_i)$ **then next**$_i$; {Monotonicity test}
  ii. $[f_u] := (f(c) + f'([u]_i) \cdot ([u]_i - c)) \cap f([u]_i)$; {Centered form}
  iii. **if** $\tilde{f} < \underline{f_u}$ **then next**$_i$;
  iv. $[h] := f''([u])$;
  v. **if** $\overline{h} < 0$ **then next**$_i$; {Concavity test}
  vi. NewtonStep$(f, [u]_i, [v], p)$; {Extended interval Newton step}
  vii. **for** $j := 1$ **to** p **do**
    A. **if** $0 \notin f'([v]_j)$ **then next**$_j$; {Monotonicity test}
    B. $c_v := m([v]_j)$;
    C. $[f_v] := (f(c_v) + f'([v]_j) \cdot ([v]_j - c_v)) \cap f([v]_j)$; {Centered form}
    D. **if** $\tilde{f} \geq \underline{f_v}$ **then** $L := L + ([v]_j, \underline{f_v})$; {Store $[v]_j$}
(c) *Bisect* := *false*;
(d) **while** $(L \neq \{\})$ **and** (**not** *Bisect*) **do**
  i. $([y], \underline{f_y})$:= Head(L); $L := L - ([y], \underline{f_y})$; $c := m([y])$;
  ii. $\tilde{f} := \min\{\tilde{f}, f(c)\}$; MultiDelete$(L, \tilde{f})$; {Midpoint test}
  iii. $[f^*] := [\underline{f_y}, \tilde{f}]$;
  iv. **if** $(d_{\text{rel}}([f^*]) < \varepsilon)$ **or** $(d_{\text{rel}}([y]) < \varepsilon)$ **then**
    $L_{\text{res}} := L_{\text{res}} + ([y], \underline{f_y})$; {Accept this subinterval}
  else
    Bisect := *true*;
until (**not** *Bisect*);
5. $([y], \underline{f_y})$:= Head(L_{res}); $[f^*] := [\underline{f_y}, \tilde{f}]$;
6. **return** $L_{\text{res}}, [f^*]$;

Algorithm 7.3 describes a verification step that checks the uniqueness of a local minimizer enclosed in the interval $[y]$. The procedure tries to do a "zero check" for the derivative f' according to condition 3 of Theorem 6.1 by applying interval Newton steps including an epsilon inflation of the iterates $[y]$. We also check the condition $f''(y) > 0$ for all $y \in [y]$ to verify that f is convex in $[y]$. We use $k_{\max} = 10$ as the maximum number of iterations, $\varepsilon = 0.25$ as the starting value for the epsilon inflation, and a factor of 8 to increase ε within the iterations. It turned out that these are good values for minimizing the effort if no verification is possible (see also [46]).

Algorithm 7.3: VerificationStep $(f, [y], yUnique)$ {Procedure}

1. $yUnique := (\underline{y} = \overline{y})$; {Point interval $\Rightarrow$ uniqueness of local minimizer}
2. **if** $yUnique$ **then return** $[y], yUnique$;

$k_{\max} := 10$; $k := 0$; $[y_{\text{in}}] := [y]$; $\varepsilon := 0.25$; {Initializations}

4. **while** (**not** $yUnique$) **and** ($k < k_{\max}$) **do** {Do $k_{\max}$ loops to achieve inclusion}
   (a) $[y_{\text{old}}] := [y] \bowtie \varepsilon$; {Epsilon inflation of $[y]$}
   (b) $[h] := f''([y_{\text{old}}])$;
   (c) **if** $\underline{h} \leq 0$ **then exit**$_{\text{while-loop}}$; {No verification possible}
   (d) $k := k + 1$; $c := m([y_{\text{old}}])$;
   (e) $[y] := c - f'_{\Diamond}(c)/f''([y_{\text{old}}])$; {Interval Newton step}
   (f) **if** $[y] \cap [y_{\text{old}}] = \emptyset$ **then exit**$_{\text{while-loop}}$; {No verification possible}
   (g) $yUnique := ([y] \overset{\circ}{\subset} [y_{\text{old}}])$; {Inner inclusion $\Rightarrow$ local uniqueness}
   (h) $[y] := [y] \cap [y_{\text{old}}]$; {Intersection with old value}
   (i) **if** $[y] = [y_{\text{old}}]$ **then** $\varepsilon := \varepsilon \cdot 8$; {Increase ε}
5. **if not** $yUnique$ **then** $[y] := [y_{\text{in}}]$;
6. **return** $[y], yUnique$;

Algorithm 7.4 now combines these procedures to compute enclosures for all global minimizers x^* of the function f and for the global minimum value f^* within the input interval $[x]$ and tries to prove the uniqueness of the local minimizers within the computed enclosures. The desired accuracy (relative diameter) of the interval enclosures is specified by the input parameter ε. 1 ulp accuracy is chosen if the specified value of ε is too small (for example 0). The enclosures for the global minimizers of f are returned in the interval vector $[Opt]$, the corresponding information on the local uniqueness of the optimizers is returned in the Boolean vector *Info*. The number of enclosures computed is returned in the integer variable N.

We use a function called **CheckParameters** as an abbreviation for the error checks for the parameters of **AllGOp1** which are necessary in an implementation. If no error occurs, **AllGOp1** delivers the N enclosures $[Opt]_i$, $i = 1, 2, \ldots, N$, and for each $i = 1, 2, \ldots, N$,

if $Info_i$ = *true*, then $[Opt]_i$ encloses unique local minimizer of f,
if $Info_i$ = *false*, then $[Opt]_i$ may enclose a local or global minimizer of f.

Algorithm 7.4: AllGOp1 $(f, [x], \varepsilon, [Opt], Info, N, [f^*], Err)$ {Procedure}

1. $Err :=$ CheckParameters;
2. **if** $Err \neq$ "No Error" **then return** Err;
3. Set ε to "1 ulp accuracy" if ε is too small;
4. GlobalOptimize $(f, [x], \varepsilon, L_{\text{res}}, [f^*])$;
5. $N :=$ Length (L_{res});
6. **for** $i := 1$ **to** N **do**

$[Opt]_i := \text{Head}(L_{res})$;

VerificationStep $(f, [Opt]_i, Info_i)$;

7. **return** $[Opt], Info, N, [f^*], Err$;

Applicability of the Algorithm

We have assumed that f is twice continuously differentiable. However, we can apply our algorithm to functions that are only once continuously differentiable or to functions that are not differentiable if we leave out some parts of the algorithm. If we do not use the interval Newton step (Step 4(b)vi in our algorithm), i.e. if we leave out Steps 4(b)iv, 4(b)v, and 4(b)vi, and replace them by the sequence

$$[v]_1 := [u]_i; \quad p := 1;$$

then our method works for functions that are only once differentiable. Algorithm 7.3 cannot be applied in this case. We can even do problems in nonsmooth global optimization. If we replace Step 4(b) by the sequence

for $i := 1$ **to** 2 **do**
  $[f_u] := f([u]_i)$;
  if $\widetilde{f} \geq \underline{f_u}$ **then** $L := L + ([u]_i, \underline{f_u})$;

then we can apply our method, consisting of subdividing and the midpoint test, to functions that are *not* differentiable.

The closer the upper bound $\widetilde{f}$ is to the global minimum value f^*, the more intervals we can delete in the midpoint test (Step 4(d)ii of Algorithm 7.2). Thus, the method can be improved by incorporating a floating-point approximate local search procedure in an attempt to decrease the value $\widetilde{f}$. See [28] or [72] for the description of such local search procedures. We explore this extension in the Exercises.

For a multiple zero x^* of f', the algorithm cannot verify the existence and the uniqueness of x^* in the enclosing result interval. Nevertheless, the zero of f', which is possibly a global minimizer, is bounded to the desired accuracy specified by ε. In this case, the corresponding component of the *Info*-vector has the value *false*.

As a consequence of the splitting (bisecting) of the intervals, it may happen that a minimizer lying exactly at the splitting point is enclosed in two different intervals. A sophisticated supplement to our method avoiding this can be found in [73].

The algorithm is well suited for parallel computers due to its natural parallelism given by the splitting of the intervals by the extended interval Newton step or by the bisection.

7.3 Implementation and Examples

7.3.1 PASCAL–XSC Program Code

We begin by describing our implementation of the operations needed for handling lists and list elements. Then, we describe the implementation of Algorithm 7.4 and its subalgorithms.

7.3.1.1 Module lst1_ari

This module supplies the type *Pair* representing a pair of an interval and a real value and the type *PairPtr* representing a list of such pairs. The local variable *FreeList* and the procedures *NewPP*, *Free*, and *FreeAll* generate and free list elements (pairs) and prevent creation of garbage in memory. *MakePair*, *GetInt*, and *GetFyi* are transfer and access functions for pairs.

The global function *EmptyList* represents an empty list. The operator + enters a new list element *P* in the list *List* according to condition (7.10). The procedure *MultiDelete* deletes all elements *P* in *List* for which *P.fyi* > *fmax*. This procedure assumes that the list elements are ordered according to condition (7.10). Function *Next* sets the list pointer *List* to the next list element. *Head* delivers the first pair of *List*, whereas *DelHead* deletes the first pair of *List*. Function *Length* delivers the number of elements in *List*.

```
{----------------------------------------------------------------------------}
{ Purpose: Definition of a list arithmetic used in connection with an       }
{    interval bisection method in global optimization for storing pairs of  }
{    an interval and a real value.                                          }
{ Method: Overloading of functions and operators of data types 'Pair' (list }
{    element) and 'PairPtr' (list).                                         }
{ Global types, operators, functions, and procedures:                       }
{    types      Pair              : list elements (pair of interval and real) }
{               PairPtr, PairElmt: list of pairs                            }
{    operators  +                 : adding a new element to a list          }
{    functions  MakePair          : transfer function for pairs             }
{               GetInt, GetFyi    : access functions for pairs              }
{               Next, Head        : access functions for lists              }
{               Length            : access function to length of list       }
{               EmptyList         : delivers an empty list                  }
{    procedures FreeAll           : free complete list                      }
{               MultiDelete       : deletes several elements in a list      }
{               DelHead           : deletes the first element of a list     }
{----------------------------------------------------------------------------}
module lst1_ari;

use
  i_ari;   { Interval arithmetic }

{----------------------------------------------------------------------------}
{ Global type definitions                                                    }
{----------------------------------------------------------------------------}
global type
  Pair     = record
               int : interval;              { Pair of an interval 'y' and   }
               fyi : real;                  { the real value 'inf(f(y))'.   }
             end;                           {-------------------------------}
```

```
  PairPtr  = ↑PairElmt;
  PairElmt = record                                        { List of pairs }
               P : Pair;                                   {---------------}
               N : PairPtr;
             end;

{----------------------------------------------------------------------------}
{ Local variable storing list of free elements (automatic garbage recycling) }
{----------------------------------------------------------------------------}
var
  FreeList : PairPtr;

{----------------------------------------------------------------------------}
{ Procedures for generating and freeing of list elements (pairs)             }
{----------------------------------------------------------------------------}
procedure NewPP (var pp: PairPtr);     { 'NewPP' generates a new list element }
begin                                  { or gets one from 'FreeList'.         }
  if FreeList = nil then               {-------------------------------------}
    begin
      new(pp);  pp↑.N:= nil;
    end
  else
    begin
      pp:= FreeList;  FreeList:= FreeList↑.N;  pp↑.N:= nil;
    end;
end;

procedure Free (var pp: PairPtr);            { 'Free' enters one element of a }
begin                                        { list in the 'FreeList'.        }
  if pp <> nil then                          {--------------------------------}
  begin
      pp↑.N:= FreeList;  FreeList := pp;  pp:= nil;
  end;
end;

global procedure FreeAll (var List: PairPtr);{ 'FreeAll' enters all elements }
var  H : PairPtr;                            { of 'List' in the 'FreeList'.  }
begin                                        {-------------------------------}
  if List <> nil then
  begin
    H:= List;
    while H↑.N <> nil do  H:= H↑.N;
    H↑.N:= FreeList;  FreeList:= List;  List:= nil;
  end;
end;

{----------------------------------------------------------------------------}
{ Transfer and access functions for pairs                                    }
{----------------------------------------------------------------------------}
global function MakePair (int: interval; fyi: real) : Pair;  { Generate pair }
begin                                                        {---------------}
  MakePair.int:= int;  MakePair.fyi:= fyi;
end;

global function GetInt (P: Pair) : interval;             { Get int-component }
begin                                                    {-------------------}
  GetInt:= P.int;
end;

global function GetFyi (P: Pair) : real;                 { Get fyi-component }
begin                                                    {-------------------}
  GetFyi:= P.fyi;
end;
```

```
{----------------------------------------------------------------------------}
{ Operators, functions, and procedures for lists of pairs                   }
{----------------------------------------------------------------------------}
{ Global function 'EmptyList' representing an empty list of pairs.          }
{----------------------------------------------------------------------------}
global function EmptyList : PairPtr;
begin
  EmptyList := nil;
end;

{----------------------------------------------------------------------------}
{ Operator + enters the pair 'P' the list 'List' in such a way that after   }
{ entering, one of the four following condition holds:                      }
{   1) O.fyi <= P.fyi < Q.fyi,                                              }
{   2)          P.fyi < Q.fyi   and  'P' is the first element of 'List',    }
{   3) O.fyi <= P.fyi           and  'P' is the last  element of 'List',    }
{   4)                               'P' is the only  element of 'List',    }
{ where 'O' is the preceding and 'Q' is the succeeding element of 'P' in    }
{ the resulting list.                                                       }
{----------------------------------------------------------------------------}
global operator + (List: PairPtr; P: Pair) Enter: PairPtr;
var
  H, HN             : PairPtr;
  ready, alreadyIn : boolean;
begin
  if (List = nil) then                              { List is empty, so new }
    begin                                           { list contains only P. }
      NewPP(H);  H↑.P:= P;  H↑.N:= nil;  Enter:= H; {-----------------------}
    end
  else if (List↑.P.fyi > P.fyi) then                { P becomes new first   }
    begin                                           { element of the list.  }
      NewPP(H);  H↑.P:= P;  H↑.N:= List;  Enter:= H; {-----------------------}
    end
  else
    begin
      H:= List;  HN:= H↑.N;  ready:= false;  alreadyIn:= H↑.P.int = P.int;

      while not (ready or alreadyIn) do             { Search for the right  }
      begin                                         { position to enter P.  }
        if (HN = nil) then                          {-----------------------}
          ready:= true
        else if (HN↑.P.fyi > P.fyi) then
          ready:= true
        else
          begin
            H:= HN;  HN:= H↑.N;  alreadyIn:= H↑.P.int = P.int;
          end;
      end;

      if not alreadyIn then
      begin
        NewPP(H↑.N);  H:= H↑.N;                     { Enter P between H and }
        H↑.P:= P;  H↑.N:= HN;                       { HN. Return List.      }
      end;                                          {-----------------------}

      Enter:= List;
    end;
end;

{----------------------------------------------------------------------------}
{ 'MultiDelete' deletes all elements 'P' in 'List' for which the condition  }
{ 'P.fyi > fmax' holds.  This procedure assumes that the 'fyi' components of }
{ the list elements are sorted in increasing order (see operator +).        }
{----------------------------------------------------------------------------}
```

```
global procedure MultiDelete (var List: PairPtr; fmax: real);
var
  DelPrev, Del : PairPtr;
  ready        : boolean;
begin
  if (List <> nil) then
  begin
    if (List↑.P.fyi > fmax) then                { All list elements satisfy }
      begin                                     { 'P.fyi > fmax'.           }
        Del:= List;  List:= nil;                {---------------------------}
      end
    else
      begin
        DelPrev:= List;  Del:= DelPrev↑.N;  ready:= (Del=nil);

        while not ready do
        begin
          if (Del = nil) then
            ready := true
          else if (Del↑.P.fyi > fmax) then
            begin
              ready:= true;  DelPrev↑.N:= nil;
            end
          else
            begin
              DelPrev:= Del;  Del:= Del↑.N;
            end;
        end;
      end;
    FreeAll(Del);
  end;
end;

global function Next (List: PairPtr) : PairPtr;   { Sets list pointer to the }
begin                                             { next list element        }
  Next:= List↑.N;                                 {--------------------------}
end;

global function Head (List: PairPtr) : Pair;    { Delivers first pair of the }
begin                                           { list, i.e. the pair P with }
  Head:= List↑.P;                               { the smallest value P.fyi.  }
end;                                            {----------------------------}

global procedure DelHead (var List: PairPtr);   { Deletes the first pair of  }
var                                             { the List.                  }
  Del : PairPtr;                                {----------------------------}
begin
  Del := List;  List:= List↑.N;  Free(Del);
end;

global function Length (List: PairPtr) : integer;  { 'Length' delivers the   }
var  i : integer;                                  { number of elements in   }
begin                                              { list 'List'.            }
  i:= 0;                                           {-------------------------}
  while List <> nil do
  begin
    i:= succ(i); List:= List↑.N;
  end;
  Length := i;
end;

{------------------------------------------------------------------------------}
{ Module initialization                                                        }
{------------------------------------------------------------------------------}
```

```
begin
  FreeList := nil;              { List of freed elements which can be used again  }
end.                            {-------------------------------------------------}
```

7.3.1.2 Module gop1

The module *gop1* supplies the global routines *AllGOp1* (the implementation of Algorithm 7.4) and the corresponding function *AllGOp1ErrMsg* which can be used to get an error message for the error code returned by *AllGOp1*. The procedures *NewtonStep*, *GlobalOptimize*, and *VerificationStep* are defined locally. All derivative evaluations are done with the help of the differentiation arithmetic *ddf_ari* described in Chapter 5.

The procedure *AllGOp1* uses the *DerivType* function *f* and the starting interval *Start* as input parameters and stores all computed enclosures in the interval vector *OptiVector*. If this vector is not long enough to store all the result intervals, the corresponding error code is returned together with the *OptiVector* containing all solutions it is able to store. If this error occurs, the user must increase the upper index bound of *OptiVector* to compute *all* optimizers.

The same applies to the information about the uniqueness, stored in the Boolean vector *InfoVector*. The user must declare both vectors *OptiVector* and *InfoVector* with lower index bound equal to 1 and with upper index bounds which are equal. These conditions, as well as the condition *Epsilon* $\geq$ *MinEpsilon* are checked at the beginning of procedure *AllGOp1*. *Epsilon* in our program corresponds to the parameter ε in the algorithms, and *MinEpsilon* corresponds to 1 ulp accuracy.

```
{----------------------------------------------------------------------------}
{ Purpose: Computing enclosures for all global minimizers and for the global }
{    minimum value of a twice continuously differentiable one-dimensional,   }
{    scalar valued function.                                                 }
{ Method: Bisection method combined with midpoint, monotonicity, concavity   }
{    test and extended interval Newton step.                                 }
{ Global procedures and functions:                                           }
{    procedure AllGOp1(...)       : computes enclosures for all zeros         }
{    function  AllGOp1ErrMsg(...) : delivers an error message text            }
{----------------------------------------------------------------------------}
module gop1;

use
  i_ari,      { Interval arithmetic          }
  xi_ari,     { Extended interval arithmetic }
  lst1_ari,   { List arithmetic              }
  ddf_ari,    { Differentiation arithmetic   }
  i_util,     { Interval utilities           }
  b_util;     { Boolean utilities            }

{----------------------------------------------------------------------------}
{ Error codes used in this module.                                           }
{----------------------------------------------------------------------------}
const
  NoError         = 0;  { No error occurred.                                  }
  lbOptiVecNot1   = 1;  { Lower bound of variable OptiVector is not equal to 1.}
  lbInfoVecNot1   = 2;  { Lower bound of variable InfoVector is not equal to 1.}
  VecsDiffer      = 3;  { Bounds of OptiVector and InfoVector do not match.   }
  VecTooSmall     = 4;  { OptiVector too small. Not all zeros can be stored.  }
```

```
{----------------------------------------------------------------------------}
{ Error messages depending on the error code.                                }
{----------------------------------------------------------------------------}
global function AllGOp1ErrMsg ( Err : integer ) : string;
var
  Msg : string;
begin
  case Err of
    NoError      : Msg := '';
    lbOptiVecNot1: Msg := 'Lower bound of OptiVector is not equal to 1';
    lbInfoVecNot1: Msg := 'Lower bound of InfoVector is not equal to 1';
    VecsDiffer   : Msg := 'Bounds of OptiVector and InfoVector do not match';
    VecTooSmall  : Msg := 'Not all optimizers found. OptiVector is too small';
    else         : Msg := 'Code not defined';
  end;
  if (Err <> NoError) then Msg := 'Error: ' + Msg + '!';
  AllGOp1ErrMsg := Msg;
end;

{----------------------------------------------------------------------------}
{ Purpose: Execution of one extended interval Newton step for the derivative }
{    of function 'f'.                                                        }
{ Parameters:                                                                }
{    In     : 'f'    : must be declared for the 'DerivType' to enable the    }
{                      internal use of the differentiation  arithmetic       }
{                      'ddf_ari'.                                            }
{             'Y'    : specifies the starting interval.                      }
{             'ddfY' : f''(Y), already computed outside of 'NewtonStep'.     }
{    Out    : 'V'    : Pair of intervals V[1] and V[2].                      }
{             'p'    : Number of valid intervals in V (0, 1, or 2).          }
{ Description:                                                               }
{    One extended interval Newton step for 'Y' is executed resulting in 'p'  }
{     interval(s) 'V[1]' and 'V[2]' which can be empty.                      }
{----------------------------------------------------------------------------}
procedure NewtonStep (function f (x: DerivType) : DerivType;
                                         Y, ddfY  : interval;
                           var                 V  : ivector;
                           var                 p  : integer);
var
  c       : real;
  fC, dfC : interval;
begin
  c:= mid(Y);  dfEval(f, intval(c), fC, dfC);    { Midpoint evaluation of 'f' }
  V:= Y ** (c - dfC div ddfY);                   { Execution of Newton step   }
  if V[1] <> EmptyIntval then p:= 1 else p:= 0;  { Fix number of non-empty    }
  if V[2] <> EmptyIntval then p:= p+1;           { intervals                  }
end;                                             {----------------------------}

{----------------------------------------------------------------------------}
{ Purpose: Execution of the global optimization method including a bisection }
{    method, midpoint test, monotonicity test, concavity test, and extended  }
{    interval Newton steps.                                                  }
{ Parameters:                                                                }
{    In     : 'f'          : must be declared for the 'DerivType' to enable  }
{                            the internal use of the differentiation         }
{                            arithmetic 'ddf_ari'.                           }
{             'Start:      : specifies the starting interval.                }
{             'Epsilon'    : specifies the desired relative accuracy         }
{    Out    : 'ResultList' : stores the candidates for enclosure of a global }
{                            minimizer.                                      }
{             'Minimum'    : stores the enclosure of the global minimum      }
{                            value.                                          }
{ Description:                                                               }
{   The procedure manages the list 'L' of pending subintervals that may      }
```

```
{   contain global minimizers. Subintervals are removed from the list and    }
{   placed in the accepted list 'ResultList' when they satisfy relative      }
{   error acceptance criteria. Subintervals are also removed from the list   }
{   by the midpoint, monotonicity, concavity tests, or by the interval       }
{   Newton steps. Subintervals are added to the pending list when an element }
{   from the list is bisected or when the extended interval Newton step      }
{   yields two candidate intervals.                                          }
{   'ResultList' returns the list of enclosures of the global minimizers,    }
{   'Minimum' returns the enclosure of the global minimum value.             }
{----------------------------------------------------------------------------}
procedure GlobalOptimize (function f (x: DerivType) : DerivType;
                                              Start : interval;
                                            Epsilon : real;
                          var            ResultList : PairPtr;
                          var               Minimum : interval);

var
  PairY                    : Pair;            { Pair ( Y, inf(f(Y) )          }
  Y, fY, dfY, ddfY, BdP    : interval;        { Initial box, f(Y), f'(Y), f''(Y) }
  U, V                     : ivector[1..2];   { Subboxes of Y and U            }
  fU, fV, fC, fCV, fBdP    : interval;        { Function evaluations of f      }
  dfU, dfV, ddfU           : interval;        { Derivative evaluations of f    }
  fmax                     : real;            { Upper bound for minimum value  }
  c, cV                    : real;            { Midpoints of intervals Y and V }
  WorkList                 : PairPtr;         { List of pairs                  }
  i, j, p                  : integer;         { Control variables              }
  Bisect                   : boolean;         { Flag for iteration and algorithm }

begin
  c:= mid(Start);  fEval(f, intval(c), fC);{ Compute upper bound for minimum }
  fmax:= sup(fC);                          {---------------------------------}

  if not UlpAcc(Start,1) then                                { Start method }
    begin                                                    {--------------}
      Y:= Start;  WorkList:= EmptyList;  ResultList:= EmptyList;

      BdP:= inf(Y);                          { Treat boundary points separately }
      for i:=1 to 2 do                       {----------------------------------}
      begin
        fEval(f, BdP, fBdP);
        if sup(fBdP) < fmax then fmax:= sup(fBdP);
        if fmax >= inf(fBdP) then
          WorkList:= WorkList + MakePair(BdP,inf(fBdP));
        BdP:= sup(Y);
      end;

      repeat                                               { Start iteration }
        U[1]:= intval(inf(Y),c);  U[2]:= intval(c,sup(Y)); { Bisect 'Y'      }
        for i:= 1 to 2 do                                  {-----------------}
        begin
          dfEval(f, U[i], fU, dfU);{ Compute dfu = f'(U)                      }
                                   {------------------------------------------}
          if (0 in dfU) then       { Monotonicity test: if not 0 in dfU stop }
          begin                    {------------------------------------------}
            fU:= (fC + dfU * (U[i] - c)) ** fU;      { Centered form 'f(U)' }
                                                     {----------------------}
            if (fmax >= inf(fU)) then
            begin
              ddfEval(f, U[i], fU, dfU, ddfU);  { Compute ddfU = f''(U).     }
                                                {----------------------------}
              if (sup(ddfU) >= 0) then     { Concavity test: if 0 > ddfU stop }
              begin                        {----------------------------------}
                NewtonStep(f,U[i],ddfU,V,p); { Extended interval Newton step }
                                             {-------------------------------}
```

```
          for j:=1 to p do
          begin
            dfEval(f, V[j], fV, dfV);        { Compute first derivative }
                                             {--------------------------}
            if (0 in dfV) then { Monotonic. test: if not 0 in dfV stop }
            begin              {---------------------------------------}
              cV:= mid(V[j]);                     { Try centered form  }
              fEval(f, intval(cV), fCV);          { to get better en-  }
              fV:= (fCV + dfV*(V[j] - cV)) ** fV; { closure of 'f(U)'  }
                                                  {--------------------}
              if (fmax >= inf(fV)) then                      { Store V }
                WorkList:= WorkList + MakePair(V[j],inf(fV));{---------}
            end;
          end; { for j ... }
        end;  { if sup(ddfU) >= 0 }
      end; { if fmax >= ... }
    end; { if 0 in dfU ... }
  end; { for i ... }

  Bisect:= false;                                    { Get next 'Y' of }
  while (WorkList <> EmptyList) and (not Bisect) do  { the work list   }
  begin                                              {-----------------}
    PairY:= Head(WorkList);  DelHead(WorkList);

    Y:= GetInt(PairY);  c:= mid(Y);  fEval(f, intval(c), fC);
    if sup(fC) < fmax then fmax:= sup(fC);
    MultiDelete(WorkList,fmax);

    Minimum:= intval(GetFyi(PairY),fmax);
    if (RelDiam(Minimum) < Epsilon) or (RelDiam(Y) < Epsilon) then
      ResultList:= ResultList + PairY            { Checking termination }
    else                                         { criterion            }
      Bisect:= true;                             {----------------------}
  end; { while }

 until (not Bisect);

 end { if not UlpAcc(Start,1) }
else
 begin                                         { Store starting interval }
   fEval(f, Start, fY);                        {-------------------------}
   ResultList:= EmptyList + MakePair(Start,inf(fY));
 end;
                                               { Compute good enclosure }
Minimum:= intval(GetFyi(Head(ResultList)),fmax); { of the global minimum  }
end;                                           {------------------------}

{----------------------------------------------------------------------------}
{ Purpose: Execution of a verification step including the use of an epsilon }
{    inflation.                                                              }
{ Parameters:                                                                }
{    In     : 'f'       : function of 'DerivType'.                           }
{    Out    : 'yUnique' : returns 'true' if the verification is successful.  }
{    In/Out : 'y'       : interval enclosure to be verified.                 }
{ Description: This procedure checks the uniqueness of the local minimizer  }
{    enclosed in the interval variable 'y' by a verification step including }
{    the use of an epsilon inflation of the iterates.                        }
{----------------------------------------------------------------------------}
procedure VerificationStep (function f (x: DerivType) : DerivType;
                            var               y : interval;
                            var         yUnique : boolean);
const
  kmax = 10;  { Maximum number of iterations }
```

```
var
  yIn, yOld, fm, dfm, fy, dfy, ddfy : interval;
  m, eps                            : real;
  k                                 : integer;
begin
  yUnique := (y.inf = y.sup);                          { y is a point interval }
  if not yUnique then                                  {-----------------------}
  begin
    yIn := y;  k := 0;  eps:= 0.25;     { Initializations                      }
  end;                                  {--------------------------------------}

  while (not yUnique) and (k < kmax) do { Do kmax loops to achieve inclusion }
  begin                                 {--------------------------------------}
    yOld:= blow(y, eps);                { Epsilon inflation of 'y'             }
    ddfEval(f, yOld, fy, dfy, ddfy);    {--------------------------------------}

    if 0 >= inf(ddfy) then            { No verification of a minimum possible }
      k := kmax                       {----------------------------------------}
    else
      begin                                   { Perform interval Newton step }
        k:= k+1;  m:= mid(yOld);              {------------------------------}
        dfEval(f, intval(m), fm, dfm);
        y := (m - dfm / ddfy);

        if y >< yOld then                          { No verification possible }
          k:= kmax                                 {--------------------------}
        else
          begin
            yUnique:= y in yOld;    { Inner inclusion ===> unique zero of f' }
            y      := y ** yOld;    { Intersection with old value            }
            if y = yOld then        {------------------------------------------}
              eps := eps * 8        { Increase the value of 'eps'            }
          end;                      {------------------------------------------}
      end;
  end;
  if not yUnique then y := yIn;
end;

{----------------------------------------------------------------------------}
{ Purpose: Computation of enclosures for all global minimizers and for the   }
{    global minimum value of a twice continuously differentiable one-dimen-   }
{    sional, scalar-valued function.                                          }
{ Parameters:                                                                 }
{    In    : 'f'             : objective function, must be declared for the }
{                              'DerivType' to enable the internal use of     }
{                              the differentiation arithmetic 'ddf_ari'.     }
{            'Start',        : specifies the starting interval.              }
{            'Epsilon'.      : specifies the desired relative accuracy       }
{                              (interval diameter) of the result intervals. }
{    Out   : 'OptiVector'    : stores and returns the enclosures for the     }
{                              global optimizers of 'f'.                     }
{            'InfoVector'    : stores the corresponding information on the   }
{                              uniqueness of the local optimizers in these   }
{                              enclosures.                                   }
{            'NumberOfOptis' : returns the number of enclosures computed.    }
{            'Minimum'     ' : returns the enclosure for the minimum value. }
{            'Err'           : returns an error code.                        }
{ Description:                                                                }
{    The enclosures for the global minimizers of 'f' are computed by calling }
{    procedure 'GlobalOptimize'. Then a verification step is applied.        }
{    The enclosures for the global minimizers of 'f' are stored in the       }
{    interval vector 'OptiVector', the corresponding information on the      }
{    uniqueness of the local minimizers in these enclosures is stored in the }
{    Boolean vector 'InfoVector'. The number of enclosures computed is       }
```

```
{    returned in the integer variable 'NumberOfOptis'. The enclosure for the }
{    global minimum value is returned in the variable 'Minimum'.             }
{    If an error occurs, the value of 'Err' is different from 0.             }
{----------------------------------------------------------------------------}
global procedure AllGOp1 (function f (x: DerivType) : DerivType;
                                                Start : interval;
                                              Epsilon : real;
                              var          OptiVector : ivector;
                              var          InfoVector : bvector;
                              var       NumberOfOptis : integer;
                              var             Minimum : interval;
                              var                 Err : integer);
var
  i               : integer;
  MinEpsilon      : real;
  ResultList, L   : PairPtr;
begin
  NumberOfOptis:= 0;
  if (lb(OptiVector) <> 1) then        { Check index bounds of result vectors }
    Err:= lbOptiVecNot1                {-------------------------------------}
  else if (lb(InfoVector) <> 1) then
    Err:= lbInfoVecNot1
  else if (ub(InfoVector) <> ub(OptiVector)) then
    Err:= VecsDiffer
  else
    begin                                { Start global optimization method  }
      Err:= NoError;                     {-----------------------------------}
      MinEpsilon:= succ(1.0) - 1.0;      { Relative machine accuracy (1 ulp) }
                                         {-----------------------------------}
      if (Epsilon < MinEpsilon) then Epsilon := MinEpsilon;  { Set 'Epsilon' }
                                                             { to 1 ulp acc. }
      GlobalOptimize(f,Start,Epsilon,ResultList,Minimum);    {---------------}

      NumberOfOptis:= Length(ResultList);
                                            { Check if there are more opti- }
      if ub(OptiVector) < NumberOfOptis then { mizers than storage space     }
      begin                                 {-------------------------------}
        Err:= VecTooSmall;  NumberOfOptis:= ub(OptiVector);
      end;

      L:= ResultList;
                                                  { Verification step for the }
      for i:=1 to NumberOfOptis do                { enclosure intervals       }
      begin                                       {---------------------------}
        OptiVector[i]:= GetInt(Head(L));  L:= Next(L);
        VerificationStep(f, OptiVector[i], InfoVector[i]);
      end;

      FreeAll(ResultList);
    end;
end;

{----------------------------------------------------------------------------}
{ Module initialization part                                                 }
{----------------------------------------------------------------------------}
begin
  { Nothing to initialize }
end.
```

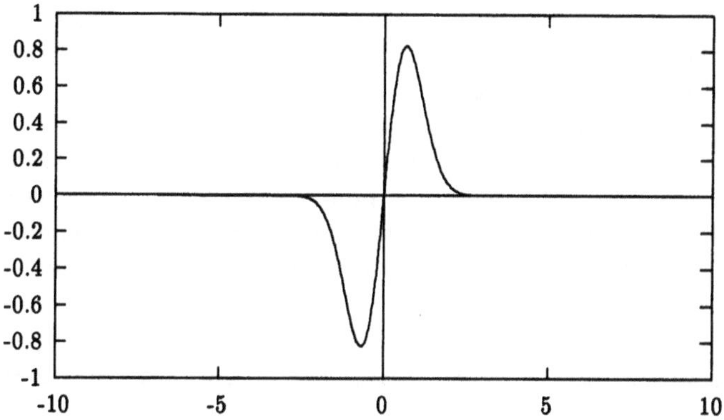

Figure 7.4: Function f (test function f_4 in [85])

7.3.2 Examples

As a sample program, we wish to use *AllGOp1* to compute all global minimizers and the global minimum of the function

$$f(x) = (x + \sin x) \cdot e^{-x^2},$$

which is the test function f_4 in [85], and of the function of Shubert

$$g(x) = -\sum_{k=1}^{5} k \sin((k+1)x + k),$$

which is the test function f_3 in [85]. Figures 7.4 and 7.5 show the plots of these two functions. Within the interval $[-10, 10]$, f has the global minimizer $-6.7957...$, and the global minimizers of g occur at $-6.7745...$, $-0.49139...$, and $5.7917...$. According

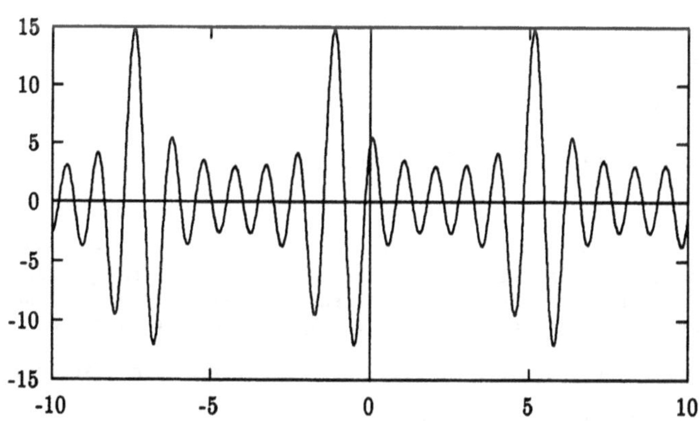

Figure 7.5: Shubert's function g (test function f_3 in [85])

to [85], both functions are difficult to minimize and represent practical problems. The large number of local optimizers of g make it very difficult for an approximation method to find the global optimizers.

We first have to define the functions f and g for the type *DerivType*. Then we use a procedure *compute* with the function f and a string (a description of the function) as parameters. This procedure prompts you to enter the necessary data for the call of procedure *AllGOp1*. If we add further function definitions in our program, we can apply this procedure in the same way using another function as parameter.

```
{----------------------------------------------------------------------------}
{ This program uses module 'gop1' to compute the global optimizers of the    }
{ functions                                                                  }
{                                                                            }
{    f(x) = (x + sin(x)) * exp(-sqr(x)).                                     }
{                                                                            }
{ and                                                                        }
{                 5                                                          }
{    g(x) = - sum ( k * sin((k+1)*x + k) )                                   }
{                k=1                                                         }
{                                                                            }
{ A starting interval and a tolerance must be entered.                       }
{----------------------------------------------------------------------------}
program gop1_ex;
use
  i_ari,     { Interval arithmetic                  }
  ddf_ari,   { Differentiation arithmetic           }
  i_util,    { Interval utilities                   }
  b_util,    { Boolean utilities                    }
  gop1;      { Global optimization one-dimensional }
const
  n = 20;  { Maximum number of optimizers to be computed }

function f (x : DerivType) : DerivType;
begin
  f := (x + sin(x)) * exp(-sqr(x));
end;

function g (x : DerivType) : DerivType;
var
 s : DerivType;
 k : integer;
begin
  s := DerivConst(0);
  for k:=1 to 5 do
    s := s + k * sin ( (k+1) * x + k );
  g := - s;
end;

{----------------------------------------------------------------------------}
{ Procedure for printing and reading information to call the procedure      }
{ 'AllGOp1'. This procedure must be called with the function 'f' and a      }
{ string 'Name' containing a textual description of that function.          }
{----------------------------------------------------------------------------}
procedure compute (function f(x:DerivType): DerivType; Name: string);
var
  SearchInterval, Minimum   : interval;
  Tolerance                 : real;
  Opti                      : ivector[1..n];
  Unique                    : bvector[1..n];
```

```
  NumberOfOptis, i, ErrCode : integer;
begin
  writeln('Computing all global minimizers of the function  ', Name);
  write('Search interval      : '); read(SearchInterval);
  write('Tolerance (relative) : '); read(Tolerance);
  writeln;
  AllGOp1(f, SearchInterval, Tolerance,
             Opti, Unique, NumberOfOptis, Minimum, ErrCode);
  for i:=1 to NumberOfOptis do
  begin
    writeln(Opti[i]);
    if unique[i] then
      writeln('encloses a locally unique candidate for a global minimizer!')
    else
      writeln('may contain a local or global minimizer!')
  end;
  writeln;
  if (NumberOfOptis <> 0) then
  begin
    writeln(Minimum);  writeln('encloses the global minimum value!');
  end;
  if ErrCode <> 0 then writeln(AllGOp1ErrMsg(ErrCode));
  writeln;  writeln(NumberOfOptis:1, ' interval enclosure(s)');  writeln;
  if (NumberOfOptis = 1) and (unique[1]) then
    writeln('We have validated that there is a unique global optimizer!');
end;

begin { Main program }
  compute(f,  '(x + SIN(x))*EXP(-x↑2)');  writeln;  writeln;
  compute(g, '-SUM(k*SIN((k+1)*x+k),k,1,5)');
end.
```

If we execute this program, we get the following runtime output:

```
Computing all global minimizers of the function  (x + SIN(x))*EXP(-x^2)
Search interval      : [-10,10]
Tolerance (relative) : 1e-12

[ -6.795786600198818E-001, -6.795786600198812E-001 ]
encloses a locally unique candidate for a global minimizer!

[ -8.242393984760771E-001, -8.242393984760764E-001 ]
encloses the global minimum value!

1 interval enclosure(s)

We have validated that there is a unique global optimizer!


Computing all global minimizers of the function  -SUM(k*SIN((k+1)*x+k),k,1,5)
Search interval      : [-10,10]
Tolerance (relative) : 1e-12

[  5.791794470920271E+000,  5.791794470920272E+000 ]
encloses a locally unique candidate for a global minimizer!
[ -6.774576143438902E+000, -6.774576143438900E+000 ]
encloses a locally unique candidate for a global minimizer!
[ -4.913908362593147E-001, -4.913908362593144E-001 ]
encloses a locally unique candidate for a global minimizer!

[  -1.20312494421672E+001,  -1.20312494421671E+001 ]
encloses the global minimum value!

3 interval enclosure(s)
```

Thus, we know that there is only one global minimizer of f and three locally unique minimizers which are good candidates for global minimizers of g within the specified starting interval $[-10, 10]$.

7.3.3 Restrictions and Hints

The objective function f must be expressible in PASCAL–XSC code as a finite sequence of arithmetic operations and elementary functions supported by the differentiation arithmetic module *ddf_ari*.

As already mentioned, the procedure *AllGOp1* stores all enclosures in the interval vector *OptiVector* which must be of sufficient length. If the first run of *AllGOp1* is not able to compute all minimizers because *OptiVector* is not long enough, then the routine must be called again with an increased index range for *OptiVector*.

The method is not very fast if a very small value of ε (*Epsilon*) is used, if the interval Newton step does not improve the actual iterates, *and* if the different tests do not discard intervals any more because of rounding and the overestimation effects of the machine interval arithmetic. Under these circumstances, the method is equivalent with a bisection method.

In *GlobalOptimize*, the evaluation of the function in differentiation arithmetic can cause a runtime error if the interval argument of an elementary function does not lie in the domain specified for this interval function (see [65]) or if a division by an interval containing zero occurs. This also may be due to the known overestimation effects of interval arithmetic (see Section 3.1). To get rid of these errors, the user may try to split the starting interval in several parts and call *AllGOp1* for these parts.

Note that the rules for getting true enclosures in connection with conversion errors (see Section 3.7) also apply here. For example, the user who wishes to compute enclosures for the minimizers of the function $f(x) = (x - \frac{1}{3})^2 \cdot (x - 0.1)^2$, must be careful that the values $\frac{1}{3}$ and 0.1 are enclosed in machine intervals when computing interval evaluations of f. An appropriate coding of f is

```
function f (x: DerivType) : DerivType;
begin
  f:= sqr(x - intval(1) / 3) * sqr(x - intval(1) / 10);
end;
```

7.4 Exercises

Exercise 7.1 Use our procedure *AllGOp1* to compute the global minimizer and the global minimum value of the function

$$h(x) = \sin x + \sin \frac{10x}{3} + \ln x - \frac{84x}{100}.$$

This function is the test function f_1 from [85] shown in Figure 7.6 with starting interval $[2.7, 7.5]$. It is a rather simple problem with several local minima.

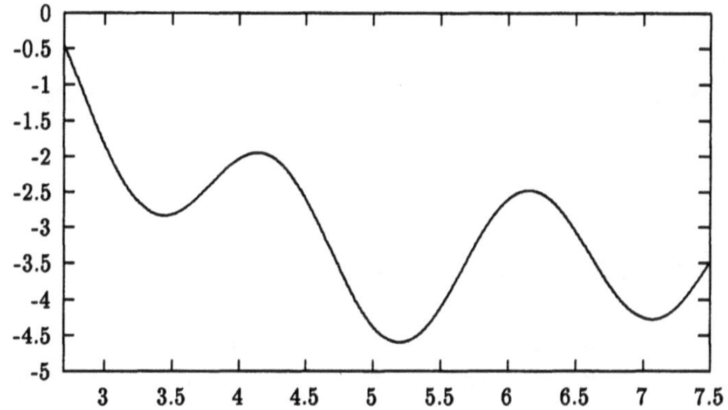

Figure 7.6: Function h (test function f_1 in [85]) with several local minima

Find an interval $[a, b]$ such that you can prove that the unique global (on $[a, b]$) minimizer located by *AllGOp1* is also the global minimizer in $\mathbb{R}$.

Exercise 7.2 Use our procedure *AllGOp1* to compute the global minimizer and the global minimum value of the function

$$r(x) = 2x^2 - \frac{3e^{-(a(x-b))^2}}{c}$$

with starting interval $[-10, 10]$. The global minimum of r is at a sharp valley for $a = 200$, $b = 0.0675$, and $c = 100$ as shown in Figure 7.7. Find a set of values a, b, and c for which an approximate optimization routine available to you fails. Hint: Make the valley very sharp, rather far from the local minimizer near $x = 0$, and deep enough to cut below the x-axis. Can you trick *AllGOp1* similarly?

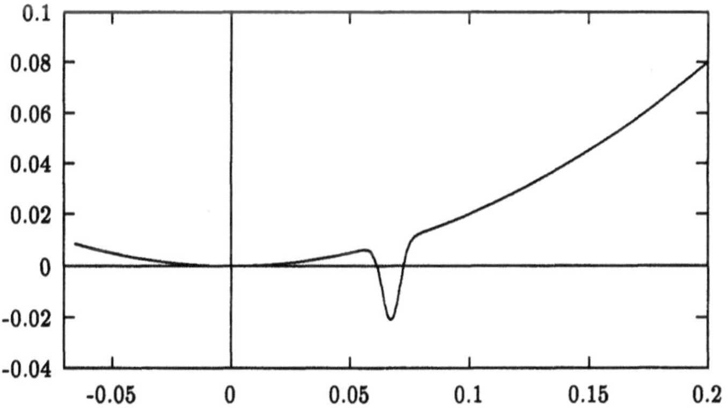

Figure 7.7: Test function r with a sharp valley

Exercise 7.3 Modify module *go1* to use a local approximate search strategy to locate a point c that may be more effective than the midpoint in eliminating subintervals from the list L (cf. Section 7.1.1). Be careful that your strategy does not leave the subinterval $[y]$. Is the resulting algorithm faster or slower than the one presented here?

7.5 References and Further Reading

The method we discussed in this chapter is an *a priori* method because the iteration starts with a (possibly large) interval enclosing all the solutions which have to be found. Here, the iterates of the method are subintervals of the previous iterates. There are also methods for finding (and bounding) one single *local* optimizer called *a posteriori* methods. These methods start with an approximation of a minimizer of f and apply a test procedure for a neighborhood interval of the approximation to verify that a zero of f' lies within that interval.

Our global optimization method can benefit from an approximation of a global minimizer in the way that we use that approximation to compute the value $\tilde{f}$ incorporated in the midpoint test. But we have to to do the same work for validating that the global minimizer has been found as for computing its value by our method. That is, a global optimization method with automatic result verification cannot be an *a posteriori* method.

For the approximation of a minimizer, there are also a huge number of classical optimization methods for local and global optimization available without any verification of the result. For an overview on such approximation methods see [85] or [88], for example.

The method presented in this chapter can easily be extended to become more efficient and faster. For more sophisticated extensions, see [25], [26], [27], [28], [35], [72], [73], and [74].

Chapter 8

Evaluation of Arithmetic Expressions

The evaluation of arithmetic expressions using floating-point arithmetic may lead to unpredictable results due to an accumulation of roundoff errors. As an example, the evaluation of $x + 1 - x$ for $x > 10^{20}$ using the standard floating-point format on almost every digital computer yields the wrong result 0. Since the evaluation of arithmetic expressions is a basic task in digital computing, we should have a method to evaluate a given expression to an arbitrary accuracy. We will develop such a method for real expressions composed of the operations $+$, $-$, $\cdot$, $/$, and $\uparrow$, where $\uparrow$ denotes exponentiation by an integer. The method is based on the principle of iterative refinement as discussed in Section 3.8.

8.1 Theoretical Background

There is a very simple method for evaluating a real arithmetic expression to a desired accuracy. Supposed we have a multiple-precision interval arithmetic. We could start evaluating the expression with single precision and check the relative diameter of the resulting interval. If the width is greater than the desired accuracy, we would repeat the evaluation with doubled precision, then in a triple-precision format, and so on. When we finally succeed, we would return the enclosing interval rounded to the target format specified by the user. Here, we suppose that the target format is the standard floating-point format. The method holds two principal disadvantages. First, we would have to implement a multiple-precision interval arithmetic. Secondly, an already computed approximation would be discarded instead of being used for the next step when restarting the evaluation with an increased precision. In the sequel, we will discuss a more sophisticated approach avoiding these disadvantages. See Fischer, Haggenmüller, and Schumacher [17] for a detailed description of the method.

8.1.1 A Nonlinear Approach

Let $f(x) = f(x_1, \ldots, x_n)$ be an arithmetic expression composed of n operands $x_1, \ldots, x_n$ and m operations. We define the set of legal operations by $\{+, -, \cdot, /, \uparrow\}$, where $\uparrow$ denotes exponentiation by an integer. We show how the evaluation of $f(x)$ is equivalent to finding the solution of a special nonlinear system of equations, which is similar to the *code lists* used in automatic differentiation techniques (cf. Rall [66]).

The procedure of transforming $f(x)$ into such a system is straight forward. Every operation of f defines a separate equation. The order of these equations is defined by the conventional rules of algebra for left-to-right evaluation.

Example 8.1 Let $f(x) = \frac{(x_1+x_3)^4(x_2-x_4)^{-3}}{x_5x_5}$. Thus, we have $n = 5$ and $m = 7$. If we denote the intermediate results of the left-to-right evaluation by x_{n+1} to x_{n+m}, then we get the following system of nonlinear equations

$$\begin{array}{lcl} x_6 &=& x_1 + x_3 \\ x_7 &=& x_6^4 \\ x_8 &=& x_2 - x_4 \\ x_9 &=& x_8^{-3} \\ x_{10} &=& x_7 \cdot x_9 \\ x_{11} &=& x_5 \cdot x_5 \\ x_{12} &=& x_{10}/x_{11} \end{array} \quad \Longleftrightarrow \quad \begin{array}{lcl} x_6 - (x_1 + x_3) &=& 0 \\ x_7 - x_6^4 &=& 0 \\ x_8 - (x_2 - x_4) &=& 0 \\ x_9 x_8^3 - 1 &=& 0 \\ x_{10} - x_7 \cdot x_9 &=& 0 \\ x_{11} - x_5 \cdot x_5 &=& 0 \\ x_{11} \cdot x_{12} - x_{10} &=& 0 \end{array}$$

with $x_{12} = x_{n+m} = f(x)$. Note that $x_{n+1}, \ldots, x_{n+m}$ are the unknowns of the system.

Once an approximate solution $(\widetilde{x}_{n+1}, \ldots, \widetilde{x}_{n+m})^{\mathrm{T}}$ of the nonlinear system is computed, the absolute error of any intermediate result is given by $\Delta x_\nu = x_\nu - \widetilde{x}_\nu$, $\nu = n+1, \ldots, n+m$. Central to our approach is the computation of enclosures of all Δx_ν. Thus, we get

$$x_\nu - \widetilde{x}_\nu \in [\Delta x]_\nu, \quad \nu = n+1, \ldots, n+m.$$

In particular, this yields $f(x) = x_{n+m} \in \widetilde{x}_{n+m} + [\Delta x]_{n+m}$.

To find enclosures of the absolute error of the intermediate results, we use the special form of the nonlinear equations. Each of the equations has the form

$$g(x_i, x_j, x_k) = 0, \quad i \le j < k, \tag{8.1}$$

where x_k is the new intermediate result. In fact, g has five special forms that will be individually considered shortly. If g has continuous partial derivatives (ours do), then expanding g about $\widetilde{x}_k$ yields

$$g(x_i, x_j, x_k) = g(x_i, x_j, \widetilde{x}_k) + \frac{\partial g}{\partial x_k}(x_i, x_j, \xi_k)\Delta x_k,$$

with ξ_k lying in the hull of x_k and $\widetilde{x}_k$. Similarly, we expand $g(x_i, x_j, \widetilde{x}_k)$ about $\widetilde{x}_j$

$$g(x_i, x_j, \widetilde{x}_k) = g(x_i, \widetilde{x}_j, \widetilde{x}_k) + \frac{\partial g}{\partial x_j}(x_i, \xi_j, \widetilde{x}_k)\Delta x_j,$$

and then $g(x_i, \widetilde{x}_j, \widetilde{x}_k)$ about $\widetilde{x}_i$ to get

$$0 = g(\widetilde{x}_i, \widetilde{x}_j, \widetilde{x}_k) + \frac{\partial g}{\partial x_k}(x_i, x_j, \xi_k)\Delta x_k + \frac{\partial g}{\partial x_j}(x_i, \xi_j, \widetilde{x}_k)\Delta x_j + \frac{\partial g}{\partial x_i}(\xi_i, \widetilde{x}_j, \widetilde{x}_k)\Delta x_i.$$

Here, we have $\xi_\nu \in x_\nu \underline{\cup} \tilde{x}_\nu$ for $\nu = i, j, k$. Hence, we obtain the following representation of the absolute error of the k-th intermediate result:

$$\Delta x_k = \frac{-g(\tilde{x}_i, \tilde{x}_j, \tilde{x}_k) - \frac{\partial g}{\partial x_j}(x_i, \xi_j, \tilde{x}_k)\Delta x_j - \frac{\partial g}{\partial x_i}(\xi_i, \tilde{x}_j, \tilde{x}_k)\Delta x_i}{\frac{\partial g}{\partial x_k}(x_i, x_j, \xi_k)}. \tag{8.2}$$

We will now use (8.2) to derive explicit rules for computing enclosures of Δx_k. We will see that each of the partial derivatives in Equation (8.2) is easily computable for each of the five special forms of g we consider below. For this, we assume that enclosures $[\Delta x]_i$ and $[\Delta x]_j$ of Δx_i and Δx_j are already known. Actually, this is no restriction, since $[\Delta x]_\nu = 0$ for the exactly known input parameters i.e. for $\nu = 1, \ldots, n$. See Exercise 8.6 for the treatment of interval input parameters.

Addition and Subtraction: Here, we have $g(x_i, x_j, x_k) = x_k - (x_i \pm x_j)$. Applying Equation (8.2), we get the following enclosure of Δx_k

$$\Delta x_k = (\tilde{x}_i \pm \tilde{x}_j - \tilde{x}_k) \pm \Delta x_j + \Delta x_i \in \underbrace{\Diamond(\tilde{x}_i \pm \tilde{x}_j - \tilde{x}_k) \pm [\Delta x]_j + [\Delta x]_i}_{[\Delta x]_k}. \tag{8.3}$$

Multiplication: For $g(x_i, x_j, x_k) = x_k - x_i x_j$, Equation (8.2) gives an expression where the unknown x_i is replaced by its enclosure $\tilde{x}_i + [\Delta x]_i$. Thus, we get

$$\begin{aligned} \Delta x_k &= (\tilde{x}_i \tilde{x}_j - \tilde{x}_k) + x_i \Delta x_j + \tilde{x}_j \Delta x_i \\ &\in \underbrace{\Diamond(\tilde{x}_i \tilde{x}_j - \tilde{x}_k) + (\tilde{x}_i + [\Delta x]_i)[\Delta x]_j + \tilde{x}_j [\Delta x]_i}_{[\Delta x]_k}. \end{aligned} \tag{8.4}$$

Division: Here, we have $g(x_i, x_j, x_k) = x_j x_k - x_i$. Applying Equation (8.2) and replacing x_j by $\tilde{x}_j + [\Delta x]_j$ yields

$$\Delta x_k = \frac{(\tilde{x}_i - \tilde{x}_j \tilde{x}_k) - \tilde{x}_k \Delta x_j + \Delta x_i}{x_j} \in \underbrace{\frac{\Diamond(\tilde{x}_i - \tilde{x}_j \tilde{x}_k) - \tilde{x}_k \Delta x_j + \Delta x_i}{\tilde{x}_j + [\Delta x]_j}}_{[\Delta x]_k}. \tag{8.5}$$

Exponentiation: We consider exponentiation by an integer in three cases. For a zero exponent, we have $x_k = 1$. Hence, we get $[\Delta x]_k = 0$. For positive exponents $n > 0$, we have $g(x_j, x_k) = x_k - x_j^n$. Applying Equation (8.2) and enclosing ξ_j by $x_j \underline{\cup} \tilde{x}_j \in (\tilde{x}_j + [\Delta x]_j) \underline{\cup} \tilde{x}_j = \tilde{x}_j + (0 \underline{\cup} [\Delta x]_j)$, we get the following enclosure of Δx_k

$$\Delta x_k = (\tilde{x}_j^n - \tilde{x}_k) + n\xi_j^{n-1}\Delta x_j \in \underbrace{\Diamond(\tilde{x}_j^n - \tilde{x}_k) + n(\tilde{x}_j + (0 \underline{\cup} [\Delta x]_j))^{n-1}[\Delta x]_j}_{[\Delta x]_k}. \tag{8.6}$$

For negative exponents $n < 0$, we use $g(x_j, x_k) = x_k x_j^{-n} - 1$. In a similar way as for positive exponents, we get

$$\begin{aligned} \Delta x_k &= \frac{\Diamond(1 - \tilde{x}_k \tilde{x}_j^{-n}) + n\tilde{x}_k \xi_j^{-(n+1)}[\Delta x]_j}{(\tilde{x}_j + [\Delta x]_j)^{-n}} \\ &\in \underbrace{\frac{\Diamond(1 - \tilde{x}_k \tilde{x}_j^{-n}) + n\tilde{x}_k(\tilde{x}_j + (0 \cup [\Delta x]_j))^{-(n+1)}[\Delta x]_j}{(\tilde{x}_j + [\Delta x]_j)^{-n}}}_{[\Delta x]_k}. \end{aligned} \tag{8.7}$$

Elementary functions: We give the enclosure (8.8) for an elementary function $s(t) : I\!R \to I\!R$ for completeness, but our implementation does not support the accurate evaluation of expressions involving elementary functions. For this special case, we have $g(x_j, x_k) = x_k - s(x_j)$. Thus, we get

$$\Delta x_k = s(\tilde{x}_j) - \tilde{x}_k + s'(\xi_j)[\Delta x]_j \in \underbrace{\Diamond(s(\tilde{x}_j) - \tilde{x}_k) + s'(\tilde{x}_j + (0 \cup [\Delta x]_j))[\Delta x]_j}_{[\Delta x]_k}. \tag{8.8}$$

We point out that it is essential for the convergence of the algorithms described below to enclose the terms for $-g(\tilde{x}_i, \tilde{x}_j, \tilde{x}_k)$ in intervals as tight as possible. We have indicated this by the $\Diamond(\ldots)$ notation. That is, these terms should be evaluated with only one final interval rounding.

8.2 Algorithmic Description

The explanations of the previous section motivate the following simple iterative refinement scheme. Let f be an arithmetic expression depending on n real-valued parameters $x_1, \ldots, x_n$. Initially, set $\tilde{x}_\nu = x_\nu$ for $\nu = 1, \ldots, n$, and compute an approximation of $f(x)$ saving the m intermediate results $\tilde{x}_{n+1}, \ldots, \tilde{x}_{n+m}$. The order of the intermediate results is given by the conventional rules for left-to-right evaluation. Set $[\Delta x]_\nu = 0$ for the exactly known parameters $x_1, \ldots, x_n$. Start the iteration by computing enclosures of the absolute error of each intermediate result using Equations (8.3)–(8.8). The procedure is finished if the relative diameter of $[y] = \tilde{x}_{n+m} + [\Delta x]_{n+m}$ is less than the desired accuracy $\varepsilon > 0$. Otherwise, improve the approximation of each intermediate result by adding the midpoint of the corresponding error interval. Repeat the procedure to get a, hopefully, better $[y]$. The iteration fails if the desired accuracy ε can not be satisfied after $k_{\max}$ steps, where $k_{\max}$ is only restricted by the memory available on an actual machine.

Algorithm 8.1: NaiveEval $(f, x, \varepsilon, [y], k, Err)$ {Procedure}

1. {Initialization: Max. number of iteration steps, error code, input parameters}
   $k_{\max} := 10$; $Err :=$ "No error";
   $\tilde{x}_\nu^{(0)} := x_\nu$; $[\Delta x]_\nu := 0$; $(\nu = 1, \ldots, n)$

{Computation of a first approximation}
Compute $\widetilde{x}_\nu^{(0)}$; $(\nu = n+1,\ldots,n+m)$
3. {Iterative refinement}
$k := 0$;
repeat
Compute $[\Delta x]_\nu$; $(\nu = n+1,\ldots,n+m)$
$[y] := \widetilde{x}_{n+m}^{(k)} + [\Delta x]_{n+m}$;
Success := $(d_{\text{rel}}([y]) \leq \varepsilon)$;
if (**not** Success) **then** {Add correction terms}
$\widetilde{x}_\nu^{(k+1)} := \widetilde{x}_\nu^{(k)} + m([\Delta x]_\nu)$; $(\nu = n+1,\ldots,n+m)$
$k := k+1$;
until *Success* **or** $(k \geq k_{\max})$;
4. {Return: Enclosure, corrections needed, and error message}
if (**not** *Success*) **then**
Err := "Cannot satisfy the desired accuracy, $k_{\max}$ is to small!";
return $[y], k, Err$;

In general, Algorithm 8.1 may not yield an enclosure that is sufficiently tight if a finite precision arithmetic is used. To satisfy a desired accuracy for the result, it is often necessary to compute some or all intermediate results with an increased precision. For this reason, we will introduce a special kind of multi-precision arithmetic. Instead of adding the correction term $m([\Delta x]_\nu)$ to get a new approximation of an intermediate result (Step 3), we will store its value separately. An intermediate result is stored in the staggered correction form $\widetilde{x}_\nu = \sum_{\mu=0}^{p} \widetilde{x}_\nu^{(\mu)}$, where p is the actual degree of precision. See Stetter [82] for more details on the staggered correction representation. For our purposes, every component of this representation is assumed to be an element of a number screen $R(b, l, e_{\min}, e_{\max})$ as introduced in Section 3.5. Some of these components may have overlapping mantissae, so a staggered representation is not unique as Example 8.2 demonstrates.

Example 8.2 Let $t = 0.123456$ and $R = R(10, 3, -5, +5)$ be a decimal screen with three significant digits. There are different staggered representations of t using components of R. For instance, we have

$$\begin{aligned} t &= 0.103 \cdot 10^0 + 0.203 \cdot 10^{-1} + 0.156 \cdot 10^{-3} \\ &= 0.133 \cdot 10^0 - 0.961 \cdot 10^{-2} + 0.660 \cdot 10^{-4}. \end{aligned}$$

We will now modify Algorithm 8.1 to employ a staggered representation of the intermediate results. For reasons of simplicity, we also use a staggered representation for the input parameters. Since these values are exactly known, we set $\widetilde{x}_\nu^{(0)} = x_\nu$ and $\widetilde{x}_\nu^{(\mu)} = 0$ for $\mu > 0$ and $\nu = 1,\ldots,n$. We could extend this algorithm to support user entry of the parameters $x_1,\ldots,x_n$ in a staggered correction format, but we do not do so here (see Exercise 8.5). In Algorithm 8.2, the actual degree of precision is denoted by p. We use the same p for each intermediate result. When the algorithm has finished, then p is a kind of condition number for the problem. Large values of

p signal that the evaluation problem was poorly conditioned. The representation of the intermediate results in the staggered format has its effects on the computation of the enclosures for the absolute error. Thus, we will refer to Algorithms 8.3–8.6 for a detailed description of how to compute these intervals.

Algorithm 8.2: Eval $(f, x, \varepsilon, [y], p, Err)$ {Procedure}

1. {Initialization: Max. number of corrections, error code, input parameters}
   $p_{\max} := 10$; $Err :=$ "No error";
   $\tilde{x}_\nu^{(0)} := x_\nu$; $\tilde{x}_\nu^{(\mu)} := 0$; $[\Delta x]_\nu := 0$; $(\mu = 1, \ldots, p_{\max}, \ \nu = 1, \ldots, n)$

2. {Computation of a first approximation}
   Compute $\tilde{x}_\nu^{(0)}$; $(\nu = n+1, \ldots, n+m)$

3. {Iterative refinement}
   $p := 0$;
   repeat
   Compute $[\Delta x]_\nu$; $(\nu = n+1, \ldots, n+m)$ using Algorithms 8.3–8.6
   $[y] := \left(\sum_{\mu=0}^{p} \tilde{x}_{n+m}^{(\mu)} + [\Delta x]_{n+m} \right)$;
   $Success := (d_{\text{rel}}([y]) \leq \varepsilon)$;
   if (**not** Success) **and** $(p < p_{\max})$ **then** {Store next staggered component}
   $\tilde{x}_\nu^{(p+1)} := m([\Delta x]_\nu)$; $(\nu = n+1, \ldots, n+m)$
   $p := p + 1$;
   until *Success* **or** $(p \geq p_{\max})$;

4. {Return: Enclosure, corrections needed, and error message}
   if (**not** *Success*) **then**
   $Err :=$ "Cannot satisfy the desired accuracy, $p_{\max}$ is to small!";
   return $[y], p, Err$;

As mentioned earlier, we do not support elementary functions. This is because would need elementary functions that accept arguments in a staggered correction format. An implementation of such functions is beyond the scope of this book (see Krämer [47]). This is also the reason for we do not give an algorithmic description of the exponentiation by an integer for arguments in a staggered correction format. To implement Equations (8.6) and (8.7) accurately, we would have to implement powers of operands in a staggered correction format. To avoid this, we will consider an integer exponentiation as a sequence of multiplications or divisions until the end of this chapter (see Algorithm A.1 in the appendix for details). Hence, the number of intermediate results increases by formally substituting these operations.

Let us now describe the algorithms for computing $[\Delta x]_\nu$ for the basic operations $+, -, \cdot$, and $/$. In order to simplify the notation slightly, we identify the parameters x_i, x_j, and x_k as x, y, and z, respectively.

Algorithm 8.3: $\mathsf{Add}(\widetilde{x},\widetilde{y},\widetilde{z},[\Delta x],[\Delta y],[\Delta z])$ {Procedure}

1. {Evaluation of (8.3) for operands in stagg. corr. format of precision p}

   $$[\Delta z] := \Diamond\Big(\sum_{\mu=0}^{p}\widetilde{x}^{(\mu)} + \sum_{\mu=0}^{p}\widetilde{y}^{(\mu)} - \sum_{\mu=0}^{p}\widetilde{z}^{(\mu)} + [\Delta x] + [\Delta y]\Big);$$

2. **return** $[\Delta z]$;

Algorithm 8.4: $\mathsf{Sub}(\widetilde{x},\widetilde{y},\widetilde{z},[\Delta x],[\Delta y],[\Delta z])$ {Procedure}

1. {Evaluation of (8.3) for operands in stagg. corr. format of precision p}

   $$[\Delta z] := \Diamond\Big(\sum_{\mu=0}^{p}\widetilde{x}^{(\mu)} - \sum_{\mu=0}^{p}\widetilde{y}^{(\mu)} - \sum_{\mu=0}^{p}\widetilde{z}^{(\mu)} + [\Delta x] - [\Delta y]\Big);$$

2. **return** $[\Delta z]$;

Algorithm 8.5: $\mathsf{Mul}(\widetilde{x},\widetilde{y},\widetilde{z},[\Delta x],[\Delta y],[\Delta z])$ {Procedure}

1. {Evaluation of (8.4) for operands in stagg. corr. format of precision p}

   $$[\Delta z] := \Diamond\Big(\sum_{\mu=0}^{p}\sum_{\nu=0}^{p}\widetilde{x}^{(\mu)}\cdot\widetilde{y}^{(\nu)} - \sum_{\mu=0}^{p}\widetilde{z}^{(\mu)} + \sum_{\mu=0}^{p}\widetilde{x}^{(\mu)}\cdot[\Delta y] + \sum_{\mu=0}^{p}\widetilde{y}^{(\mu)}\cdot[\Delta x] + [\Delta x]\cdot[\Delta y]\Big);$$

2. **return** $[\Delta z]$;

Algorithm 8.6: $\mathsf{Div}(\widetilde{x},\widetilde{y},\widetilde{z},[\Delta x],[\Delta y],[\Delta z])$ {Procedure}

1. {Evaluation of (8.5) for operands in stagg. corr. format of precision p}

   $$[u] := \Diamond\Big(\sum_{\mu=0}^{p}\widetilde{x}^{(\mu)} - \sum_{\mu=0}^{p}\sum_{\nu=0}^{p}\widetilde{z}^{(\nu)}\cdot\widetilde{y}^{(\mu)} - \sum_{\mu=0}^{p}\widetilde{z}^{(\mu)}\cdot[\Delta y] + [\Delta x]\Big);$$

   $$[v] := \Diamond\Big(\sum_{\mu=0}^{p}\widetilde{y}^{(\mu)} + [\Delta y]\Big);$$

2. **return** $[\Delta z] := [u]/[v]$;

8.3 Implementation and Examples

8.3.1 PASCAL–XSC Program Code

In this section, we give the implementation of the module *expreval* for the accurate evaluation of real arithmetic expressions. The module is based on a data type

Staggered that holds components for both the approximation and the error interval. We give the definition of a special arithmetic including the operators $+$, $-$, $\cdot$, $/$, and $\uparrow$. A little programming trick implicitly transforms an expression into a system of nonlinear equations so that evaluating the expression using this special arithmetic automatically yields a user-specified accuracy for the result. We stress that those parts of the algorithms of the previous section marked by $\diamond(\ldots)$ are implemented using the accurate expressions of PASCAL-XSC.

8.3.1.1 Module expreval

The module provides the operators $+$, $-$, $\cdot$, and $/$ for both operands of type *Staggered* and for mixed operands of type *Staggered* and *real*. The function *Power* is an implementation of the operator $\uparrow$ using the binary shift method as described in Algorithm A.1 of the appendix. The module also includes an overloaded assignment operator for converting *real* data to data of type *Staggered*. The main procedure *Eval* is an implementation of Algorithm 8.2. The function *EvalErrMsg* may be used to get an explicit error message text.

As mentioned above, we use a programming trick to transform an arithmetic expression implicitly into an equivalent system of nonlinear equations. We first convert all parameters in an expression f to operands of type *Staggered*. When f is evaluated for the first time using the special staggered arithmetic, every operator appends its result into a linear linked list. We may access these values if we restart the evaluation of f a second time, this time for computing enclosures of the error intervals according to the Algorithms 8.3–8.6. This list now may be used to update the staggered components of the intermediate results as described in Algorithm 8.2. This trick works because the code list for f is always evaluated in exactly the same order each time f is evaluated. All procedures needed for allocation and management of the internal list are declared locally, so they are hidden for programs which use the module.

Another simple trick is used to avoid an abnormal program termination caused by a division by zero. If a zero denominator occurs while executing the operator $/$, the flag *DivByZero* is set. Since all operators check this flag before starting execution, this terminates the evaluation without forcing a runtime exception. An update of those intermediate results computed before terminating the evaluation will result in an improved approximation for the denominator. Thus, it is our hope to overcome this critical point of evaluation with the next iteration step. Any other exceptions are handled by the standard exception handler of PASCAL–XSC.

```
{----------------------------------------------------------------------------}
{ Purpose: Computation of an enclosure of the value of a real arithmetic    }
{    expression composed of the operations +, -, *, /, and ↑, where ↑       }
{    denotes exponentiation by an integer.                                   }
{ Method: Iterative refinement using a defect correction mechanism. The     }
{    successively computed corrections for the result are stored in a       }
{    staggered correction format.                                            }
{ Global types, operators, functions, and procedures:                        }
{   type       Staggered   :  Data type for staggered data representation   }
{   type       StaggArray  :  Array of type 'Staggered'                     }
{   operators :=           :  Assignment of 'Staggered' := 'real'           }
```

```
{   operators +, -, *, / :  Both operands of type 'Staggered' or one of       }
{                           type 'Staggered' and one of type 'real'            }
{   function  Power(...) :  Argument of type 'Staggered', exponent of type    }
{                           'integer' (Exponentiation by an integer)           }
{   procedure Eval(...)  :  Main procedure for evaluation of an expression    }
{----------------------------------------------------------------------------}
module expreval;   { Expression evaluation }
                   { ----       ----       }

use
  x_real,    { Needed if a signaling NaN is to be returned }
  i_util,    { Utilities of type interval                  }
  i_ari;     { Interval arithmetic                         }

const
  MaxStaggPrec = 10;     { Maximum number of staggered corrections which can  }
                         { be stored in a variable of type 'Staggered'.       }
                         { Recompile this module if its value is changed.     }
global type
  Staggered  = record
                 Val : array[0..MaxStaggPrec] of real;  { Stagg. corrections }
                 Err : interval;                         { Error enclosure    }
               end;
  StaggArray = global dynamic array [*] of Staggered;

type
  IntResPtr = ↑IntRes;    { Pointer to an intermediate result }
  IntRes    = record
                Entry : Staggered;        { Staggered entry }
                Next  : IntResPtr;        { Next entry      }
              end;

{----------------------------------------------------------------------------}
{ Private variables which are globally used within other functions and      }
{ procedures of this module.                                                  }
{----------------------------------------------------------------------------}
var
  DivByZero    : boolean;      { Error flag                                }
  InitFlag     : boolean;      { Signals the initialization process        }
  ActStaggPrec : integer;      { Actual length of the staggered format     }
  HeadPtr      : IntResPtr;    { Head of the list of intermediate results  }
  ActPtr       : IntResPtr;    { Pointer to the actual intermediate result }
  FreePtr      : IntResPtr;    { Pointer to freed intermediate results     }

{----------------------------------------------------------------------------}
{ Error codes used in this module.                                            }
{----------------------------------------------------------------------------}
const
  NoError   = 0;   { No error occurred.                                }
  ItFailed  = 1;   { Maximum number of staggered corrections exceeded. }
  ZeroDiv   = 2;   { Division by zero that could not be                }
                   { removed by iterative refinement.                  }

{----------------------------------------------------------------------------}
{ Error messages depending on the error code.                                 }
{----------------------------------------------------------------------------}
global function EvalErrMsg ( ErrNo : integer ) : string;
var
  Msg : string;
begin
  case ErrNo of
    NoError  : Msg := '';
    ItFailed : Msg := 'Maximum number of staggered corrections (=' +
                      image(MaxStaggPrec) + ') exceeded';
    ZeroDiv  : Msg := 'Division by zero occurred';
```

```
    else      : Msg := 'Code not defined';
  end;
  if (ErrNo <> NoError) then Msg := 'Error: ' + Msg + '!';
  EvalErrMsg := Msg;
end;

{----------------------------------------------------------------------------}
{ Since the intermediate results of operations for staggered data must be   }
{ iteratively updated, it is necessary to store these data. For this        }
{ purpose, a linear linked list is used. Its head is accessed via 'HeadPtr', }
{ whereas its actual entry is accessed via 'ActPtr'. 'FreePtr' is a pointer  }
{ to a list of already allocated but actually unused entries. To prevent    }
{ creation of garbage in memory, the entries of this list are used first    }
{ when allocating a new intermediate result. All procedures for the handling }
{ of the list of intermediate results are locally defined.                  }
{----------------------------------------------------------------------------}
procedure InitList;                { Initialize list of intermediate results }
begin                              {-----------------------------------------}
  DivByZero := false;
  if (HeadPtr <> nil) then       { Use list of freed entries }
    FreePtr := HeadPtr
  else                           { A list was not yet allocated }
    FreePtr := nil;
  HeadPtr := nil;
  ActPtr := nil;
end;

procedure ResetList;                   { Reset error flag and set the actual }
begin                                  { pointer to the head of the list.    }
  DivByZero := false;                  {-------------------------------------}
  ActPtr := HeadPtr;
end;

procedure AllocEntry ( var p : IntResPtr );    { Allocate memory for a new   }
begin                                          { entry. Use a previously     }
  if (FreePtr <> nil) then                     { created entry from the list }
    begin                                      { of freed entries if any.    }
      p := FreePtr;                            {-----------------------------}
      FreePtr := FreePtr↑.next;
    end
  else
    new(p);
end;

procedure InitEntry ( Approx : real );  { Get a new list entry and initial- }
var                                     { ize its first staggered component }
  p : IntResPtr;                        { with the value 'Approx'.          }
begin                                   {-----------------------------------}
  AllocEntry(p);
  p↑.Entry.Val[0] := Approx;
  p↑.Next := nil;
  if (HeadPtr = nil) then
    begin  HeadPtr := p;  ActPtr := p;  end
  else
    begin  ActPtr↑.Next := p;  ActPtr := p;  end;
end;

procedure UpdateError ( Error : interval );     { Update the error component }
begin                                           { of the actual entry.       }
  ActPtr↑.Entry.Err := Error;                   {----------------------------}
  ActPtr := ActPtr↑.Next;
end;

procedure UpdateStaggComp ( i : integer );      { Update the i-th staggered  }
```

```
begin                                                     { component of all list      }
  ActPtr := HeadPtr;                                      { entries by the midpoint of }
  while (ActPtr <> nil) do                                { the error interval.        }
  begin                                                   {----------------------------}
    with ActPtr↑.Entry do
    begin
      Val[i] := mid(Err);  Err := 0;
    end;
    ActPtr := ActPtr↑.Next;
  end;
end;

{-----------------------------------------------------------------------------}
{ Assignment operator used to convert a real input parameter 'r' to the       }
{ staggered type. A real operand is assumed to be exact! It must not have     }
{ been rounded. In particular, 'r' cannot be                                  }
{    - a real value such as 0.1 which is not exactly representable in the     }
{      internal binary format,                                                }
{    - rounded by conversion to the internal binary data format during        }
{      input,                                                                 }
{    - any arithmetic expression which cannot be evaluated exactly.           }
{-----------------------------------------------------------------------------}
global operator := ( var x : Staggered; r : real );
var
  i : integer;
begin
  x.Val[0] := r;
  for i := 1 to MaxStaggPrec do x.Val[i] := 0;
  x.Err := 0;
end;

{-----------------------------------------------------------------------------}
{ Arithmetic operators +, -, *, and / for operands of type 'Staggered'. A     }
{ division by an interval containing zero will be avoided and is marked by    }
{ setting the 'DivByZero' flag. If 'DivByZero' is set, all succeeding opera-  }
{ tions are not executed. The evaluation is stopped at this point but may be  }
{ restarted after updating a new staggered component by the midpoints of the  }
{ error enclosures computed so far.                                           }
{-----------------------------------------------------------------------------}
global operator + ( x, y : Staggered ) AddStaggStagg : Staggered;
var
  i, p : integer;
  z    : Staggered;
begin
  if (not DivByZero) then
  begin
    if (InitFlag) then   { Initialize first staggered component }
      begin
        z.Val[0] := x.Val[0] + y.Val[0];
        InitEntry(z.Val[0]);
      end
    else
      begin
        z := ActPtr↑.Entry;  p := ActStaggPrec;  { Get actual values of z }

        { Error:  dz := ##( x + y - z + dx + dy ) }
        z.Err := ##( for i:=0 to p sum( x.Val[i] + y.Val[i] - z.Val[i] ) +
                     x.Err + y.Err );
        UpdateError(z.Err);
      end;
    AddStaggStagg := z;
  end;
end; {operator +}
```

```
global operator - ( x, y : Staggered ) SubStaggStagg : Staggered;
var
  i, p : integer;
  z    : Staggered;
begin
  if (not DivByZero) then
  begin
    if (InitFlag) then   { Initialize first staggered component }
      begin
        z.Val[0] := x.Val[0] - y.Val[0];
        InitEntry(z.Val[0]);
      end
    else
      begin
        z := ActPtr↑.Entry;  p := ActStaggPrec;  { Get actual values of z }

        { Error:  dz := ##( x - y - z + dx - dy ) }
        z.Err := ##( for i:=0 to p sum( x.Val[i] - y.Val[i] - z.Val[i] ) +
                     x.Err - y.Err );
        UpdateError(z.Err);
      end;
    SubStaggStagg := z;
  end;
end; {operator -}

global operator * ( x, y : Staggered ) MulStaggStagg : Staggered;
var
  i, j, p : integer;
  z       : Staggered;
begin
  if (not DivByZero) then
  begin
    if (InitFlag) then   { Initialize first staggered component }
      begin
        z.Val[0] := x.Val[0] * y.Val[0];
        InitEntry(z.Val[0]);
      end
    else
      begin
        z := ActPtr↑.Entry;  p := ActStaggPrec;  { Get actual values of z }

        { Error:  dz := ##( x*y - z + y*dx + x*dy + dx*dy ) }
        z.Err := ##( for i:=0 to p sum( for j:=0 to p sum(
                       x.Val[i]*y.Val[j] ) - z.Val[i] ) +
                     for i:=0 to p sum(
                       y.Val[i]*x.Err + x.Val[i]*y.Err ) +
                     x.Err*y.Err );
        UpdateError(z.Err);
      end;
    MulStaggStagg := z;
  end;
end; {operator *}

global operator / ( x, y : Staggered ) DivStaggStagg : Staggered;
var
  i, j, p     : integer;
  z           : Staggered;
  num, denom  : interval;
begin
  if (not DivByZero) then
  begin
    if (InitFlag) then   { Initialize first staggered component }
      if (y.Val[0] <> 0) then
        begin
```

```
          z.Val[0] := x.Val[0] / y.Val[0];
          InitEntry(z.Val[0]);
        end
      else
        begin
          DivByZero := true;
          InitEntry(0);
        end
    else
      begin
        z := ActPtr↑.Entry;  p := ActStaggPrec;  { Get actual values of z }

        { Error:  dz := ##( x - z*y + dx - z*dy )  /  ##( y + dy ) }
        num := ##( for i:=0 to p sum( x.Val[i] - for j:=0 to p sum(
                        y.Val[j]*z.Val[i] ) ) +
                    x.Err - for i:=0 to p sum( z.Val[i]*y.Err ) );
        denom := ##( for i:=0 to p sum( y.Val[i] ) + y.Err );

        if (0 in denom) then
          begin  z.Err := 0;  DivByZero:= true;  end
        else
          z.Err := num / denom;

        UpdateError(z.Err);
      end;
    DivStaggStagg := z;
  end;
end; {operator /}

{----------------------------------------------------------------------------}
{ Arithmetic operators for different operands the one of type 'real' and the }
{ other one of type 'Staggered'. All these operators are implemented by      }
{ first coercing both operands to 'Staggered' type and then calling the      }
{ corresponding operators for the type 'Staggered'.                          }
{----------------------------------------------------------------------------}
global operator + ( x : real; y : Staggered ) AddRealStagg : Staggered;
var
  z : Staggered;
begin
  z := x;  AddRealStagg := z + y;
end;

global operator + ( x : Staggered; y : real ) AddStaggReal : Staggered;
var
  z : Staggered;
begin
  z := y;  AddStaggReal := x + z;
end;

global operator - ( x : real; y : Staggered ) SubRealStagg : Staggered;
var
  z : Staggered;
begin
  z := x;  SubRealStagg := z - y;
end;

global operator - ( x : Staggered; y : real ) SubStaggReal : Staggered;
var
  z : Staggered;
begin
  z := y;  SubStaggReal := x - z;
end;

global operator * ( x : real; y : Staggered ) MulRealStagg : Staggered;
```

```
var
  z : Staggered;
begin
  z := x;  MulRealStagg := z * y;
end;

global operator * ( x : Staggered; y : real ) MulStaggReal : Staggered;
var
  z : Staggered;
begin
  z := y;  MulStaggReal := x * z;
end;

global operator / ( x : real; y : Staggered ) DivRealStagg : Staggered;
var
  z : Staggered;
begin
  z := x;  DivRealStagg := z / y;
end;

global operator / ( x : Staggered; y : real ) DivStaggReal : Staggered;
var
  z : Staggered;
begin
  z := y;  DivStaggReal := x / z;
end;

{----------------------------------------------------------------------------}
{ Power function for integer exponents using the binary shift method. Thus,  }
{ the number of successive multiplications is reduced from n to log(2,n).    }
{ Since x↑n is considered to be a monomial, we define x↑0 := 1.              }
{----------------------------------------------------------------------------}
global function Power ( x : Staggered; n : integer ) : Staggered;
var
  m    : integer;
  p, z : Staggered;
begin
  if (not DivByZero) then
  begin
    if (n = 0) then
      Power := 1
    else
      begin
        if (n > 0) then m := n  else  m := -n;

        p := 1;  z := x;                        { Binary shift method }
        while (m > 0) do                        {---------------------}
        begin
          if (m mod 2 = 1) then p := p * z;
          m := m div 2;
          if (m > 0) then z := z * z;
        end;

        if (n > 0) then  Power := p  else  Power := 1/p;
      end;
  end;
end; {function Power}

{----------------------------------------------------------------------------}
{ Purpose: The procedure 'Eval' may be used for the computation of an        }
{    enclosure of the value of a real arithmetic expression composed of the }
{    operations +, -, *, /, and ↑, where ↑ denotes exponentiation by an      }
{    integer.                                                                }
{ Parameters:                                                                }
```

```
{    In  : 'f'          : a function of type 'Staggered' whose arguments are  }
{                         passed in an array of type 'Staggered'.             }
{          'Arg'        : the real-valued arguments of 'f' stored as          }
{                         components of a real vector.                        }
{          'Eps'        : desired accuracy.                                   }
{    Out : 'Approx'     : result computed with standard floating-point        }
{                         arithmetic.                                         }
{          'Encl'       : verified enclosure of the result.                   }
{          'StaggPrec'  : number of corrections needed.                       }
{          'Err'        : error code.                                         }
{ Description:                                                                }
{   The expression 'f' is evaluated for the real arguments which are stored   }
{   sequentially in the vector 'Arg'. Initially, the real arguments are       }
{   converted to arguments of type 'Staggered'. When 'f' is evaluated for     }
{   the first time using the special staggered arithmetic, the list of        }
{   intermediate results is initialized. Each time 'f' is evaluated again,    }
{   the error of every intermediate result is enclosed. The midpoints of      }
{   these enclosures are used to update the intermediate results. The         }
{   iteration is finished if the error of the last intermediate result        }
{   (= value of 'f') is less than the desired accuracy. Otherwise, the        }
{   iteration is halted after 'MaxStaggPrec' steps.                           }
{-----------------------------------------------------------------------------}
global procedure Eval ( function f ( v : StaggArray ) : Staggered;
                            Arg       : rvector;
                            Eps       : real;
                        var Approx    : real;
                        var Encl      : interval;
                        var StaggPrec : integer;
                        var Err       : integer);
var
  i         : integer;
  x         : StaggArray[1..ub(Arg)-lb(Arg)+1];
  StaggRes  : Staggered;
  Success   : boolean;
begin
  for i := 1 to ub(Arg)-lb(Arg)+1 do                 { Initialize arguments }
    x[i] := Arg[lb(Arg)+i-1];

  InitList;                     { Initialize list for intermediate results.  }
  InitFlag := true;             { So far, no staggered corrections computed. }
  StaggRes := f(x);             { Compute first staggered component.         }

  { In general, the first component of 'StaggRes' will now hold the usual    }
  { floating-point approximation, except when a division by zero occurred.   }
  { In the latter case, a signaling NaN (= Not a Number, for more details    }
  { see the PASCAL-XSC user's guide) is returned.                            }
  {--------------------------------------------------------------------------}
  if (DivByZero) then
    Approx := x_value(x_sNaN)
  else
    Approx := StaggRes.Val[0];
                                                { Initial approximations are }
  InitFlag := false;  ActStaggPrec := 0;        { already computed.          }
  repeat
    ResetList;                 { Compute new enclosures of the absolute error }
    StaggRes := f(x);          { of any intermediate result.                  }

    Encl := ##(for i:=0 to ActStaggPrec sum(StaggRes.Val[i]) + StaggRes.Err);

    Success := (not DivByZero and (RelDiam(Encl) <= Eps) );

    { Increment actual staggered precision and store the next component of   }
    { the intermediate results by updating the list of intermediate results. }
    {------------------------------------------------------------------------}
```

```
      if ( (not Success) and (ActStaggPrec < MaxStaggPrec) ) then
      begin
        ActStaggPrec := ActStaggPrec + 1;
        UpdateStaggComp(ActStaggPrec);
      end;
    until ( Success or (ActStaggPrec = MaxStaggPrec) );
    StaggPrec := ActStaggPrec;

    if (Success) then           { Set error code }
      Err := NoError
    else if (DivByZero) then
      Err := ZeroDiv
    else
      Err := ItFailed;
  end; {procedure Eval}

  {----------------------------------------------------------------------------}
  { Module initialization part                                                 }
  {----------------------------------------------------------------------------}
  begin
    { Nothing to initialize }
  end.
```

8.3.2 Examples

The examples of this section demonstrate how to use the procedure *Eval* for an accurate black box evaluation of arithmetic expressions. The only thing we have to do is to define a function of type *Staggered* including the expression to be evaluated. By passing the parameters of the expression in a dynamic array of type *StaggArray*, we are able to define functions with a variable number of arguments. Note that the arguments are accessed starting with index 1.

Example 8.3 In our first example, we want to evaluate the real rational function

$$f(x,y) = \frac{1}{y^6 - 3xy^5 + 5x^3y^3 - 3x^5y - x^6}.$$

Let a_i and a_{i+1} be two successive Fibonacci numbers defined by

$$a_0 = 0, \quad a_1 = 1, \quad a_i = a_{i-1} + a_{i-2}, \; i = 2, 3, \ldots$$

Then we always get $f(a_i, a_{i+1}) = (-1)^i$. In the program listed below, f is defined as a function of type *Staggered*. To shorten the definition of its body, we first copy its parameters to auxiliary variables x and y. The main program accepts the parameters and the desired accuracy, evaluates f, and reports the results. If *Eval* succeeds, an enclosure of the function value and, for reasons of comparison, the value of a naive floating-point evaluation are listed.

```
{----------------------------------------------------------------------------}
{ Example for the evaluation of an arithmetic expression using procedure     }
{ 'Eval()'. Evaluate the following expression:                               }
{                                                                            }
{    f(x,y) := 1 / ( y↑6 - 3*x*y↑5 + 5*x↑3*y↑3 - 3*x↑5*y - x↑6 ).            }
```

```
{                                                                              }
{ With x and y being successive Fibonacci numbers defined by                  }
{                                                                              }
{     a_0 := 0,  a_1 := 1,  a_i := a_i-1 + a_i-2, i >= 2,                      }
{                                                                              }
{ we always get f(a_i,a_i+1) = (-1)↑i. For instance a_66 = 27777890035288     }
{ and a_67 := 44945570212853 are such two successive Fibonacci numbers.       }
{------------------------------------------------------------------------------}
program ExprEval_Example_Fibonacci ( input, output );
use
  i_ari,        { Interval arithmetic   }
  expreval;     { Expression evaluation }

function f ( v : StaggArray ) : Staggered;
var
  x, y : Staggered;
begin
  x := v[1];
  y := v[2];
  f := 1 / ( Power(y,6) - 3*x*Power(y,5) + 5*Power(x,3)*Power(y,3)
                        - 3*Power(x,5)*y - Power(x,6) );
end;

var
  Eps, Approx         : real;
  StaggPrec, ErrCode  : integer;
  Arg                 : rvector[1..2];
  Encl                : interval;
begin
  writeln('Evaluation of 1 / ( y↑6 - 3*x*y↑5 + 5*x↑3*y↑3 - 3*x↑5*y - x↑6 )');
  writeln;
  writeln('Enter the arguments:');
  write  ('   x = '); read(Arg[1]);
  write  ('   y = '); read(Arg[2]);
  writeln;
  writeln('Desired accuracy:');
  write  ('   Eps = '); read(Eps);
  writeln;

  Eval(f,Arg,Eps,Approx,Encl,StaggPrec,ErrCode);
  if (ErrCode = 0) then
    begin
      writeln('Floating-point evaluation:  ', Approx);
      writeln('Interval enclosure:         ',  Encl);
      writeln('Corrections needed:         ',StaggPrec);
    end
  else
    writeln(EvalErrMsg(ErrCode));
end.
```

A sample output is given below. The floating-point format used for computation has a precision of about 16 decimals. Thus, it does not make sense to demand a value less than 10^{-16} for the relative accuracy ε. We also point out that the result computed by the naive floating-point evaluation has neither the same order of magnitude nor any correct digits in the mantissa!

```
Evaluation of 1 / ( y^6 - 3*x*y^5 + 5*x^3*y^3 - 3*x^5*y - x^6 )

Enter the arguments:
   x = 27777890035288
   y = 44945570212853
```

```
Desired accuracy:
   Eps = 1e-20

Floating-point evaluation:    3.165189248586599E-066
Interval enclosure:        [  1.000000000000000E+000,  1.000000000000000E+000 ]
Corrections needed:           5
```

Example 8.4 Our second example deals with the evaluation of a second order difference quotient. Let $f(x)$ be a real function, twice differentiable at x. Then its second order difference quotient, defined by

$$D_f(x,h) = \frac{f(x-h) - 2f(x) + f(x+h)}{h^2}, \quad h \neq 0, \tag{8.9}$$

is a good approximation of $f''(x)$ for small values of h. It is well known that the evaluation of D_f using floating-point arithmetic is unstable. The instability is caused by cancellation while evaluating the numerator of D_f. Therefore, the evaluation of D_f for small values of h is a good example for demonstrating the benefits of our algorithm. The sample program listed below may be used to evaluate D_f for a function f defined by

$$f(x) = 540\frac{x^4 - 23x^3 + 159x^2 - 2x + 45}{x^3 + 18x^2 + 501x + 20}.$$

The exact value at $x = 1$ is $f''(1) = 36$. The coding of the sample program is similar to Example 8.3, except that the definition of D_f is realized by nested function calls.

```
{----------------------------------------------------------------------------}
{ Example for the evaluation of an arithmetic expression using procedure     }
{ 'Eval()'. The second order difference quotient for a real function f(x) is }
{ defined by Df(x,h) := (f(x-h) - 2f(x) + f(x+h)) / h↑2. This quotient is    }
{ used to approximate the second derivative of                               }
{                                                                            }
{   f(x) := 540 (x↑4 - 23x↑3 + 159x↑2 - 2x + 45) / (x↑3 + 18x↑2 + 501x + 20) }
{                                                                            }
{ at x = 1. Since f is twice differentiable, Df(1,h) should tend to          }
{ f''(1) = 36 if h tends to zero.                                            }
{----------------------------------------------------------------------------}
program ExprEval_Example_DiffQuot ( input, output );
use
  i_ari,          { Interval arithmetic   }
  expreval;       { Expression evaluation }

function f ( x : Staggered ) : Staggered;
begin
  f := 540 * (Power(x,4) - 23*Power(x,3) + 159*Power(x,2) - 2*x + 45) /
             (Power(x,3) + 18*Power(x,2) + 501*x + 20);
end;

function Df ( v : StaggArray ) : Staggered;
var
  x, h : Staggered;
begin
  x := v[1];
  h := v[2];
  Df := (f(x-h) - 2*f(x) + f(x+h)) / (h*h);
end;
```

```
var
  Eps, Approx          : real;
  StaggPrec, ErrCode : integer;
  Arg                  : rvector[1..2];
  Encl                 : interval;
begin
  writeln('Evaluation of the second order difference quotient ');
  writeln('   Df(x,h) = (f(x-h) - 2f(x) + f(x+h)) / h↑2');
  writeln('for the function');
  write  ('   f(x) = 540(x↑4 - 23x↑3 + 159x↑2 - 2x + 45) / ');
  writeln('(x↑3 + 18x↑2 + 501x + 20)');
  writeln('Note: f''''(1) = 36.');
  writeln;
  writeln('Enter the arguments:');
  write  ('   x = '); read(Arg[1]);
  write  ('   h = '); read(Arg[2]);
  writeln;
  writeln('Desired accuracy:');
  write  ('   Eps = '); read(Eps);
  writeln;

  Eval(Df,Arg,Eps,Approx,Encl,StaggPrec,ErrCode);
  if (ErrCode = 0) then
    begin
      writeln('Floating-point evaluation:  ', Approx);
      writeln('Interval enclosure:         ',  Encl);
      writeln('Corrections needed:           ',StaggPrec);
    end
  else
    writeln(EvalErrMsg(ErrCode));
end.
```

An output of the sample program is given below. Again, the result computed by naive floating-point evaluation has neither the same order of magnitude nor any correct digits in the mantissa. We stress that the result interval is verified to be an enclosure of the value of the second order difference quotient, but it is *not* proved to be an enclosure of the value of the second derivative at $x = 1$. To compute a verified enclosure of f'', we should add an enclosure for the truncation error in Equation (8.9).

```
Evaluation of the second order difference quotient
   Df(x,h) = (f(x-h) - 2f(x) + f(x+h)) / h^2
for the function
   f(x) = 540(x^4 - 23x^3 + 159x^2 - 2x + 45) / (x^3 + 18x^2 + 501x + 20)
Note: f''(1) = 36.

Enter the arguments:
   x = 1
   h = 1e-8

Desired accuracy:
   Eps = 1e-15

Floating-point evaluation:   2.842170943040400E+002
Interval enclosure:        [ 3.600000000000000E+001,  3.600000000000002E+001 ]
Corrections needed:          1
```

8.3.3 Restrictions, Hints, and Improvements

Since our method is a method for evaluating *real* expressions, we must be careful that the parameters used for evaluation are exactly representable on the computer's number screen. Actually, this is no restriction if the method is used to evaluate an expression whose input parameters come from previous computations. If an exact representation of real constants such as 0.1 is desired, one might use an appropriate representation as a fraction of integers. For more details about conversion of input data, see the remarks in Section 3.7.

There is another problem we should be wary of when defining an expression of type *Staggered*. Remember that the type of an operator is defined by the types of its operands. As an example, consider the expression $(r_1+r_2)\cdot x$, where r_1 and r_2 are operands of type *real* and x is an operand of type *Staggered*. Since both operands r_1 and r_2 are of type *real*, the sum is computed using the standard operator for a real addition instead of using the operator for addition of type *Staggered*. To be sure that for any operation the corresponding operator of type *Staggered* is used, we recommend converting each real operand of an expression by introducing auxiliary variables of type *Staggered*. In the examples of the previous section, any operators within the definition part of the functions are operators of type *Staggered*.

For the implementation of Algorithm 8.2, we made extensive use of the accurate expressions of PASCAL–XSC. The algorithm can also be applied if a multi-precision interval arithmetic is available and is used instead.

8.4 Exercises

Exercise 8.1 Check by paper and pencil how Algorithm 8.2 works for $x + y - x$ with $x = 10^{100}$ and $y = 1$. How many corrections p are needed?

Exercise 8.2 Let $f(x,y) = (x^2)^2-(2y^2)^2-(2y)^2$ and $g(x,y) = (x^2)^2-(2y)^2(y^2+1)$. Both expressions are equivalent. Use the module *expreval* to get enclosures of f and g at $x = 665857$ and $y = 470832$. Compare the results with the values computed by ordinary floating-point arithmetic. For these special arguments, the exact result is 1.

Exercise 8.3 Evaluate the expression

$$f(x,y) = \frac{1}{107751}(1682xy^4 + 3x^3 + 29xy^2 - 2x^5 + 832)$$

at $x = 192119201$ and $y = 35675640$. Use the module *expreval* to get an enclosure of $f(x,y)$. Compare the result with the value computed by ordinary floating-point arithmetic. For these special arguments the exact result is 1783.

Exercise 8.4 Use the module *expreval* to evaluate the sample polynomials of Chapter 4 on page 64.

Exercise 8.5 Extend procedure *Eval* to accept arguments to f in a staggered correction form and to return the value of f as a *Staggered* type.

Exercise 8.6 Extend the module *expreval* to support interval-valued expressions. For an interval parameter $[x]_\nu$, $\nu \in \{1, \ldots, n\}$, use $\widetilde{x}_\nu := m([x]_\nu)$ and $[\Delta x]_\nu := [x]_\nu - \widetilde{x}_\nu$ for initialization. The method is effective for small intervals, but does not necessarily yield the desired accuracy. See Fischer, Haggenmüller, and Schumacher [17] for more details.

8.5 References and Further Reading

A first algorithm for the verified evaluation of arithmetic expressions was given by Böhm [10]. Compared to the method described above, this algorithm is based on a transformation to an equivalent system of linear equations, so it could not handle elementary functions. A summary of the nonlinear approach is found in the papers of Fischer, Haggenmüller, and Schumacher [17], [18]. A more sophisticated approach which works by computing an upper bound for the condition of the problem, and then delivers the precision necessary to achieve the desired accuracy was given by Hammer [23]. It includes an implementation in PASCAL–XSC including standard functions and a method for avoiding the problem of data conversion. A different approach which is based on the principles of reverse methods was given by Fischer [16]. In contrast to our algorithm, this approach yields different degrees of precision for computing different intermediate results. Finally, we remark that there are commercial implementations of the nonlinear approach discussed above which come with the subroutine libraries ACRITH of IBM and ARITHMOS of SIEMENS (see [32], [80]).

Chapter 9

Zeros of Complex Polynomials

We consider the complex polynomial $p : \mathbb{C} \to \mathbb{C}$ defined by

$$p(z) = \sum_{i=0}^{n} p_i z^i, \qquad p_i \in \mathbb{C}, \quad i = 0, \ldots, n, \qquad p_n \neq 0. \tag{9.1}$$

The Fundamental Theorem of algebra asserts that this polynomial has n zeros counted by multiplicity. Finding these roots is a non trivial problem in numerical mathematics. Most algorithms deliver only approximations of the exact zeros without any or with only weak statements concerning the accuracy.

In this section, we describe an algorithm that computes verified enclosures of the roots of a complex polynomial by enclosing the zeros in narrow bounds. The coefficients of the deflated polynomial also are enclosed.

9.1 Theoretical Background

9.1.1 Description of the Problem

The algorithm described in Section 9.2 is based on the fact that the roots of the complex polynomial (9.1) of degree n match the eigenvalues of the companion matrix

$$A = \begin{pmatrix} 0 & \ldots & 0 & -p_0/p_n \\ 1 & & & -p_1/p_n \\ & \ddots & & \vdots \\ & & 1 & -p_{n-1}/p_n \end{pmatrix} \in \mathbb{C}^{n \times n} \tag{9.2}$$

since $(-1)^n \cdot p_n \cdot \det(A - zI) = p(z)$, where I is the identity matrix of dimension n. Hence, the problem of finding a zero of the complex polynomial p is equivalent to finding an eigenvalue z^* of the matrix $A \in \mathbb{C}^{n \times n}$. We solve the eigenvalue problem

$$Aq^* = z^* q^* \quad \text{or} \quad (A - z^* I)q^* = 0, \qquad z^* \in \mathbb{C}, \qquad q^* \in \mathbb{C}^n, \tag{9.3}$$

where q^* is an eigenvector corresponding to the eigenvalue z^* consisting of the coefficients $q_0^*, q_1^*, \ldots, q_{n-1}^*$ of the deflated polynomial $q^*(z) = \sum_{i=0}^{n-1} q_i^* z^i = \frac{p(z)}{z - z^*}$.

We get a verified enclosure of an eigenvalue of the companion matrix and therefore of a zero of the complex polynomial by using Schauder's fixed-point theorem (see [30]), which is a generalization of Brouwer's fixed-point theorem described in Section 3.8. Additionally, the coefficients of the deflated polynomial are enclosed.

The coefficients of the deflated polynomial may be determined recursively by Horner's evaluation of the polynomial p at the point z^*:

$$\begin{aligned} q^*_{n-1} &:= p_n \\ q^*_{i-1} &:= q^*_i z^* + p_i, \quad i = n-1, \ldots, 1. \end{aligned}$$

From the definition of q^*, we can scale the vector $q^* = (q^*_0, q^*_1, \ldots, q^*_{n-1})^T$ by fixing $q^*_{n-1} := p_n$, thus avoiding the division by p_n in the matrix A.

We have a system of nonlinear equations in the n unknowns $q^*_0, q^*_1, \ldots, q^*_{n-2}$, and z^*. Let the vector q be the first $n-1$ components of the desired eigenvector $q := (q^*_0, q^*_1, \ldots, q^*_{n-2})^T$. The eigenvalue z^* often is stored as the nth component of a vector $(q, z)^T$.

9.1.2 Iterative Approach

We introduce the following *a posteriori* method to determine the zeros of a complex polynomial. Let $f : \mathbb{C}^n \to \mathbb{C}^n$ be a nonlinear, differentiable function. A well known strategy to solve a nonlinear system $f(x) = 0$ is the simplified Newton iteration using the fixed-point form of the problem. Let a starting approximation $x^{(0)}$ be given, let $R = f'(x^{(0)})^{-1}$, and iterate according to

$$x^{(k+1)} = x^{(k)} - R \cdot f(x^{(k)}) =: g(x^{(k)}), \quad k = 0, 1, 2, \ldots \tag{9.4}$$

If $x^{(0)}$ is close to the fixed-point x^*, the sequence of $x^{(k)}$ for $k \to \infty$ approaches the fixed-point $x^* = g(x^*)$ with $f(x^*) = 0$.

To apply the simplified Newton iteration to the eigenvalue problem (9.3), we let $x = (q, z)^T = (q_0, q_1, \ldots, q_{n-2}, z)^T$, $q_{n-1} := p_n = const$, and define the function $f : \mathbb{C}^n \to \mathbb{C}^n$ as

$$f(x) = f\left(\begin{pmatrix} q \\ z \end{pmatrix}\right) = (A - zI)\begin{pmatrix} q \\ p_n \end{pmatrix}.$$

Then matrix $R \approx J_f^{-1}$ is an approximation to the inverse of the Jacobian matrix

$$\begin{aligned} J_f &= f'(x) = \left[(A - zI)\begin{pmatrix} q \\ p_n \end{pmatrix}\right]' \\ &= (A - zI)\begin{pmatrix} 1 & & 0 & 0 \\ & \ddots & & \vdots \\ & & 1 & \\ 0 & & & 0 \end{pmatrix} - \begin{pmatrix} 0 & \cdots & 0 & q_0 \\ \vdots & & \vdots & \vdots \\ \vdots & & \vdots & q_{n-2} \\ 0 & \cdots & 0 & p_n \end{pmatrix} \qquad (9.5) \\ &= \begin{pmatrix} -z & & & -q_0 \\ 1 & \ddots & & \vdots \\ & \ddots & -z & -q_{n-2} \\ & & 1 & -p_n \end{pmatrix}. \end{aligned}$$

Because of the special shape of J_f, its inverse R can be determined directly by eliminating the diagonal elements according to Gauss:

$$\begin{pmatrix} 1 & z & & \\ & 1 & & \\ & & \ddots & \\ & & & 1 \end{pmatrix} \cdots \begin{pmatrix} 1 & & & \\ & \ddots & & \\ & & 1 & z \\ & & & 1 \end{pmatrix} \cdot J_f = \begin{pmatrix} 0 & & & -w_0 \\ 1 & & & -w_1 \\ & \ddots & & \\ & & 1 & -w_{n-1} \end{pmatrix} \tag{9.6}$$

$$\text{with } \begin{aligned} w_{n-1} &:= p_n \\ w_i &:= q_i + z \cdot w_{i+1}, \quad i = n-2, \dots, 0. \end{aligned}$$

With the known inverse of the right-hand side, we immediately get

$$R = J_f^{-1} = \begin{pmatrix} -w_1/w_0 & 1 & & \\ & & \ddots & \\ -w_{n-1}/w_0 & & & 1 \\ -1/w_0 & & & 0 \end{pmatrix} \cdot \begin{pmatrix} 1 & z & & \\ & 1 & & \\ & & \ddots & \\ & & & 1 \end{pmatrix} \cdots \begin{pmatrix} 1 & & & \\ & \ddots & & \\ & & 1 & z \\ & & & 1 \end{pmatrix}. \tag{9.7}$$

For numerical stability reasons, it is better to perform a residual correction instead of iterative improvement of the complete complex vector $(q, z)^T$. We use the following notation for the residual correction:

$$x = \widetilde{x} + \Delta_x, \quad \text{that is} \quad \begin{pmatrix} q \\ z \end{pmatrix} = \begin{pmatrix} \widetilde{q} + \Delta_q \\ \widetilde{z} + \Delta_z \end{pmatrix}$$

with $\widetilde{x}$ (resp. $\widetilde{q}$ and $\widetilde{z}$) being approximations of the exact values x (resp. q and z). Then the Newton iteration has the form

$$\begin{aligned} x^{(k+1)} - \widetilde{x} &= x^{(k)} - \widetilde{x} - R \cdot f(x^{(k)}), \quad \text{or} \\ \Delta_x^{(k+1)} &= \Delta_x^{(k)} - R \cdot f(\widetilde{x} + \Delta_x^{(k)}). \end{aligned}$$

For our special eigenvalue problem, we get

$$\begin{pmatrix} \Delta_q^{(k+1)} \\ \Delta_z^{(k+1)} \end{pmatrix} = \begin{pmatrix} \Delta_q^{(k)} \\ \Delta_z^{(k)} \end{pmatrix} - R \cdot f(\begin{pmatrix} \widetilde{q} + \Delta_q^{(k)} \\ \widetilde{z} + \Delta_z^{(k)} \end{pmatrix}) =: g(\begin{pmatrix} \Delta_q^{(k)} \\ \Delta_z^{(k)} \end{pmatrix}) = g(\Delta_x^{(k)})$$

with $g : \mathbb{C}^n \to \mathbb{C}^n$, $q, \Delta_q \in \mathbb{C}^{n-1}$, and $z, \Delta_z \in \mathbb{C}$. All further calculations depend only on the residual values $(\Delta_q, \Delta_z)^T$ denoted as Δ_x.

Some transformations of the right-hand side (omitting the iteration index k) yield

$$\begin{aligned} g(\Delta_x) &= \Delta_x - R \cdot \Big(A - (\widetilde{z} + \Delta_z) I \Big) \begin{pmatrix} \widetilde{q} + \Delta_q \\ p_n \end{pmatrix} \\ &= \Delta_x - R \cdot \Bigg[(A - \widetilde{z} I) \begin{pmatrix} \widetilde{q} \\ p_n \end{pmatrix} - \Delta_z \begin{pmatrix} \widetilde{q} \\ p_n \end{pmatrix} \\ &\qquad + (A - \widetilde{z} I) \begin{pmatrix} \Delta_q \\ 0 \end{pmatrix} - \Delta_z \begin{pmatrix} \Delta_q \\ 0 \end{pmatrix} \Bigg] \end{aligned}$$

$$= \Delta_x - R\cdot(A-\widetilde{z}I)\begin{pmatrix}\widetilde{q}\\ p_n\end{pmatrix} + R\cdot\Delta_z\begin{pmatrix}\Delta_q\\ 0\end{pmatrix}$$

$$-R\cdot\left((A-\widetilde{z}I)\begin{pmatrix}1 & & 0 & 0\\ & \ddots & & \vdots\\ & & 1 & \\ 0 & & & 0\end{pmatrix} - \begin{pmatrix}0 & \cdots & 0 & \widetilde{q}_0\\ \vdots & & \vdots & \vdots\\ \vdots & & \vdots & \widetilde{q}_{n-2}\\ 0 & \cdots & 0 & p_n\end{pmatrix}\right)\cdot\Delta_x\,.$$

We do some additional substitutions for a more algorithmic representation:

$$g(\Delta_x) := -R\cdot d + R\cdot\Delta_z\begin{pmatrix}\Delta_q\\ 0\end{pmatrix} + (I - R\cdot J_f)\Delta_x, \tag{9.8}$$

with

$$d := (A-\widetilde{z}I)\begin{pmatrix}\widetilde{q}\\ p_n\end{pmatrix}. \tag{9.9}$$

In summary, we have a formulation of the Newton iteration with

$$\Delta_x^{(k+1)} = g(\Delta_x^{(k)}) \quad\text{denoting}\quad \begin{pmatrix}\Delta_q^{(k+1)}\\ \Delta_z^{(k+1)}\end{pmatrix} = g(\begin{pmatrix}\Delta_q^{(k)}\\ \Delta_z^{(k)}\end{pmatrix}) \tag{9.10}$$

to solve the eigenvalue problem (9.3).

The Approximate Iteration

We must determine good approximations of the exact eigenvector q^* and eigenvalue z^* to avoid inflation effects using the interval version of the Newton iteration (9.10). For this purpose, we first use a non-interval residual iteration algorithm starting with an arbitrary starting approximation $\widetilde{z}$ for a root of $p(z)$. The initial eigenvector $\widetilde{q}$ corresponding to that eigenvalue $\widetilde{z}$ is computed recursively by Horner's evaluation of the polynomial p at the point $\widetilde{z}$. We use the iteration function g of (9.8) (in floating-point arithmetic) to improve the corresponding residual vector $\Delta_x = (\Delta_q, \Delta_z)^T$ until sufficient accuracy is achieved.

The Verification Step

We define the interval function $g_{[]} : I\mathbb{C}^n \to I\mathbb{C}^n$ as an interval extension of (9.8) for the interval version of Newton's iteration algorithm to get

$$[\Delta_x]^{(k+1)} = g([\Delta_x]^{(k)}) \quad\text{denoting}\quad \begin{pmatrix}[\Delta_q]^{(k+1)}\\ [\Delta_z]^{(k+1)}\end{pmatrix} = g_{[]}(\begin{pmatrix}[\Delta_q]^{(k)}\\ [\Delta_z]^{(k)}\end{pmatrix}) \tag{9.11}$$

with $[\Delta_q] \in I\mathbb{C}^{n-1}$, and $[\Delta_z] \in \mathbb{C}$. Omitting the iteration index k,

$$g_{[]}([\Delta_x]) := -R\cdot d + R\cdot[\Delta_z]\begin{pmatrix}[\Delta_q]\\ 0\end{pmatrix} + (I - R\cdot J_f)[\Delta_x],$$

with the fixed values d from (9.9), J_f from (9.5) and R from (9.7).

Schauder's fixed-point theorem says: If we have the enclosure

$$[\Delta_x]^{(k+1)} = g_{[]}([\Delta_x]^{(k)}) \overset{\circ}{\subset} [\Delta_x]^{(k)}, \tag{9.12}$$

then there exists a (not necessarily unique) fixed-point of $g_{[]}$, and with this a solution $x^* \in \widetilde{x} + [\Delta_x]^{(k+1)}$ of the eigenvalue problem (9.4).

If we choose R as an exact inverse of J_f, i.e. $R \equiv J_f^{-1}$, we get $I - R \cdot J_f = 0$, and a simplified iteration function

$$g_{[]}([\Delta_x]) := -R \cdot d + R \cdot [\Delta_z] \begin{pmatrix} [\Delta_q] \\ 0 \end{pmatrix}.$$

This simplification is also valid if we replace R by an interval matrix $[R]$ that is a verified enclosure of the exact inverse J_f^{-1} (see [10]). Using complex interval arithmetic, we get

$$J_f^{-1} \in [R] = \begin{pmatrix} -[w]_1/[w]_0 & 1 & & \\ & & \ddots & \\ -[w]_{n-1}/[w]_0 & & & 1 \\ -1/[w]_0 & & & 0 \end{pmatrix} \cdot \begin{pmatrix} 1 & \widetilde{z} & & \\ & 1 & & \\ & & \ddots & \\ & & & 1 \end{pmatrix} \cdots \begin{pmatrix} 1 & & & \\ & \ddots & & \\ & & 1 & \widetilde{z} \\ & & & 1 \end{pmatrix},$$

which allows us to simplify the interval function even on a computer to

$$g_{[]}([\Delta_x]) := -[R] \cdot [d] + [R] \cdot [\Delta_z] \begin{pmatrix} [\Delta_q] \\ 0 \end{pmatrix}. \tag{9.13}$$

Subsequently, we start a new iteration step by evaluating the function $g_{[]}$ for a complex interval vector argument until we achieve enclosure (9.12). For computational reasons (cf. Section 3.6), we start with a slightly inflated approximation. The Schauder fixed-point theorem guarantees that there exists a solution of the fixed-point problem (9.12) in $([\Delta_q]^{(k+1)}, [\Delta_z]^{(k+1)})^T$. If $0 \notin [w_0]$, it also follows that $[R]$ is non-singular. This yields

$$\left(A - (\widetilde{z} + [\Delta_z])I \right) \begin{pmatrix} \widetilde{q} + [\Delta_q] \\ p_n \end{pmatrix} \supseteq (A - z^* I) \begin{pmatrix} q^* \\ p_n \end{pmatrix} = 0.$$

That is, $(\widetilde{z} + [\Delta_z])$ is a verified enclosure of an eigenvalue z^*, which is a root of the complex polynomial p, and $(\widetilde{q} + [\Delta_q])$ is a verified enclosure of a corresponding eigenvector q^*, the components of which are the coefficients of the deflated polynomial.

9.2 Algorithmic Description

First, we present the algorithm Approximation to improve an initial approximation $\widetilde{z}$ of a root of a complex polynomial $p(z) = \sum_{i=0}^{n} p_i z^i$. Approximation works by transforming problem (9.1) to the equivalent problem of finding an eigenvalue and its corresponding eigenvector for the companion matrix (9.2). It improves the approximations of a root and the coefficients of the corresponding deflated polynomial to avoid overestimations during the floating-interval calculations. A residual iteration method is used to improve the starting approximation until the accuracy necessary to start the interval algorithm is achieved. No guarantee is claimed for the correctness of the approximation computed by **Approximation**.

In the following algorithm, the vector $\Delta \in \mathbb{C}^n$ denotes the combined vector $\Delta_x = (\Delta_q, \Delta_z)^T$ mentioned in the preceding section. The matrix R is not computed explicitly since the components w_i represent the entire matrix. With this the complexity of the inversion is reduced to $O(n)$. The iteration $\Delta^{(k+1)} := g(\Delta^{(k)})$ is done directly using some loops that are equivalent to the mathematical formulation in (9.8). We use $k_{\max} = 50$ as the maximum number of iterations, and $\varepsilon = 10^{-10}$ as the value for the relative error. If the condition number of the inverse R is extremely large, then the convergence of the residual iteration is slow. To avoid the possibility of an unbounded number of iterations at Step 3, we halt after $k_{\max}$ iterations. It turned out that $k_{\max}$ and ε are good values for minimizing the effort to get sufficiently accurate approximations $\widetilde{z}$ and $\widetilde{q}$.

Algorithm 9.1: Approximation $(p, \widetilde{q}, \widetilde{z}, Err)$ {Procedure}

1. {Constant for termination criteria}
   $\varepsilon := 10^{-10}$; $k_{\max} := 50$;
2. {Determination of an approximate eigenvector $\widetilde{q}$}
   {for the initial polynomial zero $\widetilde{z}$ }
   $\widetilde{q}_{n-1} := p_n$;
   $\widetilde{q}_i := \widetilde{q}_{i+1} * \widetilde{z} + p_{i+1}$; $(i = n-2, \cdots, 0)$
3. {Floating-point iteration of an eigenvalue $\widetilde{z}$ and an eigenvector $\widetilde{q}$}
   $k := 0$;
   repeat
   (a) {Compute the defect $d = (A - \widetilde{z}I)\widetilde{q}$ (see (9.9))}
   $d_0 := \square(-\widetilde{z} * \widetilde{q}_0 - p_0)$;
   $d_i := \square(\widetilde{q}_{i-1} - \widetilde{z} * \widetilde{q}_i - p_i)$; $(i = 1, \cdots, n-1)$
   (b) {Compute the components of R, the inverse of the Jacobian J_f}
   {(see (9.7)) }
   $w_{n-1} := \widetilde{q}_{n-1}$;
   $w_i := \square(\widetilde{q}_i + \widetilde{z} * w_{i+1})$; $(i = n-2, \cdots, 0)$
   if $w_0 = 0$ **then** $Err :=$ "Inversion failed";
   (c) {Compute $\Delta^{(k+1)} = g(\Delta^{(k)})$ (see (9.8)}
   $\Delta_{n-1}^{(k+1)} := d_{n-1}$;
   $\Delta_i^{(k+1)} := \square(d_i + \widetilde{z} * \Delta_{i+1}^{(k)})$; $(i = n-2, \cdots, 0)$
   $t := \Delta_0^{(k+1)} / w_0$;
   $\Delta_i^{(k+1)} := \square(-\Delta_{i+1}^{(k+1)} + t * w_{i+1})$; $(i = 0, \cdots, n-2)$

$:= t;$

(d) {Determine the new iterates $\widetilde{q}$ and $\widetilde{z}$}

$$\widetilde{q}_i := \widetilde{q}_i + \Delta_i^{(k+1)}; \quad (i = 0, \cdots, n-2)$$
$$\widetilde{z} := \widetilde{z} + \Delta_{n-1}^{(k+1)};$$

(e) $k := k + 1$;

until $\|\Delta\|_\infty / \max(\|(\widetilde{q}_i)_{i=0}^{n-2}\|_\infty, |\widetilde{z}|) \leq \varepsilon$ **or** $(k = k_{\max})$;

4. **return** $\widetilde{q}, \widetilde{z}, Err$;

The algorithm **IntervalIteration** computes a verified enclosure of a root of a complex polynomial using an interval iteration. Starting with good approximations $\widetilde{z}$ for a root of the complex polynomial $p(z)$ and $\widetilde{q}$ for the coefficients of the deflated polynomial $q(z) = \frac{p(z)}{z-\widetilde{z}}$, a verification strategy based on Schauder's fixed-point theorem is used to determine (if possible) a guaranteed enclosure of a polynomial root z^*. In addition, guaranteed enclosures of the coefficients of the deflated polynomial $q^*(z) = \frac{p(z)}{z-z^*}$ are returned as a by-product. We use $k_{\max} = 10$ as the maximum number of iterations, and $\varepsilon = 0.1$ as the value for the epsilon inflation. It turned out that these are good values for minimizing the effort if no verification is possible (see also [10]).

Algorithm 9.2: IntervalIteration $(p, \widetilde{q}, \widetilde{z}, [q], [z], Err)$ {Procedure}

1. {Computation of the interval evaluation}

(a) {Compute an enclosure of the defect $[d] = (A - \widetilde{z}I)\widetilde{q}$ (see 9.9)}

$$[d]_0 := \Diamond(-\widetilde{z} * \widetilde{q}_0 - p_0);$$
$$[d]_i := \Diamond(\widetilde{q}_{i-1} - \widetilde{z} * \widetilde{q}_i - p_i); \quad (i = 1, \cdots, n-1)$$

(b) {Compute an enclosure of $[R]$, the inverse of the Jacobian J_f} {(see (9.7))}

$$[w]_{n-1} := \widetilde{q}_{n-1};$$
$$[w]_i := \Diamond(\widetilde{q}_i + \widetilde{z} * [w]_{i+1}); \quad (i = n-2, \cdots, 0)$$

if $0 \in [w]_0$ **then** $Err :=$ "Verified enclosure failed";

(c) {Compute an enclosure $[\Delta]^{(0)} = \begin{pmatrix} [\Delta_q] \\ [\Delta_z] \end{pmatrix} \supset \begin{pmatrix} q^* - \widetilde{q} \\ z^* - \widetilde{z} \end{pmatrix}$}

$$[\Delta]_{n-1}^{(0)} := [d]_{n-1};$$
$$[\Delta]_i^{(0)} := \Diamond([d]_i + \widetilde{z} * [\Delta]_{i+1}^{(0)}); \quad (i = n-2, \cdots, 0)$$
$$[t] := [\Delta]_0^{(0)} / [w]_0;$$
$$[\Delta]_i^{(0)} := \Diamond(-[\Delta]_{i+1}^{(0)} + [t] * [w]_{i+1}); \quad (i = 0, \cdots, n-2)$$
$$[\Delta]_{n-1}^{(0)} := [t];$$

2. {Interval iteration}

$k := 0$; $\varepsilon := 10^{-1}$; $k_{\max} := 10$;

repeat

(a) {Slightly enlarge the enclosure interval}
$[\Delta]^{(k)} := [\Delta]^{(k)} \bowtie \varepsilon$;

(b) {Determine a new enclosure interval $[\Delta]^{(k+1)} = g_{\tilde{u}}([\Delta]^{(k)})$ (see (9.13))}
$[v]_{n-1} := 0$;
$[v]_i := \Diamond([\Delta]_{n-1}^{(k)} * [\Delta]_i^{(k)} + \tilde{z} * [v]_{i+1})$; $(i = n-2, \cdots, 0)$
$[v]_0 := [v]_0/[w]_0$;
$[\Delta_i]^{(k+1)} := \Diamond([\Delta]_i^{(0)} + [v]_{i+1} - [v]_0 * [w]_{i+1})$; $(i = 0, \cdots, n-2)$
$[\Delta]_{n-1}^{(k+1)} := [\Delta]_{n-1}^{(0)} - [v]_0$;

(c) $k := k + 1$;

until $([\Delta]^{(k)} \overset{\circ}{\subset} [\Delta]^{(k-1)})$ **or** $(k = k_{\max})$;

3. {Verification of the result (see (9.12))}
if $([\Delta]^{(k)} \overset{\circ}{\subset} [\Delta]^{(k-1)})$ **then**
$[q]_i := \Diamond(\tilde{q}_i + [\Delta]_i^{(k)})$; $(i = 0, \cdots, n-2)$
$[q]_{n-1} := p_n$;
$[z] := \Diamond(\tilde{z} + [\Delta]_{n-1}^{(k)})$;
else *Err* := "Inclusion failed"

4. **return** $[q], [z]$, *Err*;

The algorithm **CPolyZero** uses the algorithms **Approximation** and **IntervalIteration** presented above. Complex polynomials $p(z) = \sum_{i=0}^{n} p_i z^i$ of degree one can be solved directly using interval operations. To solve complex polynomials of degree $n > 1$ Algorithms 9.1 and 9.2 are used to determine guaranteed enclosures for a polynomial zero z^* and for the coefficients of the deflated polynomial $q(z) = \frac{p(z)}{z - z^*}$.

Algorithm 9.3: CPolyZero $(p, \tilde{z}, [z], [q], Err)$ {Procedure}

1. {Initialization and treatment of the special cases}
Err := "No Error";
if $(n = 0)$ **or** $((n = 1)$ **and** $(p_1 = 0))$ **then** *Err* := 'Constant polynomial';
else if $(n = 1)$ **then**
$[z] := -(p_0 \oslash p_1)$;
$[q]_0 := p_1$;

2. {Usual case}
else
Approximation $(p, \tilde{q}, \tilde{z}, Err)$;
IntervalIteration $(p, \tilde{q}, \tilde{z}, [z], [q], Err)$;

3. **return** $[q], [z]$, *Err*;

Applicability of the Algorithm

Simple Zeros

This algorithm can only be used to enclose simple zeros of the given polynomial p. If we try to enclose a multiple zero by starting the routine IntervalIteration with a good approximation of a multiple zero, we will get an error message because computing a verified enclosure of the inverse R of the matrix J_f fails. From (9.6), we get

$$|\det(J_f)| = |w_0| = |\sum_{i=0}^{n-1} q_i z^i| = |p'(z)|, \tag{9.14}$$

where z is an eigenvalue, and q is an eigenvector of the eigenproblem (9.3). If we have a multiple zero z of the polynomial, (9.14) yields

$$|\det(J_f)| = |p'(z)| = 0.$$

Thus, the matrix J_f is singular. The inverse R cannot be enclosed because it does not exist.

Improvements

If two or more zeros lie very close together, it may happen that the enclosure of the inverse of J_f fails, because J_f is in some numerical sense too close to a singular matrix, i.e. it is too ill conditioned. To overcome this problem in many cases, the enclosure of the inverse has to be computed with double mantissa length using #-expressions (cf. Chapter 10).

As a further refinement, we can compute the approximate solutions $\tilde{z}$ and $\tilde{q}$ in the routine Approximation with double length and make use of the double length in the routine IntervalIteration. Both improvements require the use of #-expressions.

With these improvements, we can separate and verify two zeros differing only in the last digit of a mantissa of single length (cf. [10], [19]).

Real polynomials

If we have a real polynomial $p(z) = \sum_{i=0}^{n} p_i z^i$ ($p_i \in I\!R$) (imaginary part of each coefficient is zero), and we have a real starting approximation $\tilde{z} \in I\!R$, we will never get enclosures of a pair of conjugate complex zeros of $p(z)$, because all complex arithmetic operations deliver a real result, as the imaginary parts of all operands will stay zero. At any rate, the algorithm described here also works on real polynomials. If we want to find a complex conjugate zero of a real polynomial, we must choose a non-real starting approximation $\tilde{z}$.

9.3 Implementation and Examples

9.3.1 PASCAL–XSC Program Code

We list the program code for enclosing a root of a complex polynomial. In the listing, interval data are named with double characters, e.g. dd[i] denotes the interval $[d]_i$.

The algorithm described above has been implemented in separate modules. First, the arithmetic modules for complex polynomials *cpoly* and complex interval polynomials *cipoly* are listed. The module *cpzero* containing all the routines necessary to compute the enclosures of a zero of the complex polynomial $p(z) = \sum_{i=0}^{n} p_i z^i$ and of the coefficients of the deflated polynomial follows.

9.3.1.1 Module cpoly

This module supplies a global type definition named *CPolynomial* representing a complex polynomial $p(z) = \sum_{i=0}^{n} p_i z^i$ $(p_i, z \in \mathbb{C})$.The routines *read* and *write* for the input and output of complex polynomials are defined and exported. Since no operations on polynomials are requested by Algorithm 9.3, no operators have been implemented in this module.

```
{----------------------------------------------------------------------------}
{ Purpose: Declaration of data type for representation of a complex poly-    }
{    nomial by its coefficients, and of I/O procedures for this data type.   }
{ Global types, operators and procedures:                                    }
{    type CPolynomial      : representation of complex polynomials           }
{    operator :=           : assignment of complex polynomials by a real value}
{    procedure read(...)   : input of data type CPolynomial                  }
{    procedure write(...)  : output of data type CPolynomial                 }
{ Remark: Variables of type 'CPolynomial' should be declared with lower      }
{    bound 0 (zero).                                                         }
{----------------------------------------------------------------------------}
module cpoly;

use
  iostd,  { Needed for abnormal termination with 'exit' }
  c_ari;  { Complex arithmetic }

global type
  CPolynomial  = global dynamic array[*] of complex;

global operator := ( var p : CPolynomial; r : real );
var
  i : integer;
begin
  for i := 0 to ub(p) do p[i] := r;
end;


global procedure read ( var t : text; var p : CPolynomial );
var
  i    : integer;
begin
  if (lb(p) <> 0) then
  begin
    write('Error: Variable of type CPolynomial was declared with ');
    writeln('lower bound <> 0!');
```

```
    exit(-1); { Abnormal program termination }
  end;
  write('  x↑0 * '); read(t,p[0]);
  for i := 1 to ub(p) do
    begin  write('+ x↑',i:0,' * '); read(t,p[i])  end;
end;

global procedure write ( var t : text; p : CPolynomial );
var
  i          : integer;
  PolyIsZero : boolean;    { Signals 'p' is a zero polynomial }
begin
  PolyIsZero := true;
  for i := 0 to ub(p) do
    if (p[i] <> 0) then
    begin
      if PolyIsZero then write(t,'   ')  else  write(t,'+ ');
      writeln(t,p[i],' * x↑',i:1);
      PolyIsZero := false;
    end;
  if PolyIsZero then writeln(t,'   0 (= zero polynomial)');
end;

{----------------------------------------------------------------------------}
{ Module initialization part                                                 }
{----------------------------------------------------------------------------}
begin
  { Nothing to initialize }
end.
```

9.3.1.2 Module cipoly

This module supplies the global type definition for variables representing a complex interval polynomial. We define the I/O-routines *read* and *write*, an operator *in* to decide whether all coefficients of one polynomial are inside the coefficients of another polynomial, and the function *blow* described in Section 3.6.

```
{----------------------------------------------------------------------------}
{ Purpose: Declaration of data type for representation of a complex interval }
{    polynomial by its coefficients, and of I/O procedures for this data     }
{    type. Boolean operator in and function for epsilon inflation are        }
{    supplied                                                                }
{ Global types, operators and procedures:                                    }
{    type CIPolynomial     : representation of complex interval polynomials  }
{    operator :=           : assignment of complex interval polynomials by a }
{                            real value                                      }
{    procedure read(...)   : input of data type CIPolynomial                 }
{    procedure write(...)  : output of data type CIPolynomial                }
{    operator in           : boolean-valued operator determining whether a   }
{                            complex interval polynomial lies (componentwise) }
{                            in another complex interval polynomial          }
{    function blow         : componentwise epsilon inflation of a complex    }
{                            interval polynomial                             }
{ Remark: Variables of type 'CIPolynomial' should be declared with lower     }
{    bound 0 (zero).                                                         }
{----------------------------------------------------------------------------}
module cipoly;

use
  iostd,    { Needed for abnormal termination with 'exit' }
  i_ari,    { Interval arithmetic         }
```

```
  ci_ari;  { Complex interval arithmetic }

global type
  CIPolynomial  = global dynamic array[*] of cinterval;

global operator := ( var p : CIPolynomial; r : real );
var
  i : integer;
begin
  for i := lb(p) to ub(p) do p[i] := r;
end;

global procedure read ( var t : text; var p : CIPolynomial );
var
  i : integer;
begin
  if (lb(p) <> 0) then
  begin
    write('Error: Variable of type CIPolynomial was declared with ');
    writeln('lower bound <> 0!');
    exit(-1); { Abnormal program termination }
  end;
  write('  x↑0 * '); read(t,p[0]);
  for i := 1 to ub(p) do
    begin  write('+ x↑',i:0,' * '); read(t,p[i])  end;
end;

global procedure write ( var t : text; p : CIPolynomial );
var
  i          : integer;
  PolyIsZero : boolean;   { Signals 'p' is a zero polynomial }
begin
  PolyIsZero := true;
  for i := 0 to ub(p) do
    if (p[i] <> 0) then
    begin
      if PolyIsZero then write(t,'  ')  else  write(t,'+ ');
      writeln(t,'(', re(p[i]),',');
      writeln(t,'   ',im(p[i]),' ) * x↑',i:1);
      PolyIsZero := false;
    end;
  if PolyIsZero then writeln(t,'   0 (= zero polynomial)');
end;

global operator in ( p, q : CIPolynomial ) res : boolean;
var
  i    : integer;
  incl : boolean;
begin
  incl := true;  i := 0;
  while (incl = true) and (i <= ub(p)) do
    begin  incl := p[i] in q[i];  i := i+1   end;
  res  :=  incl;
end;

global function blow ( p   : CIPolynomial;
                       eps : real          ) : CIPolynomial[0..ub(p)];
var
  i : integer;
begin
  for i := 0 to ub(p) do blow[i] := blow(p[i],eps);
end;
```

```
{----------------------------------------------------------------------------}
{ Module initialization part                                                 }
{----------------------------------------------------------------------------}
begin
  { Nothing to initialize }
end.
```

9.3.1.3 Module cpzero

The module *cpzero* supplies the global routine *CPolyZero* that computes the enclosure of a root of the complex polynomial $p(z) = \sum_{i=0}^{n} p_i z^i$ according to Algorithm 9.3. The function *CPolyZeroErrMsg* returns an error message for the error code returned by *CPolyZero*.

If no error occurs during the calculation, *CPolyZero* returns true enclosures of a root $[z]$ of $p(z)$ and of the coefficients of the deflated polynomial $[q]$ computed by the routines Approximation and IntervalIteration.

```
{----------------------------------------------------------------------------}
{ Purpose: Determination and enclosure of a root of a complex polynomial,   }
{    and of the coefficients of the deflated polynomial.                     }
{ Method: The root of a complex polynomial matches with the eigenvalue of   }
{    its companion matrix and the coefficients of the deflated polynomial   }
{    match to the components of the corresponding eigenvector. The eigen-   }
{    vector and eigenvalue are determined using the simplified Newton       }
{    iteration with iterative refinement.                                   }
{ Global procedures and functions:                                           }
{   procedure CPolyZero(...)       : computes an enclosure for a root and for }
{                                    the deflated complex polynomial        }
{   function CPolyZeroErrMsg(...) : delivers an error message text          }
{----------------------------------------------------------------------------}
module cpzero;        { Complex polynomial zero }
                      { -       -          ---- }
use
  r_util,             { Utilities of type real         }
  c_ari,              { Complex arithmetic             }
  ci_ari,             { Complex interval arithmetic    }
  cpoly,              { Complex polynomials            }
  cipoly;             { Complex interval polynomials   }

{----------------------------------------------------------------------------}
{ Error codes used in this module. In the comments below p[0],..., p[n] are  }
{ the coefficients of a polynomial p.                                        }
{----------------------------------------------------------------------------}
const
  NoError   = 0;      { No error occurred.                              }
  ZeroPoly  = 1;      { Zero polynomial, i.e. n = 0 and p[0] = 0.       }
  ConstPoly = 2;      { Constant polynomial, i.e. n = 0 and p[0] <> 0.  }
  InvFailed = 3;      { Inversion of the Jacobian failed.               }
  VerFailed = 4;      { Verified inversion of the Jacobian failed.      }
  IncFailed = 5;      { Inclusion failed.                               }
  IllProb   = 6;      { Illegal Problem, i.e. polynomials with lower    }
                      { bound <> 0 passed to the main procedure.        }

{----------------------------------------------------------------------------}
{ Error messages depending on the error code.                                }
{----------------------------------------------------------------------------}
global function CPolyZeroErrMsg(Err : integer) : string;
var
  Msg : string;
begin
```

```
  case Err of
    NoError  : Msg := '';
    ZeroPoly : Msg := 'Zero polynomial occurred';
    ConstPoly: Msg := 'Constant polynomial <> 0 occurred';
    InvFailed: Msg := 'Inversion of the Jacobian failed';
    VerFailed: Msg := 'Verified inversion of the Jacobian failed';
    IncFailed: Msg := 'Inclusion failed';
    IllProb  : Msg := 'Illegal polynomial with least index <> 0 occurred';
    else       : Msg := 'Code not defined';
  end;
  if (Err <> NoError) then Msg := 'Error: ' + Msg + '!';
  CPolyZeroErrMsg := Msg;
end;

function MaxNorm(p : CPolynomial) : real; { Function used to get the maximum }
var                                       { norm of 'p', where 'p' is        }
  i        : integer;                     { interpreted as a complex vector  }
  max, tmp : real;                        {----------------------------------}
begin
  max := abs(p[0]);
  for i := 1 to ub(p) do
  begin
    tmp := abs(p[i]);
    if (tmp > max) then max := tmp;
  end;
  MaxNorm := max;
end;

{----------------------------------------------------------------------------}
{ Purpose: Determination of a floating-point approximation for a root of     }
{    polynomial p and for the deflated polynomial.                           }
{ Parameters:                                                                }
{    In     : 'p'  : represents a complex polynomial by its coefficients     }
{    In/Out: 'z'   : starting approximation (in) for a root and is returned  }
{                    as an improved floating-point approximation             }
{    Out    : 'q'  : returns an approximation (for the coefficients of) the  }
{                    deflated polynomial                                     }
{             'Err': error flag                                              }
{ Description: For a given starting approximation 'z', a floating-point      }
{    Newton iteration is performed to get improved approximations for a root }
{    'z' of polynomial 'p', as well as for the coefficients 'q' of the       }
{    deflated polynomial.                                                    }
{----------------------------------------------------------------------------}
procedure Approximation(    p : CPolynomial; var z   : complex;
                        var q : CPolynomial; var Err : integer );
const
  kmax  = 50;                           { Maximum number of iteration steps }
  eps   = 10E-10;                       { Relative error of approximation   }
var
  k, i, n      : integer;
  t            : complex;
  delta, d, w  : CPolynomial[0..ub(p)-1];
  qHlp         : CPolynomial[0..ub(p)-2];
begin
  Err := NoError;  n := ub(p);                          { Initialization }
                                                        {----------------}
  q[n-1] := p[n];                        { Determination of an approximate }
  for i := n-2 downto 0 do               { eigenvector q                   }
    q[i] := q[i+1]*z + p[i+1];           {---------------------------------}

  k := 0;                                     { Floating-point iteration }
  repeat                                      { of z and q               }
                                              {--------------------------}
```

```
    d[0] := #*( -z*q[0] - p[0] );                    { Computation of the defect }
    for i := 1 to n-1 do                             { d = (A - z*I)*q           }
      d[i] := #*( q[i-1] - z*q[i] - p[i] );          {---------------------------}

    w[n-1] := q[n-1];                            { Computation of R, the inverse }
    for i := n-2 downto 0 do                     { of the Jacobian J             }
      w[i] := #*( q[i] + z*w[i+1] );             {-------------------------------}

    if (w[0] = 0) then
      Err := InvFailed
    else
      begin
        delta[n-1] := d[n-1];                     { Computation of delta↑(k+1) = }
        for i := n-2 downto 0 do                  { g(delta↑(k))                 }
          delta[i] := d[i] + z*delta[i+1];        {------------------------------}
        t := delta[0]/w[0];
        for i := 0 to n-2 do
          delta[i] := -delta[i+1] + t*w[i+1];
        delta[n-1] := t;

        for i := 0 to n-2 do            { Determine the new iterates of q and z }
        begin                           { and store the first n-1 components of }
          q[i] := q[i] + delta[i];      { q in qHlp                             }
          qHlp[i] := q[i];              {---------------------------------------}
        end;
        z := z + delta[n-1];

        k := k+1;
      end;

  { Stop iteration if the relative round-off error of the coefficients and }
  { of the zero is approximately <= a given epsilon. Stop also if the      }
  { maximum number of iteration steps is exceeded, or if an error occurred.}
  until ( Err <> NoError)
        or ( k = kmax )
        or ( (MaxNorm(delta) / max(MaxNorm(qHlp),abs(z))) <= eps);

end; { procedure Approximation }

{----------------------------------------------------------------------------}
{ Purpose: Determination of enclosures for a root of polynomial p and for    }
{    the deflated polynomial.                                                }
{ Parameters:                                                                }
{    In    : 'p'   : represents a complex polynomial by its coefficients     }
{            'q'   : approximation (of the coefficients of) the deflated     }
{                    polynomial                                              }
{            'z'   : floating-point approximation of a root                  }
{    Out   : 'qq'  : returns an enclosure (for the coefficients of) the      }
{                    deflated polynomial                                     }
{            'zz'  : returns an enclosure of the root                        }
{            'Err' : error flag                                              }
{ Description: For starting floating-point approximations 'z' of a root      }
{    and 'q' of the deflated polynomial an interval residual iteration is    }
{    performed to get enclosures 'zz' for the root of polynomial 'p', as     }
{    well as for the coefficients of the deflated polynomial.                }
{----------------------------------------------------------------------------}
procedure IntervalIteration( p, q   : CPolynomial;       z   : complex;
                             var qq  : CIPolynomial; var zz  : cinterval;
                             var Err : integer                          );
const
  kmax   = 10;                               { Maximum number of iteration steps }
var
  i, k, n          : integer;
  eps              : real;
```

```
  tt               : cinterval;
  dd, vv, ww       : CIPolynomial[0..ub(p)-1];
  ddelta, ddelta_0,
  ddelta_old       : CIPolynomial[0..ub(p)-1];
begin
  Err := NoError;  n := ub(p);                          { Initialization }
                                                        {----------------}
  dd[0] := ##( -p[0] - z*q[0] );                   { Compute an inclusion of the }
  for i := 1 to n-1 do                             { defect [d] = (A - z*I)*q    }
    dd[i] := ##( q[i-1] - z*q[i] - p[i]);          {-----------------------------}

  ww[n-1] := q[n-1];                           { Compute an inclusion of [R], the }
  for i := n-2 downto 0 do                     { inverse of the Jacobian J        }
    ww[i] := ##( q[i] + z*ww[i+1] );           {----------------------------------}
  if (0 in ww[0]) then
    Err := VerFailed
  else
    begin
      ddelta_0[n-1] := dd[n-1];                         { Compute a starting }
      for i := n-2 downto 0 do                          { inclusion [y]↑(0)  }
        ddelta_0[i] := ##( dd[i] + z*ddelta_0[i+1] );   {--------------------}
      tt := ddelta_0[0]/ww[0];
      for i := 0 to n-2 do
        ddelta_0[i] := ##( -ddelta_0[i+1] + tt*ww[i+1] );
      ddelta_0[n-1] := tt;

      { Interval iteration }
      k := 0;  ddelta := ddelta_0;                      { Interval iteration }
      repeat                                            {--------------------}
        ddelta_old := ddelta;

        case k of                                      { Slightly enlarge the }
          0..3:  eps := 0.125;                         { inclusion interval   }
          4..6:  eps := 0.5;                           {----------------------}
          else:  eps := 5;
        end;
        ddelta_old := blow(ddelta,eps);

        vv[n-1] := 0;                       { Determine a new inclusion interval }
        for i := n-2 downto 0 do            {------------------------------------}
          vv[i] := ##( ddelta_old[n-1]*ddelta_old[i] + z*vv[i+1] );
        vv[0] := vv[0]/ww[0];
        for i := 0 to n-2 do
          ddelta[i] := ##( ddelta_0[i] + vv[i+1] - vv[0]*ww[i+1] );
        ddelta[n-1] := ddelta_0[n-1] - vv[0];

        k := k+1;
      until (ddelta in ddelta_old) or (k = kmax);

      if (ddelta in ddelta_old) then               { Verification of the result }
        begin                                      {----------------------------}
          for i := 0 to n-2 do qq[i] := ##( q[i] + ddelta[i] );
          qq[n-1] := p[n];
          zz := ##( z + ddelta[n-1] );
        end
      else
        Err := IncFailed;
    end;

end;  { procedure IntervalIteration }

{----------------------------------------------------------------------------}
{ Purpose: Determination of enclosures for a root of polynomial p and for    }
{    the deflated polynomial                                                 }
```

```
{ Parameters:                                                               }
{    In     : 'p'  : represents a complex polynomial by its coefficients    }
{             'z'  : floating-point approximation of a root                 }
{    Out    : 'qq' : returns an enclosure (for the coefficients of) the     }
{                    deflated polynomial                                    }
{             'zz' : returns an enclosure of the root                       }
{             'Err': error flag                                             }
{ Description: For starting approximations 'z' of a root, enclosures 'zz'  }
{    for a root of polynomial 'p', as well as 'qq' for the coefficients of  }
{    the deflated polynomial are computed. Since the root of a complex      }
{    polynomial matches with an eigenvalue of its companion matrix and the  }
{    coefficients of the deflated polynomial match to the components of the }
{    corresponding eigenvector, the eigenvector and eigenvalue are determined}
{    by iterative refinement. First, good floating-point approximations are }
{    computed, and then verified enclosures for a root and for a deflated   }
{    polynomial are returned.                                               }
{---------------------------------------------------------------------------}
global procedure CPolyZero(    p   : CPolynomial;
                               z   : complex;
                           var qq  : CIPolynomial;
                           var zz  : cinterval;
                           var Err : integer      );
var
  i, n : integer;
  q    : CPolynomial[0..ub(p)-1];
begin
  if (lb(p) <> 0) or (lb(qq) <> 0) then       { Illegal polynomial declaration }
    Err := IllProb
  else
    begin
      Err := NoError;  n := ub(p);                          { Initialization }

      if (n = 0) or (n = 1) then
        if (n = 0) then                         { Polynomial of degree n = 0 }
          if (p[0] = 0) then Err := ZeroPoly    { Zero polynomial            }
                        else Err := ConstPoly   { Constant polynomial        }
        else                                    { Polynomial of degree n = 1 }
          if (p[1] = 0) then
            if (p[0] = 0) then Err := ZeroPoly  { Zero polynomial            }
                          else Err := ConstPoly { Constant polynomial        }
          else
            begin
              zz := -intval(p[0]) / p[1];
              qq[0] := p[1];
            end
      else                                      { Polynomial of degree n > 1 }
        begin                                   { (Common case)              }
          Approximation(p,z,q,Err);
          if (Err = NoError) then IntervalIteration(p,q,z,qq,zz,Err);
        end;
    end;
end; { procedure CPolyZero }

{---------------------------------------------------------------------------}
{ Module initialization part                                                }
{---------------------------------------------------------------------------}
begin
  { Nothing to initialize }
end.
```

9.3.2 Example

We consider the following complex polynomial of 4th degree. We want to find and enclose a root for the arbitrary starting approximation $\tilde{z} = (1+i)$ using Algorithm 9.3.

$$p(z) = z^4 - (5+3i)z^3 + (6+7i)z^2 - (54-22i)z + (120+90i)$$

$$\text{exact zeros}: \quad \pm\sqrt{2}(1+2i), \quad 5, \quad \text{and} \quad 3i.$$

$$\sqrt{2} \approx 1.41421356237309\ldots$$

$$\text{starting approximation}: \quad (1+i)$$

The following main program calls *CPolyZero* to calculate an enclosure $[z]$ of an exact root z^*. That is, it is guaranteed that $z^* \in [z]$. Guaranteed enclosures $[q]$ of the coefficients of the deflated polynomial $q(z) = \frac{p(z)}{z-[z]}$ are determined, too.

```
program cpzero_example(input, output);
use
  c_ari,                                { Complex arithmetic              }
  ci_ari,                               { Complex interval arithmetic     }
  cpoly,                                { Complex polynomials             }
  cipoly,                               { Complex interval polynomials    }
  cpzero;                               { Complex polynomial zero finder  }

procedure main(n : integer);
var
  ErrCode : integer;
  z       : complex;
  zz      : cinterval;
  p       : CPolynomial[0..n];
  qq      : CIPolynomial[0..n-1];
begin
  writeln('Enter the coefficients in increasing order: ');  read(p);
  writeln;
  write('Enter the starting approximation: ');  read(z);
  writeln;

  CPolyZero(p, z, qq, zz, ErrCode);

  if (ErrCode = 0) then
    begin
      writeln('Polynomial:'); writeln(p);
      writeln('Zero found in:'); writeln(zz); writeln;
      writeln('The coefficients of the reduced polynomial are:'); writeln(qq);
    end
  else
    writeln(CPolyZeroErrMsg(ErrCode));
end; { procedure main }

var
  n : integer;

begin
  n := -1;
  while (n < 0) do
  begin
    write('Enter the degree of the polynomial (>=0): '); read(n); writeln;
  end;
  main(n);
end.
```

Our implementation of Algorithm 9.3 produces the following output:

```
Enter the degree of the polynomial (>=0): 4

Enter the coefficients in increasing order:
  x^0 * (120,90)
+ x^1 * (-54,22)
+ x^2 * (6,7)
+ x^3 * (-5,-3)
+ x^4 * (1,0)

Enter the starting approximation: (1,1)

Polynomial:
  (  1.200000000000000E+002,  9.000000000000000E+001 ) * x^0
+ ( -5.400000000000000E+001,  2.200000000000000E+001 ) * x^1
+ (  6.000000000000000E+000,  7.000000000000000E+000 ) * x^2
+ ( -5.000000000000000E+000, -3.000000000000000E+000 ) * x^3
+ (  1.000000000000000E+000,  0.000000000000000E+000 ) * x^4

Zero found in:
( [  1.414213562373094E+000,  1.414213562373096E+000 ],
  [  2.828427124746189E+000,  2.828427124746191E+000 ] )

The coefficients of the reduced polynomial are:
  ([ -4.242640687119286E+001, -4.242640687119284E+001 ],
   [  2.121320343559642E+001,  2.121320343559643E+001 ] ) * x^0
+ ([  1.414213562373094E+000,  1.414213562373096E+000 ],
   [ -3.384776310850236E+000, -3.384776310850235E+000 ] ) * x^1
+ ([ -3.585786437626906E+000, -3.585786437626904E+000 ],
   [ -1.715728752538100E-001, -1.715728752538098E-001 ] ) * x^2
+ ([  1.000000000000000E+000,  1.000000000000000E+000 ],
   [  0.000000000000000E+000,  0.000000000000000E+000 ] ) * x^3
```

The complex rectangular intervals are presented in the form (re, im), where re and im are real intervals represented by their lower and upper bounds, respectively.

9.3.3 Restrictions and Hints

Close Zeros

If two zeros of the polynomial are so close together that they are identical in their number representation up to the mantissa length and differ only in digits beyond the mantissa, they are called "a numerical multiple zero". Such zeros cannot be verified with our program described above because they cannot be separated by the given number representation. The program handles them just like a true multiple zero (see Section 9.2) and terminates. However, we could implement the algorithm CPolyZero of Section 9.2 using a multi-precision floating-point and floating-interval arithmetic (see [10], [55]).

Clusters of Zeros

If two or more zeros are extremely close together, i.e. they form a cluster, it is not possible to verify a zero of this cluster with the implementation given in Section 9.3.1 because we may enclose a zero of the derivative p' of the polynomial p. Hence, an enclosure of the determinant of the Jacobian matrix (see 9.2) contains zero, and the verified inversion of the Jacobian matrix fails. We may overcome this limit of the implementation by computing the inverse of the Jacobian matrix with higher accuracy, e.g. with double mantissa length, using the exact scalar product (see [54]).

9.4 Exercises

Exercise 9.1 Use program *cpzero_example* to enclose the other three roots of the polynomial given in Section 9.3.2.

By repeating the deflation of a verified zero from the reduced polynomial *pdeflated* $\in [q]$, the approximation of a new zero in the reduced polynomial, and the verification of the new zero in the original polynomial, we get all simple zeros of the polynomial. The deflated polynomial we can get from the parameter $[q]$ of the routine CPolyZero. For approximating a new zero, the deflated polynomial *pdeflated* is used. The verification of the new zero is done in the original polynomial p because the zeros of the approximate deflated polynomial are smeared out because $[q]$ has interval-valued coefficients, while p has point-valued coefficients.

Algorithm 9.4: AllCPolyZeros (p) {Procedure}

1. *pdeflated* := p;
   for $i := 1$ **to** $n - 1$ **do**
   {Verification of a new zero}
   (a) {Approximate a new zero of *pdeflated*}
   (b) {Verify the new zero for p}
   (c) {Deflate verified zero from *pdeflated*}

Exercise 9.2 Build a module containing the routines and definitions of the program of Section 9.3.1. Implement the above Algorithm 9.4 in PASCAL–XSC, using the routines and definitions of the program of Section 9.3.1 and test the implementation with the example of Section 9.3.2. Note, that Algorithm AllCPolyZeros might fail to determine *all* zeros of a polynomial if there exist e.g. multiple zeros.

Exercise 9.3 Determine enclosures for all zeros of the Wilkinson polynomial $p(z) = \prod_{i=1}^{n}(z - i)$ with $n = 5$ using Algorithm 9.4. Start computation with complex and real starting approximations.

9.5 References and Further Reading

Algorithm 9.3 is an extension of one first proposed by Böhm [10] for polynomials with real coefficients. A commercial implementation was released by IBM with the

Fortran subroutine library ACRITH [32]. The algorithm was generalized for polynomials with complex coefficients by Geörg [19] and Grüner [22]. This modification may be used to enclose simple real and complex zeros of complex polynomials, as well as simple real and complex zeros of real polynomials. Additional comments are given by Krämer in [48].

Part III

Multi-Dimensional Problems

Chapter 10

Linear Systems of Equations

Finding the solution of a linear system of equations is one of the basic problems in numerical algebra. We will develop a verification algorithm for square systems with full matrix based on a Newton-like method for an equivalent fixed-point problem.

10.1 Theoretical Background

10.1.1 A Newton-like Method

Let $Ax = b$ be a real system of equations with $A \in I\!R^{n\times n}$ and $b, x \in I\!R^n$. Finding a solution of the system $Ax = b$ is equivalent to finding a zero of $f(x) = Ax - b$. Hence, Newton's method gives the following fixed-point iteration scheme

$$x^{(k+1)} = x^{(k)} - A^{-1}(Ax^{(k)} - b), \quad k = 0, 1, \dots . \tag{10.1}$$

Here, $x^{(0)}$ is some arbitrary starting value. In general, the inverse of A is not exactly known. Thus instead of (10.1), we use the Newton-like iteration

$$x^{(k+1)} = x^{(k)} - R(Ax^{(k)} - b), \quad k = 0, 1, \dots, \tag{10.2}$$

where $R \approx A^{-1}$ is an approximate inverse of A.

Let us now replace the real iterates $x^{(k)}$ by interval vectors $[x]^{(k)} \in I\,I\!R^n$. If there exists an index k with $[x]^{(k+1)} \subseteq [x]^{(k)}$, then, by Brouwer's fixed-point theorem (see page 51), Equation 10.2 has at least one fixed-point $x \in [x]^{(k)}$. Supposed R is regular. Then this fixed-point is also a solution of $Ax = b$. However, if we consider the diameter of $[x]^{(k+1)}$, we get $d([x]^{(k+1)}) = d([x]^{(k)}) + d(R(A[x]^{(k)} - b)) \geq d([x]^{(k)})$. Thus in general, the subset relation will not be satisfied. For this reason, we modify the right-hand side of (10.2) to

$$x^{(k+1)} = Rb + (I - RA)x^{(k)}, \quad k = 0, 1, \dots, \tag{10.3}$$

where I denotes the $n \times n$ identity matrix. Rump [76] proved for Equation (10.3) that if there exists an index k with $[x]^{(k+1)} \overset{\circ}{\subset} [x]^{(k)}$, then the matrices R and A are regular, and there is a unique solution x of the system $Ax = b$ with $x \in [x]^{(k+1)}$. Moreover this result is valid for any matrix R. Note that the $\overset{\circ}{\subset}$ operator denotes the inclusion in the interior as defined in Section 3.1.

10.1.2 The Residual Iteration Scheme

It is a well-know numerical principle that an approximate solution $\tilde{x}$ of $Ax = b$ may be improved by solving the system $Ay = d$, where $d = b - A\tilde{x}$ is the residual of $A\tilde{x}$. Since $y = A^{-1}(b - A\tilde{x}) = x - \tilde{x}$, the exact solution of $Ax = b$ is given by $x = \tilde{x} + y$. Applying Equation (10.3) to the residual system leads to

$$y^{(k+1)} = \underbrace{R(b - A\tilde{x})}_{=:\, z} + \underbrace{(I - RA)}_{=:\, C} y^{(k)}, \quad k = 0, 1, \dots . \tag{10.4}$$

According to Rump's results mentioned above, the residual equation $Ay = d$ has a unique solution $y \in [y]^{(k+1)}$ if we succeed in finding an index k satisfying $[y]^{(k+1)} \overset{\circ}{\subset} [y]^{(k)}$ for the corresponding interval iteration scheme. Moreover, since $y = x - \tilde{x} \in [y]^{(k+1)}$, we then have a verified enclosure of the unique solution of $Ax = b$ given by $\tilde{x} + [y]^{(k+1)}$.

These results remain valid if we replace the exact expressions for z and C in (10.4) by interval extensions. However, to avoid overestimation effects, it is highly recommended to evaluate $b - A\tilde{x}$ and $I - RA$ without any intermediate rounding.

10.1.3 How to Compute the Approximate Inverse

We recall that the results summarized above are valid even for an arbitrary matrix R. However, it is a non-surprising empirical fact that the better R approximates the inverse of A, the faster the contained-in-the-interior relation for two successive iterates will be satisfied. We favor Crout's algorithm (cf. [83]) with partial pivoting for matrix inversion.

Let $A' = PA$ be the matrix which comes from A by row interchanges so that a factorization $A' = LU$ with a lower-triangular matrix L and an upper-triangular matrix U exists. If we normalize the diagonal entries of L to unity, and if we save the indices for the pivotal rows in a vector p, initialized by $p = (1, 2, \dots, n)^T$, the decomposition procedure is as follows:

$$\left.\begin{array}{l}
v_k = a_{ki} - \sum\limits_{j=1}^{i-1} l_{kj} a_{ji}, \quad k = i, \dots, n \\[2ex]
|v_j| = \max\limits_{i \le k \le n} |v_k| \\[2ex]
\text{if } j \ne i: \quad p_i \leftrightarrow p_j, v_i \leftrightarrow v_j, a_i \leftrightarrow a_j, l_i \leftrightarrow l_j, u_i \leftrightarrow u_j \\[2ex]
u_{ii} = v_i \\[1ex]
\left.\begin{array}{l}
u_{ik} = a_{ik} - \sum\limits_{j=1}^{i-1} l_{ij} u_{jk} \\[2ex]
l_{ki} = \dfrac{v_k}{v_i}
\end{array}\right\} k = i+1, \dots, n
\end{array}\right\} i = 1, \dots, n. \tag{10.5}$$

Here, the double arrow ($\leftrightarrow$) indicates that the values of the specified elements or rows are swapped.

Once the LU-decomposition is available, the approximate inverse R is computed column by column using simple forward/backward substitution. Let $e^{(k)}$ denote the k-th unit vector. We successively solve the systems $Ly = Pe^{(k)}$ and $Ux = y$ by

$$y_i = e_{p_i}^{(k)} - \sum_{j=1}^{i-1} l_{ij} y_j, \quad i = 1, \ldots, n, \text{ and} \tag{10.6}$$

$$x_i = \frac{1}{u_{ii}} \left(y_i - \sum_{j=i+1}^{n} u_{ij} x_j \right), \quad i = n, n-1, \ldots, 1. \tag{10.7}$$

The vector x is an approximation for the k-th column of R.

We define the condition number of a regular square matrix A by

$$\mathrm{cond}(A) := \|A\|_\infty \cdot \|A^{-1}\|_\infty.$$

For solving a linear system of equations with conventional floating-point arithmetic, it is a rule of thumb that $\log(\mathrm{cond}(A))$ is a rough indicator for the number of decimals of lost accuracy in the solution. For instance, a condition number of 10^5 signals a loss of about five decimals of accuracy in the solution. Some of the algorithms described below return the quantity $\|A\|_\infty \cdot \|R\|_\infty$ as an estimate for the condition number of the matrix A.

10.2 Algorithmic Description

First, we give the algorithm for computing an approximate inverse. Except for the special case of 2×2 matrices, which is handled by applying well-known explicit rules, Crout's method is used as described in Section 10.1.3. To save memory, the matrices L and U are not explicitly allocated. Actually, their elements are stored by overwriting the input matrix A. Thus, any appearance of components of L and U in Equations (10.5)–(10.7) is replaced by the corresponding elements of A. The inversion procedure is stopped if a pivotal element is less than *Tiny*. Otherwise, we would risk an overflow exception at runtime when executing the division by the pivotal element. The value of *Tiny* depends on the exponent range of the floating-point format being used. For the floating-point system used by PASCAL–XSC with a minimum positive element of about 10^{-324} (see Section 3.7), a value of 10^{-200} is a good choice. The forward substitution stage of the algorithm takes advantage of the leading zero elements in the unit vector $e^{(k)}$.

Algorithm 10.1: MatInv(A, R, Err) {Procedure}

1. {Initialization}
   $Tiny := 10^{-200}$; $p := (1, 2, \ldots, n)^T$; $Err :=$ "No error";
2. {Special case: $n = 2$}
   if $(n = 2)$ **then**
   $Det := \square(a_{11} \cdot a_{22} - a_{21} \cdot a_{12})$;

> **if** ($|Det| < Tiny$) **then**
> **return** $Err :=$ "Matrix is probably singular!";
> $r_{11} := \quad a_{22}/Det; \quad r_{12} := -a_{12}/Det;$
> $r_{21} := -a_{21}/Det; \quad r_{22} := \quad a_{11}/Det;$
>
> 3. {Usual case: $n \neq 2$}
> **else**
> {LU-Factorization, see (10.5)}
> **for** $i = 1$ **to** n **do**
> $$v_k := \square\left(a_{ki} - \sum_{j=1}^{i-1} a_{kj} \cdot a_{ji}\right); \quad (k = i, \ldots, n)$$
> $|v_j| := \max_{i \le k \le n} |v_k|;$
> **if** ($|v_j| < Tiny$) **then**
> **return** $Err :=$ "Matrix is probably singular!";
> **if** ($j \neq i$) **then**
> swap (p_i, p_j); swap (v_i, v_j); swap (a_i, a_j);
> $a_{ii} := v_i;$
> **for** $k = i + 1$ **to** n **do**
> $$a_{ik} := \square\left(a_{ik} - \sum_{j=1}^{i-1} a_{ij} \cdot a_{jk}\right);$$
> $a_{ki} := v_k / v_i;$
>
> 4. {Computation of the inverse column by column}
> **for** $k = 1$ **to** n **do**
> (a) {Forward substitution, see (10.6)}
> Find index l with $p_l = k$;
> $y_i := 0; \quad (i = 1, \ldots, l-1)$
> $y_l := 1;$
> $$y_i := -\square\left(\sum_{j=1}^{i-1} a_{ij} \cdot y_j\right); \quad (i = l+1, \ldots, n);$$
> (b) {Backward substitution, see (10.7)}
> $$x_i := \square\left(y_i - \sum_{j=i+1}^{n} a_{ij} \cdot x_j\right) / a_{ii}; \quad (i = n, n-1, \ldots, 1)$$
> $r_{*,k} := x; \quad$ {k-th column of R}
>
> 5. {Return: Approximate inverse R and error code Err}
> **return** R, Err;

The procedure for solving the linear system of equations $Ax = b$ has two principal steps: We compute an approximate solution $\tilde{x}$, and then we try to find an enclosure for the error of this approximation. That is, the method is an *a posteriori* method as described in Section 3.8. To get a good approximation, we use (10.2) for a simple real residual iteration (see Step 2 of Algorithm 10.5). Since this pre-iteration will be followed by a verification step, we may apply some heuristic considerations to improve the value it computes. We try to predict whether some of the components of the exact solution might vanish. Let x and y be two successive iterates. Our heuristic is that any component of y which has diminished in more than n orders of magnitude is a good candidate for a zero entry in the exact solution. In Algorithm 10.2, those components of the new iterate y are changed to zero. It turned out that $n = 5$ is a good value for practical use.

Algorithm 10.2: CheckForZeros (x, y) {Procedure}

1. {Initialization}
   $Factor := 10^5$;
2. {Updating zero-suspicious entries of y, i.e. those entries with $|y_i/x_i| \leq 10^{-5}$}
   if $(|y_i| \cdot Factor < |x_i|)$ **then** $y_i := 0$; $\quad (i = 1, \ldots, n)$
3. {Return the updated y}
   return y;

In general, the real iteration will be stopped if the relative error of any component of two successive iterates is less than δ. Otherwise, we halt the iteration after $k_{\max}$ steps. Here, we use $\delta = 10^{-12}$ because the floating-point format of PASCAL–XSC has about 15 significant decimal digits. That is, we try to compute about 12 correct digits of the mantissa of the approximate solution. Algorithm 10.3 describes the stopping criterion for the iteration. Here, we make use of another heuristic. If the components of two successive iterates differ in sign, or if one of them vanishes, we take this as an indicator for a zero entry in the exact solution. Thus, we implicitly set its relative error to zero.

Algorithm 10.3: Accurate (x, y) {Function}

1. {Initialization: Desired relative accuracy}
   $\delta := 10^{-12}$;
2. {Check if the relative error of y with respect to x is less than or equal to δ}
   $i := 1$;
   repeat
     if $(x_i \cdot y_i \leq 0)$ **then** $ok := True$ {Relative error implicitly set to 0}
     else
       $ok := (|y_i - x_i| \leq \delta \cdot |y_i|)$;
     $i := i+1$;
   until (not ok) **or** $(i > n)$;
3. {Return value of the flag ok}
   return Accurate $:= ok$;

For the verification step, we refer to Section 10.1.2. To accelerate the procedure, the iterates are inflated at the beginning of each iteration loop. For a general remark about the ε-inflation, we refer to Section 3.8. As recommended in [46], we use a constant value of $\varepsilon = 1000$ for the ε-inflation. We refer to Falcó Korn [15] for a more sophisticated inflation strategy. It is an empirical fact that the inner inclusion is satisfied nearly always after a few steps or never. Thus in Algorithm 10.4, the iteration is stopped after $p_{\max}$ steps.

Algorithm 10.4: VerificationStep $([x], [z], [C], \mathit{IsVerified})$ {Procedure}

1. {Initialization}
   $\varepsilon := 1000$; $p_{\max} := 3$; $p := 0$; $[x]^{(0)} := [z]$;
2. **repeat**
   $[x]^{(p)} := [x]^{(p)} \bowtie \varepsilon$; {$\varepsilon$-Inflation}
   $[x]^{(p+1)} := \Diamond([z] + [C] \cdot [x]^{(p)})$;
   $\mathit{IsVerified} := ([x]^{(p+1)} \overset{\circ}{\subset} [x]^{(p)})$;
   $p := p + 1$;
   until $\mathit{IsVerified}$ **or** $(p \geq p_{\max})$;
   $[x] := [x]^{(p)}$;
3. {Return enclosure $[x]$ and flag *IsVerified*}
   return $[x]$, *IsVerified*;

We now give the complete algorithm based on the Algorithms 10.2–10.4 for computing a verified enclosure of the solution of a linear system of equations. The procedure fails if the computation of an approximate inverse R fails or if the inclusion in the interior can not be established. A condition number estimate is returned by the variable *Cond*. We need a narrow enclosure $[z]$ of $z = R(b - A\tilde{x})$ to start the verification step. Thus in Step 3, we first compute the residual $d = \Box(b - A\tilde{x})$. If the computation was afflicted with a roundoff error, an enclosure of this error is given by $[d] = \Diamond(b - A\tilde{x} - d)$. So, we have $b - A\tilde{x} \in d + [d]$ and $z \in \Diamond(Rd + R[d])$. In Step 4, we remark that $\tilde{x}$ is an exact, but not necessarily unique, solution of the system if $[z] = 0$.

Algorithm 10.5: LinSolve $(A, b, [x], \mathit{Cond}, \mathit{Err})$ {Procedure}

1. {Computation of an approximate solution}
   MatInv (A, R, Err);
   if ($\mathit{Err} \neq$ "No error") **then**
   return Err := "Matrix is probably singular!"
   $\mathit{Cond} := \|A\|_\infty \cdot \|R\|_\infty$;
2. {Real residual iteration for an approximate solution}
   $k_{\max} := 10$; $k := 0$; $\tilde{x}^{(0)} := R \cdot b$;
   repeat
   $d := \Box(b - A \cdot \tilde{x}^{(k)})$;
   $\tilde{x}^{(k+1)} := \Box(\tilde{x}^{(k)} + R \cdot d)$;
   CheckForZeros $(\tilde{x}^{(k)}, \tilde{x}^{(k+1)})$;
   Success := Accurate $(\tilde{x}^{(k)}, \tilde{x}^{(k+1)})$;
   $k := k + 1$;

> **until** *Success* **or** $(k \geq k_{\max})$;
> $\tilde{x} := \tilde{x}^{(k)}$;
>
> 3. {Computation of enclosures $[C]$ and $[z]$ for $C = I - RA$ and $z = R(b - A\tilde{x})$}
> $[C] := \Diamond(I - R \cdot A)$;
> $d := \Box(b - A \cdot \tilde{x})$;
> $[d] := \Diamond(b - A \cdot \tilde{x} - d)$;
> $[z] := \Diamond(R \cdot d + R \cdot [d])$;
>
> 4. {Verification step}
> **if** $([z] = 0)$ **then**
> $[x] := \tilde{x}$; {Exact solution}
> **else**
> VerificationStep($[x], [z], [C]$, *IsVerified*);
> **if** (**not** *IsVerified*) **then**
> *Err* := "Verification failed, the system is probably ill-conditioned!";
> **return** *Cond*, *Err*;
> **else**
> $[x] := \tilde{x} + [x]$; {Approximation plus correction interval}
>
> 5. {Return: Enclosure, condition number estimate, and error code}
> **return** $[x]$, *Cond*, *Err*;

10.3 Implementation and Examples

10.3.1 PASCAL–XSC Program Code

We give the listings of the modules *matinv* for matrix inversion and *linsys* for solving a linear systems of equations. We emphasize that those parts of the algorithms of the previous section which are marked by $\Box(\ldots)$ and $\Diamond(\ldots)$ are implemented using accurate expressions. Error codes are passed by parameters of type *integer*. The error code 0 means that no error occurred.

10.3.1.1 Module matinv

The procedure *MatInv* is an implementation of Algorithm 10.1. It takes into account that the index range of the matrix parameters may start with lower indices different from one. The procedure checks whether the system is square and whether the input and the output matrices have the same shape. The function *MatInvErrMsg* may be used to get an explicit error message.

```
{----------------------------------------------------------------------------}
{ Purpose: Computation of an approximate inverse of a real square matrix.    }
{ Method:  LU decomposition applying Crout's algorithm.                      }
{ Global functions and procedures:                                           }
{   procedure MatInv(...)      : Matrix inversion                            }
{   function MatInvErrMsg(...) : To get an error message text                }
{----------------------------------------------------------------------------}
```

```
module matinv;       { Matrix inversion routines }
                     { ---    ---                }
use
  mv_ari;            { Real matrix/vector arithmetic }

{----------------------------------------------------------------------------}
{ Error codes used in this module.                                           }
{----------------------------------------------------------------------------}
const
  NoError      = 0;      { No error occurred.                             }
  NotSquare    = 1;      { Matrix to be inverted is not square.           }
  DimensionErr = 2;      { Input and output matrices differ in shape.     }
  Singular     = 3;      { Matrix to be inverted is probably singular.    }

{----------------------------------------------------------------------------}
{ Error messages depending on the error code.                                }
{----------------------------------------------------------------------------}
global function MatInvErrMsg ( Err : integer ) : string;
var
  Msg : string;
begin
  case Err of
    NoError     : Msg := '';
    NotSquare   : Msg := 'Matrix to be inverted is not square';
    DimensionErr: Msg := 'Input and output matrices differ in shape';
    Singular    : Msg := 'Inversion failed, matrix is probably singular';
    else        : Msg := 'Code not defined';
  end;
  if (Err <> NoError) then Msg := 'Error: ' + Msg + '!';
  MatInvErrMsg := Msg;
end;

{----------------------------------------------------------------------------}
{ Purpose: The procedure 'MatInv' may be used for the computation of an      }
{    approximate inverse of a real square matrix.                            }
{ Parameters:                                                                }
{    In  : 'A'   : matrix to be inverted, passed as reference parameter to   }
{                  save computation time for copying 'A'.                    }
{    Out : 'R'   : approximate inverse.                                      }
{          'Err' : error code.                                               }
{ Description:                                                               }
{   Inversion of a regular matrix A stored in 'A' using LU decomposition.    }
{   For LU decomposition, formally a permutation matrix P is determined so   }
{   that the product P*A may be decomposed into a lower-triangular matrix L  }
{   and an upper-triangular matrix U with P*A = L*U. The diagonal elements   }
{   of L are set to 1. Using Crout's algorithm, the elements of both matri-  }
{   ces L and U are stored by temporary overwriting a copy of the input      }
{   matrix 'A'. The permutation matrix P is not explicitly generated. The    }
{   indices of row interchanges are stored in the index vector 'p' instead.  }
{   The i-th element of P*b may be accessed indirectly using the p[i]-th     }
{   entry of 'b'. Accurate expressions are used to avoid cancellation errors.}
{   The k-th column of the inverse R of P*A is computed by forward/backward  }
{   substitution with the k-th unit vector e_k as the right-hand side of the }
{   system: U*y = P*e_k ==> y, L*x = y ==> x. For error codes, see above.    }
{----------------------------------------------------------------------------}
global procedure MatInv ( var A   : rmatrix;
                          var R   : rmatrix;
                          var Err : integer  );
const
  Tiny = 1E-200;       { A divisor less than 'Tiny' is handled as zero }
var
  n1, nn, m1, mm   : integer;    { For lower and upper bounds of 'A' }
  p1, pn, q1, qm   : integer;    { For lower and upper bounds of 'R' }
  i, j, k, l, n, m : integer;    { For loops      }
```

```
  Max, Temp          : real;          { Help variables }
  AA          : rmatrix[1..ub(A)-lb(A)+1,1..ub(A)-lb(A)+1];   { A copy of 'A'}
  p           : dynamic array[1..ub(A)-lb(A)+1] of integer;
  v, x        : rvector[1..ub(A)-lb(A)+1];
begin
  { Get lower and upper bounds of the rows and columns of 'A' and 'R' }
  n1 := lb(A,1);  nn := ub(A,1);  n := nn-n1+1;
  m1 := lb(A,2);  mm := ub(A,2);  m := mm-m1+1;
  p1 := lb(R,1);  pn := ub(R,1);
  q1 := lb(R,2);  qm := ub(R,2);

  { Check for correct dimensions of 'A' and 'R' }
  Err := NoError;

  if (n <> m) then                          { Error: 'A' is not square         }
    Err := NotSquare
  else if ( (n <> pn-p1+1) or
            (m <> qm-q1+1) ) then           { Error: Dimensions not compatible }
    Err := DimensionErr
  else if (n = 2) then                      { Special case: (2,2)-matrix       }
    begin
      AA := A;      { Copy 'A' to avoid computation of index offsets }
      Temp := #*( AA[1,1]*AA[2,2] - AA[2,1]*AA[1,2] );   { = Det(AA) }
      if (abs(Temp) < Tiny) then
        Err := Singular
      else
        begin
          R[p1,q1] :=  AA[2,2] / Temp;  R[p1,qm] := -AA[1,2] / Temp;
          R[pn,q1] := -AA[2,1] / Temp;  R[pn,qm] :=  AA[1,1] / Temp;
        end;
    end
  else   { Usual case: Dimension of 'A' > 2 }
    begin
      AA := A;          { Copy 'A' to avoid computation of index offsets }
      for i := 1 to n do p[i] := i;   { Initializing permutation vector }

      { Start LU factorization }
      i := 1;
      while ((Err = NoError) and (i <= n)) do
      begin
        { Compute the numerators of those elements of the i-th column }
        { of L which are not updated so far and store them in 'v'.    }
        for k := i to n do
          v[k] := #*(AA[k,i] - for j:=1 to i-1 sum(AA[k,j] * AA[j,i]));

        { Look for the column pivot }
        j := i;  Max := abs(v[i]);
        for k := i+1 to n do
        begin
          Temp := abs(v[k]);
          if (Temp > Max) then
            begin  j := k;  Max := Temp;  end;
        end;

        { Swap rows of 'AA' and 'v', store the permutation in 'p' }
        if (j <> i) then
        begin
          x := AA[i];     AA[i] := AA[j];  AA[j] := x;
          k := p[i];      p[i] := p[j];    p[j] := k;
          Temp := v[i];  v[i] := v[j];    v[j] := Temp;
        end;

        if (Max < Tiny) then      { Pivot element < 'Tiny', inversion failed }
          Err := Singular         { matrix 'A' assumed to be singular        }
```

```
      else
        begin
          Temp := v[i];
          AA[i,i] := Temp;        { Update the diagonal element of U }
        end;

      if (Err = NoError) then
      begin
        { Update U's row and L's column elements }
        for k := i+1 to n do
        begin
          AA[i,k] := #*(AA[i,k] - for j:=1 to i-1 sum(AA[i,j]*AA[j,k]));
          AA[k,i] := v[k] / Temp;
        end;
        i := i+1;
      end;
    end; {while}

    if (Err = NoError) then

      { Now 'AA' is overwritten with the subdiagonal elements of L in its  }
      { lower left triangle and with the elements of U in its diagonal and }
      { its upper right triangle. The elements of the inverse matrix are   }
      { computed column by column using forward/backward substitution.     }
      {--------------------------------------------------------------------}
      for k := 1 to n do
      begin
        { Forward substitution: L*x = P*e_k, where e_k is the k-th unit }
        { vector. Note: If P*e_k has m leading zeros, this results in   }
        { x_i = 0 for 1,..,l-1 and x_l = 1. Thus, forward substitution  }
        { starts at index l+1.                                          }
        {---------------------------------------------------------------}
        l := 1;
        while (p[l] <> k) do
          begin x[l] := 0; l := l+1;  end;
        x[l] := 1;
        for i := l+1 to n do
          x[i] := -#*(for j:=l to i-1 sum(AA[i,j]*x[j]));

        { Backward substitution: U * x = x, where the right-hand side is }
        { the result of the forward substitution. It will be overwritten }
        { by the solution of the system, the k-th column of the inverse  }
        { matrix.                                                        }
        {----------------------------------------------------------------}
        for i := n downto 1 do
          x[i] := #*(x[i] - for j:=i+1 to n sum(AA[i,j]*x[j])) / AA[i,i];

        R[*,q1+k-1] := x;    { Remember index offset! }
    end; {if (Err = NoError) ...}
  end; {Usual case}
end;

{----------------------------------------------------------------------------}
{ Module initialization part                                                 }
{----------------------------------------------------------------------------}
begin
  { Nothing to initialize }
end.
```

10.3.1.2 Module linsys

The local procedure *LinSolveMain* is an implementation of Algorithm 10.5. It takes into account that the index range of its parameters may start with a lower bound different from one. The procedure checks if the system is square and if the dimensions of the parameters are compatible. The local procedures *CheckForZeros*, *Accurate*, and *VerificationStep* are implementations of the Algorithms 10.2–10.4. The module provides two global procedures *LinSolve* one with and the other without computing a condition number estimate for the input matrix. Both procedures call *LinSolveMain* with an appropriate list of parameters. The global function *LinSolveErrMsg* gives an explicit error message.

```
{----------------------------------------------------------------------------}
{ Purpose: Computation of a verified solution of a square linear system of   }
{    equations A*x = b with full real matrix A and real right-hand side b.   }
{ Method: Transformation of A*x = b to fixed-point form and applying an      }
{    interval residual iteration.                                            }
{ Global functions and procedures:                                           }
{   procedure LinSolve(...)       : To get a verified enclosure of the       }
{                                   solution, two versions                   }
{   function LinSolveErrMsg(...) : To get an error message text              }
{----------------------------------------------------------------------------}
module linsys;          { Linear system solving }
                        { ---    ---            }
use
  mv_ari,               { Matrix/vector arithmetic          }
  mvi_ari,              { Matrix/vector interval arithmetic }
  matinv;               { Matrix inversion                  }

{----------------------------------------------------------------------------}
{ Error codes used in this module.                                           }
{----------------------------------------------------------------------------}
const
  NoError      = 0;    { No error occurred.                        }
  NotSquare    = 1;    { System to be solved is not square.        }
  DimensionErr = 2;    { Dimensions of A*x = b are not compatible. }
  InvFailed    = 3;    { System is probably singular.              }
  VerifFailed  = 4;    { Verification failed, system is probably   }
                       { ill-conditioned.                          }

{----------------------------------------------------------------------------}
{ Error messages depending on the error code.                                }
{----------------------------------------------------------------------------}
global function LinSolveErrMsg ( Err : integer ) : string;
var
  Msg : string;
begin
  case Err of
    NoError     : Msg := '';
    NotSquare   : Msg := 'System to be solved is not square';
    DimensionErr: Msg := 'Dimensions of A*x = b are not compatible';
    InvFailed   : Msg := 'System is probably singular';
    VerifFailed : Msg := 'Verification failed, system is probably ' +
                         'ill-conditioned';
    else        : Msg := 'Code not defined';
  end;
  if (Err <> NoError) then Msg := 'Error: ' + Msg + '!';
  LinSolveErrMsg := Msg;
end;
```

```
{----------------------------------------------------------------------------}
{ Computes the absolute value of a real vector component by component.       }
{----------------------------------------------------------------------------}
function abs( v : rvector ) : rvector[lb(v)..ub(v)];
var
  i : integer;
begin
  for i := lb(v) to ub(v) do abs[i] := abs(v[i]);
end;


{----------------------------------------------------------------------------}
{ Computes an upper bound for the maximum norm of a real matrix 'M'.         }
{----------------------------------------------------------------------------}
function MaxNorm( var M : rmatrix ) : real;
var
  i, j     : integer;
  Max, Tmp : real;
  AbsMi    : rvector[lb(M,2)..ub(M,2)];
begin
  Max := 0;
  for i := lb(M) to ub(M) do
  begin
    AbsMi := abs(M[i]);
    Tmp := #>( for j:=lb(M,2) to ub(M,2) sum( AbsMi[j] ) );
    if (Tmp > Max) then Max := Tmp;
  end;
  MaxNorm := Max;
end;

{----------------------------------------------------------------------------}
{ The vectors x and y are successive approximations for the solution of a    }
{ linear system of equations computed by iterative refinement. If a compo-   }
{ nent of y is diminished by more than 'Factor', it is a good candidate for  }
{ a zero entry. Thus, it is set to zero.                                     }
{----------------------------------------------------------------------------}
procedure CheckForZeros ( var x, y : rvector );
const
  Factor = 1E+5;
var
  i : integer;
begin
  for i := lb(y) to ub(y) do
    if ( abs(y[i])*Factor < abs(x[i]) ) then y[i] := 0;
end;

{----------------------------------------------------------------------------}
{ The vectors x and y are successive iterates. The function returns TRUE if  }
{ the relative error of all components x_i and y_i is <= 10↑(-12), i.e. y_i  }
{ has about 12 correct decimals. If x_i or y_i vanishes, the relative error  }
{ is implicitly set to zero.                                                 }
{----------------------------------------------------------------------------}
function Accurate ( var x, y : rvector ) : boolean;
const
  Delta     = 1E-12;   { Relative error bound }
var
  ok     : boolean;
  i, n   : integer;
  abs_yi : real;
begin
  i := lb(y); n := ub(y);
  repeat
    if (sign(x[i])*sign(y[i]) <= 0) then       { Relative error set to 0 }
      ok := true
```

```
    else
      begin
        abs_yi := abs(y[i]);                      { Relative error > Delta? }
        ok := (abs(y[i] - x[i]) <= Delta * abs_yi );
      end;
    i := i+1;
  until (not ok) or (i > n);
  Accurate := ok;
end;

{----------------------------------------------------------------------------}
{ This procedure 'VerificationStep()' performs the iteration                 }
{ [y] = blow([x],Eps), [x] = [z] + [C]*[y] for k = 1,2,... until the new     }
{ iterate [x] lies in the interior of [y] or until the maximum number of     }
{ iterations is exceeded. The flag 'IsVerified' is set if an inclusion in    }
{ the interior could be established.                                         }
{----------------------------------------------------------------------------}
procedure VerificationStep ( var xx, zz     : ivector;
                             var C          : imatrix;
                             var IsVerified : boolean );
const
  MaxIter = 3;          { Maximum number of iteration steps }
  Epsilon = 1000;       { Factor for the epsilon inflation }
var
  p  : integer;
  yy : ivector[lb(xx)..ub(xx)];
begin
  xx := zz; p := 0;                          { Initialize:  [x] := [z] }
  repeat
    yy := blow(xx,Epsilon);                  { Epsilon inflation            }
    xx := zz + C*yy;                         { New iterate: [x] := [z]+[C]*[y] }
    IsVerified := xx in yy;                  { Inclusion in the interior?    }
    p := p+1;
  until ( IsVerified or (p >= MaxIter) );
end;

{----------------------------------------------------------------------------}
{ Purpose: The procedure 'LinSolveMain()' computes a verified solution of a  }
{    square linear system of equations A*x=b.                                 }
{ Parameters:                                                                 }
{    In  : 'A'           : matrix of the system, passed as reference          }
{                          parameter to save time for copying it.             }
{          'b'           : right-hand side of the system, passed as           }
{                          reference parameter to save time for copying it.   }
{          'ComputeCond' : flag signalling whether a condition number         }
{                          estimate is to be computed.                        }
{    Out : 'xx'          : enclosure of the unique solution.                  }
{          'Cond'        : condition number estimate.                         }
{          'Err'         : error code.                                        }
{ Description: An approximate inverse 'R' of 'A' is computed by calling       }
{   procedure 'MatInv()'. Then an approximate solution 'x' is computed        }
{   applying a conventional real residual iteration. For the final verifica- }
{   tion, an interval residual iteration is performed. An enclosure of the    }
{   unique solution is returned in the interval vector 'xx'. The procedure    }
{   also returns a condition number estimate 'Cond' if the flag 'ComputeCond'}
{   is set. 'Cond' is initialized by -1. A negative value for 'Cond' signals }
{   that an estimate could not be computed.                                   }
{----------------------------------------------------------------------------}
procedure LinSolveMain ( var A          : rmatrix; var b    : rvector;
                         var xx         : ivector; var Cond : real;
                             ComputeCond : boolean; var Err  : integer );
const
  MaxResCorr = 10;    { Maximum number of real residual corrections }
```

```
var
  IsVerified : boolean;
  k, n, m    : integer;
  x, y, d    : rvector[lb(xx)..ub(xx)];
  zz         : ivector[lb(xx)..ub(xx)];
  R          : rmatrix[lb(A,1)..ub(A,1),lb(A,2)..ub(A,2)];
  C          : imatrix[lb(A,1)..ub(A,1),lb(A,2)..ub(A,2)];

begin
  { Get length of the rows and columns of 'A', initialize condition number }
  n := ub(A,1) - lb(A,1);  m := ub(A,2) - lb(A,2);  Cond := -1;

  if (n <> m) then                              { Error: 'A' is not square     }
    Err := NotSquare
  else if ( (n <> ub(b)-lb(b)) or              { Error: Dimensions of A*x = b }
            (m <> ub(xx)-lb(xx)) ) then        {        are not compatible    }
    Err := DimensionErr
  else
    Err := NoError;

  if (Err = NoError) then
  begin
    MatInv(A,R,Err);
    if (Err <> NoError) then
      Err := InvFailed
    else
      begin {Algorithm}
        if ComputeCond then                         { Compute condition number}
          Cond := MaxNorm(A) *> MaxNorm(R);         {-------------------------}

        x := R*b;  k := 0;              { Real residual iteration }
        repeat                          {-------------------------}
          y := x;
          d := #*(b - A*y);
          x := #*(y + R*d);
          CheckForZeros(y,x);
          k := k+1;
        until Accurate(y,x) or (k >= MaxResCorr);

        { Prepare verification step, i.e. compute enclosures [C] }
        { and [z] of C = (I - R*A) and z = R*(b - A*x).          }
        {--------------------------------------------------------}
        C := ##(id(A) - R*A);
        d := #*(b - A*x);
        zz := ##(b - A*x - d);      { Now b-A*x lies in d+[z]. }
        zz := ##(R*d + R*zz);       { Thus, R*(b-A*x) lies in ##(R*d+R*[z]) }

        { If R*(b-A*x) = 0, then x is an exact solution. Otherwise try to }
        { find a verified enclosure [x] for the absolute error of x.      }
        {-----------------------------------------------------------------}
        if (zz = null(zz)) then
          xx := x
        else
          begin
            VerificationStep(xx,zz,C,IsVerified);   { Attempt to compute [x] }
            if not IsVerified then
              Err := VerifFailed
            else
              xx := x + xx;   { The exact solution lies in x+[x] }
          end;
      end; {Algorithm}
  end; { if (Err = NoError) ... }
end; {procedure LinSolve}
```

```
{----------------------------------------------------------------------------}
{ Purpose: The procedure 'LinSolve()' computes a verified solution of a      }
{    square linear system of equations A*x=b without returning a condition   }
{    number estimate.                                                        }
{ Parameters:                                                                }
{    In  : 'A'          : matrix of the system, passed as reference         }
{                         parameter to save time for copying it.             }
{          'b'          : right-hand side of the system, passed as           }
{                         reference parameter to save time for copying it.   }
{    Out : 'xx'         : enclosure of the unique solution.                  }
{          'Err'        : error code.                                        }
{ Description: Calls 'LinSolveMain()' for solving the linear system with the }
{    flag 'ComputeCond' not set.                                             }
{----------------------------------------------------------------------------}
global procedure LinSolve ( var A  : rmatrix; var b   : rvector;
                            var xx : ivector; var Err : integer );
var
  DummyCond : real;   { Dummy parameter for call of 'LinSolveMain()' }
begin
  LinSolveMain(A,b,xx,DummyCond,false,Err);
end;

{----------------------------------------------------------------------------}
{ Purpose: The procedure 'LinSolve()' computes a verified solution of a      }
{    square linear system of equations A*x=b and returns a condition         }
{    number estimate.                                                        }
{ Parameters:                                                                }
{    In  : 'A'          : matrix of the system, passed as reference         }
{                         parameter to save time for copying it.             }
{          'b'          : right-hand side of the system, passed as           }
{                         reference parameter to save time for copying it.   }
{    Out : 'xx'         : enclosure of the unique solution.                  }
{          'Cond'       : condition number estimate.                         }
{          'Err'        : error code.                                        }
{ Description: Calls 'LinSolveMain()' for solving the linear system with the }
{    flag 'ComputeCond' set.                                                 }
{----------------------------------------------------------------------------}
global procedure LinSolve ( var A   : rmatrix; var b    : rvector;
                            var xx  : ivector; var Cond : real;
                            var Err : integer                    );
begin
  LinSolveMain(A,b,xx,Cond,true,Err);
end;

{----------------------------------------------------------------------------}
{ Module initialization part                                                 }
{----------------------------------------------------------------------------}
begin
  { Nothing to initialize }
end.
```

10.3.2 Example

In our sample program, we demonstrate how to use the modules defined in the previous section. As in Section 2.5, we give the definition of a control procedure *main* with a single parameter *n* of type *integer*. The dynamic arrays needed to store the matrix of the system, its right-hand side, and the enclosure of the solution, are allocated dynamically depending on *n*. After entering the system, the program tries to enclose a solution. If the algorithm succeeds, an enclosure of the solution and a naive floating-point approximation are printed. Otherwise, an error message is displayed. Finally, a condition number estimate for the input matrix is printed.

```
program linsys_example ( input, output );
use
  mv_ari,      { Matrix/vector arithmetic          }
  mvi_ari,     { Matrix/vector interval arithmetic }
  matinv,      { Matrix inversion      }
  linsys;      { Linear system solver }
{-----------------------------------------------------------------------}
procedure main ( n : integer );
var
  ErrCode : integer;
  Cond    : real;
  A, R    : rmatrix[1..n,1..n];
  b       : rvector[1..n];
  x       : ivector[1..n];
begin
  writeln('Enter matrix A:'); read(A); writeln;
  writeln('Enter vector b:'); read(b); writeln;

  LinSolve(A,b,x,Cond,ErrCode);

  if (ErrCode = 0) then
    begin
      { Compare the result to a naive floating-point approximation }
      MatInv(A,R,ErrCode);
      write('Naive floating-point approximation:');
      writeln(R*b);

      write('Verified solution found in:');
      writeln(x);
    end
  else
    writeln(LinSolveErrMsg(ErrCode));

  if (ErrCode in [0,4]) then
  begin
    writeln('Condition estimate:');
    writeln(Cond:9);
  end;

end;
{-----------------------------------------------------------------------}
var
  n : integer;
begin
  n := 0;
  while (n <= 0) do
    begin  write('Enter the dimension of the system: ');  read(n);  end;
  writeln;
  main(n);
end.
```

As sample input data, we use a Boothroyd/Dekker system (see [89]) whose elements are integers defined by

$$a_{ij} = \binom{n+i-1}{i-1} \cdot \binom{n-1}{n-j} \cdot \frac{n}{i+j-1}, \quad i,j = 1,\ldots,n.$$

With the right-hand side $b = (1, 2, \ldots, n)$, the exact solution of the system is given by $x = (0, 1, \ldots, n-1)$. For $n = 10$, our sample program leads to the following output:

```
Enter the dimension of the system: 10

Enter matrix A:
   10     45     120     210     252     210     120      45     10     1
   55    330     990    1848    2310    1980    1155     440     99    10
  220   1485    4752    9240   11880   10395    6160    2376    540    55
  715   5148   17160   34320   45045   40040   24024    9360   2145   220
 2002  15015   51480  105105  140140  126126   76440   30030   6930   715
 5005  38610  135135  280280  378378  343980  210210   83160  19305  2002
11440  90090  320320  672672  917280  840840  517440  205920  48048  5005
24310 194480  700128 1485120 2042040 1884960 1166880  466752 109395 11440
48620 393822 1432080 3063060 4241160 3938220 2450448  984555 231660 24310
92378 755820 2771340 5969040 8314020 7759752 4849845 1956240 461890 48620

Enter vector b:
1 2 3 4 5 6 7 8 9 10

Naive floating-point approximation:
 3.769740075654227E-010
 9.999999949022040E-001
-1.999999975192608E+000
 2.999999904595484E+000
-3.999999691503490E+000
 4.999999132005541E+000
-5.999997815384631E+000
 6.999994975405571E+000
-7.999989266972989E+000
 8.999978443156579E+000

Verified solution found in:
[  0.000000000000000E+000,  0.000000000000000E+000 ]
[  1.000000000000000E+000,  1.000000000000000E+000 ]
[ -2.000000000000000E+000, -2.000000000000000E+000 ]
[  3.000000000000000E+000,  3.000000000000000E+000 ]
[ -4.000000000000000E+000, -4.000000000000000E+000 ]
[  5.000000000000000E+000,  5.000000000000000E+000 ]
[ -6.000000000000000E+000, -6.000000000000000E+000 ]
[  7.000000000000000E+000,  7.000000000000000E+000 ]
[ -8.000000000000000E+000, -8.000000000000000E+000 ]
[  9.000000000000000E+000,  9.000000000000000E+000 ]

Condition estimate:
 1.1E+015
```

Note that the resulting enclosure of the solution is a thin interval vector. This is a consequence of our strategy for updating zero-suspicious entries in Algorithm 10.2. After the real residual iteration of Algorithm 10.5, the approximate solution is already exact! The verification step is used only to get a mathematical proof of this fact. In general, the enclosure of the solution is not a thin interval vector.

10.3.3 Restrictions and Improvements

The method described in this chapter demands exactly representable input data. Neither the matrix A nor the right-hand side b of the system may be disturbed by conversion errors during input. Actually, this is no restriction if A and b result from previous computations. To see why this is important, let us consider the 2×2 system with

$$A = \begin{pmatrix} 1 & 1 \\ \varepsilon & 0 \end{pmatrix}, \quad b = \begin{pmatrix} 1 \\ 1 \end{pmatrix}, \quad \varepsilon \neq 0.$$

The exact solution of the system is $x = (1/\varepsilon, 1 - 1/\varepsilon)^{\mathrm{T}}$. If we solve the system for $\varepsilon = 10^{-20}$ using the above program, we get the following result

```
Verified solution found in:
[  1.000000000000000E+020,  1.000000000000001E+020 ]
[ -1.000000000000001E+020, -1.000000000000000E+020 ]
```

which does not contain the exact solution $(10^{20}, 1 - 10^{20})^{\mathrm{T}}$. The reason for this incorrect result is that $\varepsilon = 10^{-20}$ is not exactly representable on a binary number screen (see Section 3.7). Thus after data conversion, we solved a perturbed system with a $\square(\varepsilon)$ entry for the lower left element of A.

To get rid of the problem of input conversion, we might enclose all input data in small intervals. We would have to modify our algorithm so that it works with interval data. The modifications of Algorithm 10.5 to accept interval input as well as complex and complex interval data are subject to a subsequent volume of this book. This future volume is also planned to deal about the treatment of rectangular linear systems of equations.

10.4 Exercises

Exercise 10.1 How large a Hilbert matrix can the procedure *LinSolve* handle? Hilbert matrices are very ill-conditioned for large dimensions. The elements of a Hilbert matrix H are defined by

$$h_{ij} = \frac{1}{i + j - 1} \quad i, j = 1, \ldots, n.$$

Multiply H by the common denominator of all its elements to get an exactly representable input matrix. Use an arbitrary vector as right-hand side of the system.

Exercise 10.2 Algorithm 10.5 fails for very ill-conditioned systems. For instance, we are not able to solve a Boothroyd/Dekker system of dimension 15 using procedure *LinSolve*. As mentioned earlier, the better R approximates the inverse of A, the better the algorithm works. Therefore, it is a good idea to improve the behavior of our algorithm by refining the approximate inverse.

If R is an already computed approximation of A^{-1}, we can use a little trick to improve its value. By computing

$$\begin{aligned} R_1 &= (RA)^{-1}R, \\ R_2 &= \square((RA)^{-1}R - R_1) \end{aligned}$$

we get $A^{-1} = R_1 + R_2$. R_1 is a new approximation to A^{-1}, and R_2 is its refinement. This fact can be used to modify Algorithm 10.5. If the verification step has failed, it has to be restarted after computing an improved double-length approximate inverse. To accomplish this goal, we do not want Step 4 to return with an error message. An improved version of Algorithm 10.5 has the following form:

Algorithm 10.6: NewLinSolve $(A, b, [x], Err)$ {Procedure}

1. – 3. see Algorithm 10.5
4. {Verification step}
   if $([z] = 0)$ **then**
     return $[x] := \tilde{x}$, *Cond*, *Err*; {Exact solution}
   else
     VerificationStep $([x], [z], [C], IsVerified)$;
     if $(IsVerified)$ **then**
       $[x] := \tilde{x} + [x]$; {Approximation plus correction interval}
       return $[x]$, *Cond*, *Err*;
5. {Computing a double-length approximate inverse stored in R_1 and R_2}
   MatInv $(R \cdot A, R_2, Err)$;
   if $(Err \neq$ "No error"$)$ **then**
     return $Err :=$ "Verification failed, system is probably ill-conditioned!";
   $R_1 := R_2 \cdot R$;
   $R_2 := \Box(R_2 \cdot R - R_1)$;
6. {Computation of enclosures $[C]$ and $[z]$}
   $[C] := \Diamond(I - R_1 \cdot A - R_2 \cdot A)$;
   $d := \Box(b - A \cdot \tilde{x})$;
   $[d] := \Diamond(b - A \cdot \tilde{x} - d)$;
   $[z] := \Diamond(R_1 \cdot d + R_2 \cdot d + R_1 \cdot [d] + R_2 \cdot [d])$;
7. – 8. see 4. – 5. of Algorithm 10.5

Supplement the module *linsys* by a procedure *NewLinSolve* according to the modifications described in Algorithm 10.6. Test the new procedure for a Boothroyd/Dekker system of dimension 15. The improved version should succeed in enclosing the solution.

Exercise 10.3 Another possible extension of Algorithm 10.5 is to use the staggered type arithmetic from Chapter 8 to store and to compute $\tilde{x}$ and R. Make these modifications and compare the accuracy, set of problems for which it works, and speed with those of the procedure *LinSolve*.

10.5 References and Further Reading

The basic principles of a self-verifying algorithm for linear systems of equations are due to Rump [76]. A straight-forward extension of the algorithms to interval systems and complex systems is given in Rump [78]. A compact summary of the mathematical theory is found in Rump [79]. Comprehensive discussions on interval linear systems of equations including preconditioning strategies are given in the books of Hansen [28] and Neumaier [64]. There are also commercial implementations for the verified computation of linear systems of equations which come with the subroutine libraries ACRITH of IBM and ARITHMOS of SIEMENS (see [32], [80]).

Chapter 11

Linear Optimization

A linear programming problem consists of a linear function to be maximized (or minimized) subject to linear equality and inequality constraints. Any linear program (LP) can be put by well-known transformations into standard form

$$\text{(LP)} \qquad \left(\begin{array}{ccccc} z & = & c^{\mathrm{T}}x & = & \max! \\ & & Ax & = & b \\ & & x & \geq & 0 \end{array} \right) \tag{11.1}$$

$$\Leftrightarrow$$

$$\max\{c^{\mathrm{T}}x \mid x \in X\}, \quad X := \{x \in I\!\!R^n \mid Ax = b,\ x \geq 0\},$$

where A is a real $m \times n$ matrix, $b \in I\!\!R^m$, $c \in I\!\!R^n$, and $m < n$. The input data of (11.1) are given by the triple $P = (A, b, c) \in I\!\!R^{m \cdot n + m + n}$.

Linear programming problems occur frequently in many economic, engineering, and scientific applications, so many software libraries include routines to estimate optimal solutions. Of course, such routines for computing approximate solutions are subject to rounding and cancellation errors that may result in "answers" which are not truly optimal or are not even feasible. We describe an algorithm for the self-validated solution to linear programming problems. The algorithm we present computes an enclosure of the true optimal solution set. The self-validated nature of our algorithm is especially important as the applications of linear programming become larger, more complex, and deeply embedded as sub-problems in even more complex models.

11.1 Theoretical Background

11.1.1 Description of the Problem

The function $c^{\mathrm{T}}x$ is called the *objective function*, and X is called the *set of feasible solutions* of (11.1). X is a convex polyhedron which is the intersection of the affine variety $\{x \in I\!\!R^n \mid Ax = b\}$ with the positive orthant in $I\!\!R^n$.

A point $x \in I\!\!R^n$ is called *feasible* if $x \in X$. It is called *optimal* if, in addition, $c^{\mathrm{T}}x \geq c^{\mathrm{T}}x'$ for all $x' \in X$. If such a point exists, the value $z_{\mathrm{opt}} := \max\{c^{\mathrm{T}}x \mid x \in X\}$ is called the *optimal value*. We seek an enclosure of the set of all optimal points.

A main theorem in linear programming asserts that if there exists an optimal solution, it has at most m components different from zero. Furthermore, there is only a finite number of such solutions. The *simplex method* invented by Dantzig [13] is an algorithm to compute an approximation of such an optimal solution efficiently (see [84]).

Solutions of $Ax = b$ where $n - m$ components of x are zero can be described by partitioning the rectangular matrix A consisting of the columns $(a_{*,1}, \ldots, a_{*,n})$ into two submatrices A_B and A_N, where $A_B = (a_{*,\beta_1}, \ldots, a_{*,\beta_m})$ is a nonsingular $m \times m$ submatrix with the column vectors a_{*,β_i} of A, and $A_N = (a_{*,\nu_1}, \ldots, a_{*,\nu_{n-m}})$ is the $m \times (n - m)$ submatrix with the remaining column vectors of A. The set of indices $\mathcal{B} = \{\beta_1, \ldots, \beta_m\} \subseteq \{1, \ldots, n\}$ which determines the nonsingular quadratic matrix A_B is called a *basic index set* of (11.1), and $\mathcal{N} := \{\nu_1, \ldots, \nu_{n-m}\} := \{1, \ldots, n\} \setminus \mathcal{B}$ is the set of *nonbasic indices*.

For a given partitioning $\mathcal{B}$ and $\mathcal{N}$, the linear equation $Ax = b$ can be expressed in the form

$$A_B \cdot x_B + A_N \cdot x_N = b \tag{11.2}$$

where $x = (x_B, x_N)^{\mathrm{T}}$ is partitioned analogously. Multiplying equation (11.2) by the inverse A_B^{-1} yields

$$x_B = A_B^{-1} \cdot b - H \cdot x_N \quad \text{with} \quad H := A_B^{-1} \cdot A_N.$$

Partitioning $c = (c_B, c_N)^{\mathrm{T}}$ similarly, the objective function can be expressed in the form

$$\begin{aligned} c^{\mathrm{T}} x = c_B^{\mathrm{T}} \cdot x_B + c_N^{\mathrm{T}} \cdot x_N &= c_B^{\mathrm{T}} \cdot (A_B^{-1} \cdot b - H \cdot x_N) + c_N^{\mathrm{T}} \cdot x_N \\ &= c_B^{\mathrm{T}} \cdot A_B^{-1} \cdot b - (c_B^{\mathrm{T}} \cdot H - c_N^{\mathrm{T}}) \cdot x_N. \end{aligned}$$

Therefore,

$$c^{\mathrm{T}} x = c_B^{\mathrm{T}} \cdot A_B^{-1} b - d^{\mathrm{T}} \cdot x_N \quad \text{with} \quad d := H^{\mathrm{T}} \cdot c_B - c_N. \tag{11.3}$$

Thus, both the m variables x_B and the value of the objective function are determined by the $n - m$ components of x_N.

A *basic solution* (corresponding to the basic index set $\mathcal{B}$) is a solution for which the $n - m$ components of x_N are zero.

$$x := (x_B, x_N)^{\mathrm{T}} \quad \text{with} \quad x_B := A_B^{-1} \cdot b \quad \text{and} \quad x_N := 0.$$

In addition, if $x \geq 0$ (this is the case if $x_B := A_B^{-1} b \geq 0$), then $x \in X$ is called a *basic feasible solution*. The value of the objective function is $z = c^{\mathrm{T}} x = c_B^{\mathrm{T}} A_B^{-1} b$.

Geometrically, the basic feasible solutions correspond to the vertices of the convex polyhedron X. The vector d in (11.3) is called the *vector of reduced costs*, since it indicates how the objective function depends on x_N.

11.1.2 Verification

The "results" of interest for the linear programming problem (11.1) are:

1. The *optimal value* z_{opt},

2. the *optimal basic index set* $\mathcal{B}$, and
3. the *optimal basic feasible solution* x.

The simplex method efficiently determines an optimal value z_{opt} and one optimal basic index set $\mathcal{B}$ with the corresponding optimal feasible solution x. Implementations of the simplex algorithm in software libraries are usually quite reliable, but the accumulation of rounding and cancellation errors may cause them to return a non-optimal or infeasible solution, or to fail to find all optimal solutions in case the algorithm is unable to recognize that more than one vertex of X is optimal. Therefore, we compute enclosures for the *optimal value* z_{opt}, the *set of optimal basic index sets* $\mathcal{V}_{\text{opt}}$, and the *set of optimal basic feasible solutions* $\mathcal{X}_{\text{opt}}$ for a linear programming problem P. Algorithm 11.7 calculates an interval $[z]$, a *set of index sets* $\mathcal{S} = \{\mathcal{B}^{(1)}, \ldots, \mathcal{B}^{(s)}\}$, and interval vectors $[x]^{(1)}, \ldots, [x]^{(s)} \in I\mathbb{R}^n$ such that the following three conditions are satisfied:

1. $z_{\text{opt}} \in [z]$
2. $\mathcal{B} \in \mathcal{V}_{\text{opt}} \Rightarrow \mathcal{B} \in \mathcal{S}$
3. $x \in \mathcal{X}_{\text{opt}} \Rightarrow$ there exists a $\mathcal{B}^{(i)} \in \mathcal{S}$ with $x \in [x]^{(i)}$

The following *a posteriori* method either computes such enclosures $[z]$, $\mathcal{S}$, and $[x]^{(1)}, \ldots, [x]^{(s)}$ satisfying conditions 1, 2, and 3 and guarantees that the real problem P admits optimal solutions, or else the method gives a warning that no enclosures could be calculated. The latter is the case if the set of feasible solutions X is empty or if the objective function is unbounded on X.

Let $\mathcal{B} \in \mathcal{S}$ be an index set $\{\beta_1, \ldots, \beta_m\} \subseteq \{1, \ldots, n\}$. Let $[x_B]$, $[y]$, $[h]_{*,\nu}$ with $\nu \in \mathcal{N} := \{\nu_1, \ldots, \nu_{n-m}\} := \{1, \ldots, n\} \setminus \mathcal{B}$ be vectors of inclusion for the solutions of the systems of linear equations

$$\begin{aligned} A_B \cdot x_B &= b \\ A_B^{\mathrm{T}} \cdot y &= c_B \\ A_B \cdot h_{*,\nu} &= a_{*,\nu}. \end{aligned} \tag{11.4}$$

These vector enclosures are calculated by the self-validating Algorithm 10.5 for solving square linear systems. If the algorithm succeeds, it is verified that the real matrix A_B is nonsingular.

We define

$$\begin{aligned} [d] &:= A_N^{\mathrm{T}} \cdot [y] - c_N \\ [x] &:= ([x_B], [x_N])^{\mathrm{T}}, \quad [x_N] := 0 \\ [H] &:= ([h]_{*,\nu_1}, \ldots, [h]_{*,\nu_{m-n}}) \\ [z] &:= c_B^{\mathrm{T}} \cdot [x_B] \cap b^{\mathrm{T}} \cdot [y]. \end{aligned}$$

The interval tableau $[T]$ corresponding to the basic index set $\mathcal{B} \in \mathcal{S}$ has the form

$$[T] := \left(\begin{array}{c|c} [z] & [d]^{\mathrm{T}} \\ \hline [x_B] & [H] \end{array} \right) = \left(\begin{array}{c|ccc} [z] & [d]_{\nu_1} & \cdots & [d]_{\nu_{n-m}} \\ \hline [x]_{\beta_1} & [h]_{\beta_1\nu_1} & \cdots & [h]_{\beta_1\nu_{n-m}} \\ \vdots & \vdots & & \vdots \\ [x]_{\beta_m} & [h]_{\beta_m\nu_1} & \cdots & [h]_{\beta_m\nu_{n-m}} \end{array} \right). \tag{11.5}$$

The self-validating algorithm for solving linear programming problems with input data $P = (A, b, c)$ is based on the following two theorems whose proofs appear in Krawzcyk [51] and Jansson [34].

Theorem 11.1 *Let $\mathcal{B} \in \mathcal{S}$ be a basic index set of P and let $[T]$ be the corresponding interval tableau.*

1. *If the conditions*
$$\underline{x}_B > 0 \quad \text{and} \quad \underline{d} > 0 \tag{11.6}$$
are satisfied, then the linear programming problem P has a unique optimal solution contained in $[x]$. Moreover $\mathcal{V}_{\text{opt}} = \mathcal{B}$, and $z_{\text{opt}} \in [z]$.
2. *If $\mathcal{B} \in \mathcal{V}_{\text{opt}}$, then*
$$\overline{x}_B \geq 0 \quad \text{and} \quad \overline{d} \geq 0. \tag{11.7}$$

If the condition (11.6) is satisfied, then P is called *basis stable*, i.e. the linear programming problem P has a unique optimal solution. Inequalities (11.7) give a necessary condition that $\mathcal{B}$ is an optimal basic index set for the problem P. Such basic index sets are only of interest if P is not basis stable.

The second theorem shows how to compute an enclosure for a *set of neighboring basic index sets* $\mathcal{L}$ of $\mathcal{B}$, where $\mathcal{B} \in \mathcal{V}_{\text{opt}}$ and all $\mathcal{B}' \in \mathcal{L}$ differ in exactly one index from $\mathcal{B}$. Each neighboring basic index set denotes a different neighboring vertex of the set of feasible solutions X. With Theorem 11.2 we have a sufficient criterion to determine a list of candidates that is a true superset of all neighboring basic index sets that may represent an optimal solution.

Theorem 11.2 *Let $\mathcal{B}$ be a basic index set of P with $\overline{x}_B \geq 0$ and $\overline{d} \geq 0$. Let $[T]$ be the corresponding interval tableau. Let $\widetilde{\mathcal{L}}$ be defined as the set of all index sets $\mathcal{B}' = (\mathcal{B} \setminus \{\beta\}) \cup \{\nu\}$ with $\beta \in \mathcal{B}$ and $\nu \in \mathcal{N}$ that satisfy one of the following conditions:*

$$0 \in [d]_\nu,\ \overline{h}_{\beta_\nu} > 0 \quad \text{and} \quad \frac{\underline{x}_\beta}{\overline{h}_{\beta\nu}} \leq \min\left\{\frac{\overline{x}_\beta}{\underline{h}_{\beta\nu}} \mid \underline{h}_{\beta\nu} > 0,\ \beta \in \mathcal{B}\right\} \tag{11.8}$$

$$0 \in [x]_\beta,\ \underline{h}_{\beta_\nu} < 0 \quad \text{and} \quad \frac{\overline{d}_\nu}{\underline{h}_{\beta\nu}} \geq \max\left\{\frac{\underline{d}_\nu}{\overline{h}_{\beta\nu}} \mid \overline{h}_{\beta\nu} < 0,\ \nu \in \mathcal{N}\right\}. \tag{11.9}$$

If $\mathcal{B} \in \mathcal{V}_{\text{opt}}$, then $\widetilde{\mathcal{L}} \supseteq \mathcal{L}$.

11.2 Algorithmic Description

With Algorithm 11.1 we compute the interval tableau $[T]$ of (11.5) using the algorithm LinSolve from Chapter 10 for solving the square linear systems of equations. If LinSolve fails to solve one of the linear systems, an error code is returned.

Algorithm 11.1: ComputeTableau $(A, b, c, \mathcal{B}, \mathcal{N}, [T], Err)$ {Procedure}

1. {Initialization}
   Err := "No error";
2. {Solve linear systems of equations (11.4)}
   LinSolve $(A_{\mathcal{B}}, b, [x_{\mathcal{B}}], Err)$; $[x_{\mathcal{N}}] := 0$;
   LinSolve $(A_{\mathcal{B}}^T, c_{\mathcal{B}}, [y], Err)$;
   for $\nu \in \mathcal{N}$ **do**
     LinSolve $(A_{\mathcal{B}}, a_{*,\nu}, [h]_{*,\nu}, Err)$;
   if $Err \neq$ "No error" {for any LinSolve call} **then**
     return Err := "Submatrix $A_{\mathcal{B}}$ probably singular";
3. {Compute components of interval tableau $[T]$}
   $[x] := ([x_{\mathcal{B}}], [x_{\mathcal{N}}])^T$;
   $[d] := A_{\mathcal{N}}^T \cdot [y] - c_{\mathcal{N}}$;
   $[H] := ([h]_{\nu_1}, \ldots, [h]_{\nu_{m-n}})$;
   $[z] := c_{\mathcal{B}}^T \cdot [x_{\mathcal{B}}] \cap b^T \cdot [y]$;
4. **return** $[T], Err$;

Algorithm 11.2 determines whether the interval tableau $[T]$ corresponding to the actual basic index set $\mathcal{B}$ represents a basis stable solution. According to condition (11.6) the existence of a basis stable solution indicates a unique optimal solution for the linear programming problem P, i.e. $\mathcal{B} = \mathcal{S} \supseteq \mathcal{V}_{opt}$, $[x] \supseteq \mathcal{X}_{opt}$ and $[z] \supseteq z_{opt}$. Thus, no further calculations are necessary.

Algorithm 11.2: BasisStable $([T])$ {Function}

1. {Check whether $\underline{x}_{\mathcal{B}} > 0$ (cf. (11.6))}
   for $\beta \in \mathcal{B}$ **do**
     if $\underline{x}_\beta \leq 0$ **then return** *IsStable* := *false*;
2. {Check whether $\underline{d} > 0$ (cf. (11.6))}
   for $\nu \in \mathcal{N}$ **do**
     if $\underline{d}_\nu \leq 0$ **then return** *IsStable* := *false*;
3. **return** *IsStable* := *true*;

We use Algorithm 11.3 to decide whether the interval tableau $[T]$ corresponding to the actual basic index set $\mathcal{B}$ represents a possibly optimal solution for the linear programming problem P, i.e. $\mathcal{B} \in \{\mathcal{B}^{(1)}, \ldots, \mathcal{B}^{(s)}\} = \mathcal{S} \supseteq \mathcal{V}_{opt}$, $[x] \in \{[x]^{(1)}, \ldots, [x]^{(s)}\} \supseteq \mathcal{X}_{opt}$, and $[z] \supseteq z_{opt}$. If (11.7) is satisfied, then $\mathcal{B}$, $[x]$, and $[z]$ are stored to the list of optimal solutions.

Algorithm 11.3: PossiblyOptimalSolution $([T])$ {Function}

1. {Check whether $\overline{x}_{\mathcal{B}} >= 0$ (cf. (11.7))}
   for $\beta \in \mathcal{B}$ **do**
     if $\overline{x}_\beta < 0$ **then return** *IsOptimal* := *false*;

```
    {Check whether d̄ >= 0 (cf. (11.7))}
    for ν ∈ N do
      if d̄_ν < 0 then return IsOptimal := false;
 3. return IsOptimal := true;
```

Algorithms 11.4 and 11.5 check whether the set of feasible solutions X is empty or if the objective function $c^T x$ is unbounded on X. This is done by an examination of the interval tableau $[T]$. X is probably (i.e. numerically) empty if there does not exist a $\nu \in \mathcal{N}$ with $[h]_{\beta\nu} < 0$ for all $\beta \in \mathcal{B}$ with $0 \in [x]_\beta$. The objective function is probably (i.e. numerically) unbounded if there does not exist a $\beta \in \mathcal{B}$ with $[h]_{\beta\nu} > 0$ for all $\nu \in \mathcal{N}$ with $0 \in [d]_\nu$.

Algorithm 11.4: EmptySolutionSet ([T]) {Function}

```
 1. {Initialization}
    IsEmpty := false;
 2. {Check whether for all β ∈ B with 0 ∈ [x]_β there}
    {exists a ν ∈ N with h̄_βν < 0                   }
    for β ∈ B with 0 ∈ [x]_β do
      while (not IsEmpty) do
        IsEmpty := true;
        for ν ∈ N do
          if h̄_βν < 0 then
            IsEmpty := false; exit_for-loop;
 3. return IsEmpty;
```

Algorithm 11.5: Unbounded ([T]) {Function}

```
 1. {Initialization}
    IsUnbounded := false;
 2. {Check whether for all ν ∈ N with 0 ∈ [d]_ν there}
    {exists a β ∈ B with h_βν > 0                     }
    for ν ∈ N with 0 ∈ [d]_ν do
      while (not IsUnbounded) do
        IsUnbounded := true;
        for β ∈ B do
          if h_βν > 0 then
            IsUnbounded := false; exit_for-loop;
 3. return IsUnbounded;
```

For the actual basic index set $\mathcal{B}$, Algorithm 11.6 determines a set of neighboring basic index sets $\mathcal{L}$ that are candidates for being optimal basic index sets. A true superset of all relevant candidates is determined according to Theorem 11.2.

Algorithm 11.6: NeighboringList $([T], \mathcal{B}, \mathcal{N})$ {Function}

1. {Initialization}
   $\mathcal{L} := \emptyset$;

2. {Determine a list of neighboring basic index sets that are}
   {candidates for optimal solutions (see Theorem 11.2) }

   (a) {Search for candidates $\nu \in \mathcal{N}$ and $\beta \in \mathcal{B}$ by determination}
   {of the minimum of quotients $[x]_\beta/[h]_{\beta\nu}$ (cf. (11.8)) }
   for all $\nu \in \mathcal{N}$ **with** $0 \in d_\nu$ **do**
   {Determine minimum of $\overline{x}_\beta/\underline{h}_{\beta\nu}$ for column ν}
   colmin $:= \min(\overline{x}_\beta/\underline{h}_{\beta\nu})$;
   {Determine candidates $\beta \in \mathcal{B}$ for exchange of indices}
   if $\underline{x}_\beta/\overline{h}_{\beta\nu} \leq$ *colmin* **then**
   {Determine new basic index set}
   $\mathcal{B}' := (\mathcal{B} \setminus \{\beta\}) \cup \{\nu\}$;
   $\mathcal{L} := \mathcal{L} + \mathcal{B}'$;

   (b) {Search for candidates $\beta \in \mathcal{B}$ and $\nu \in \mathcal{N}$ by determination}
   {of the maximum of quotients $[d]_\nu/[h]_{\beta\nu}$ (cf. (11.9)) }
   for all $\beta \in \mathcal{B}$ **with** $0 \in x_\beta$ **do**
   {Determine maximum of $\underline{d}_\nu/\overline{h}_{\beta\nu}$ for row β}
   rowmax $:= \max(\underline{d}_\nu/\overline{h}_{\beta\nu})$;
   {Determine candidates $\nu \in \mathcal{N}$ for exchange of indices}
   if $\overline{d}_\nu/\underline{h}_{\beta\nu} \geq$ *rowmax* **then**
   {Determine new basic index set}
   $\mathcal{B}' := (\mathcal{B} \setminus \{\beta\}) \cup \{\nu\}$;
   $\mathcal{L} := \mathcal{L} + \mathcal{B}'$;

3. **return** $\mathcal{L}$;

We now present the algorithm **LinOpt** to determine enclosures for the optimal value z_{opt}, the complete set of optimal basic index sets $\mathcal{V}_{opt}$, and the set of optimal basic feasible solutions $\mathcal{X}_{opt}$ for a linear programming problem P. The problem is given by the triple $P = (A, b, c)$. As stated in the previous section, we also need an initial optimal basic index set $\mathcal{B}_{start}$ to start the algorithm.

During the iteration in Step 2a of Algorithm 11.7, we select a basic index set $\mathcal{B}$ out of the list of candidates for optimal basic index sets $\mathcal{C}$ and update the list of examined basic index sets $\mathcal{E}$. In Step 2b we compute the interval tableau (11.5) using Algorithm 11.1. Next, we check whether the actual basic index set $\mathcal{B}$ represents a basis stable solution using Algorithm 11.2. If not, we use in Step 2d Algorithms 11.4 and 11.5 to decide if the linear programming problem is irregular, i.e. X is empty or $c^T x$ is unbounded on X. In Step 2e, we validate the optimality of the actual solution using Algorithm 11.3. If this is the case, we store it to the list of optimal solutions and then determine new candidates for optimal basic index sets, the list of neighboring basic index sets $\mathcal{L}$, using Algorithm 11.6.

Algorithm 11.7: LinOpt $(A, b, c, \mathcal{B}_{start}, z_{opt}, \mathcal{V}_{opt}, \mathcal{X}_{opt}, No, Err)$ {Proc.}

1. {Initialization}
   $Err :=$ "No error"; $k_{max} := 100$;
   $k := 0$; $\mathcal{C} := \mathcal{B}_{start}$; $\mathcal{E} := \emptyset$; $\mathcal{V}_{opt} := \emptyset$; $\mathcal{X}_{opt} := \emptyset$; $No := 0$;
2. {Iteration}
   repeat
   (a) {Determine index sets $\mathcal{B}$ and $\mathcal{N}$}
   select $\mathcal{B} \in \mathcal{C}$; $\mathcal{C} := \mathcal{C} \setminus \mathcal{B}$; $\mathcal{E} := \mathcal{E} \cup \mathcal{B}$;
   $\mathcal{N} := \{1, \ldots, n\} \setminus \mathcal{B}$;
   (b) {Compute interval tableau $[T]$}
   ComputeTableau $(A, b, c, \mathcal{B}, \mathcal{N}, [T], Err)$;
   if $Err \neq$ "No error" **then exit**$_{repeat\text{-}loop}$
   (c) {Check whether the tableau is basis stable}
   if BasisStable $([T])$ **then**
   {Store optimal value, optimal basic index set and optimal solution}
   $z_{opt} := z$; $\mathcal{X}_{opt} := ([x_B], [x_N])^T$; $\mathcal{V}_{opt} := \mathcal{B}$; $No := No + 1$;
   return $z_{opt}, \mathcal{V}_{opt}, \mathcal{X}_{opt}$;
   (d) {Check whether the set of feasible solutions is empty or}
   {if the objective function is unbounded }
   if EmptySolutionSet $([T])$ **then**
   $Err :=$ "Set of feasible solutions is empty";
   exit$_{repeat\text{-}loop}$
   if Unbounded $([T])$ **then**
   $Err :=$ "Objective function is unbounded";
   exit$_{repeat\text{-}loop}$
   (e) {Check whether the actual solution is possibly optimal}
   if PossiblyOptimalSolution $([T])$ **then**
   {Store optimal value, optimal basic index set, and optimal solution}
   $z_{opt} := z_{opt} \cup z$; $\mathcal{X}_{opt} := \mathcal{X}_{opt} \cup ([x_B], [x_N])^T$;
   $\mathcal{V}_{opt} := \mathcal{V}_{opt} \cup \mathcal{B}$; $No := No + 1$;
   {Determine the list of neighboring basic index sets $\mathcal{L}$}
   $\mathcal{L} :=$ NeighboringList $([T], \mathcal{B}, \mathcal{N})$;
   $\mathcal{L} := (\mathcal{L} \setminus \mathcal{E}) \setminus \mathcal{C}$; $\mathcal{C} := \mathcal{C} \cup \mathcal{L}$;
   (f) $k := k + 1$;

   until $(\mathcal{C} = \emptyset)$ **or** $(k = k_{max})$;
3. {Check for errors}
   if $(k = k_{max})$ **then** $Err :=$ "Maximum number of iterations exceeded";
   if $(No = 0)$ **then** $Err :=$ "Initial basic index set is not optimal";
4. **return** $z_{opt}, \mathcal{V}_{opt}, \mathcal{X}_{opt}, Err$;

Applicability of the Algorithm

Initial Index Set

A basic index set $\mathcal{B}_{start} \in \mathcal{V}_{opt}$ can be determined by computing an approximation of the optimal solution and the corresponding basic index set $\mathcal{B}$ for the linear programming problem $P = (A, b, c)$ using the standard simplex algorithm. Calculate vectors of inclusion $[x_B]$ and $[y]$ for the systems of linear equations

$$A_B \cdot x_B = b, \quad A_B^{\mathrm{T}} \cdot y = c_B.$$

If $\overline{x}_B \geq 0$ and $\overline{d} \geq 0$ with $[d] := A_N^{\mathrm{T}} \cdot [y] - c_N$, then it is proved by Theorem 11.1 that $\mathcal{B}$ is a possibly optimal basic index set of the linear programming problem $P = (A, b, c)$. Therefore, $\mathcal{B}_{start} := \mathcal{B} \in \mathcal{V}_{opt}$.

Stopping Criteria

Algorithm 11.7 stops in Step 2b or 2d and returns an error if either $[T]$ is not computable or one of the conditions in Step 2d is fulfilled.

If $[T]$ is not computable, Algorithm 11.1 has failed to solve the systems of linear equations (11.4). This failure indicates that the actual submatrix A_B is singular or nearly singular.

If in Step 2d the first or second condition is satisfied, this indicates that the set of feasible solutions X is probably empty or the objective function is unbounded on X. In this case, it is not possible to compute enclosures for z_{opt}, $\mathcal{V}_{opt}$, and $\mathcal{X}_{opt}$.

Improvements

The main work of the Algorithm 11.7 is to compute the interval tableau $[T]$, i.e. to solve the systems of linear equations (11.4). The only purpose of the interval tableaus is to compute an enclosure of the set of basic index sets $\mathcal{L}$ by Theorem 11.2. For the computation of $\mathcal{L}$, it is enough to get enclosures $[x_B]$ and $[d]$ and enclosures for the columns of H with $0 \in [d]_\nu$ and the rows of $[H]$ with $0 \in [x]_\beta$ (cf. (11.8) and (11.9)). In practical applications, $[d]_\nu > 0$ and $[x]_\beta > 0$ holds for most indices $\nu \in \mathcal{N}$ and $\beta \in \mathcal{B}$. Therefore, it is much less effort to compute only the necessary columns and rows of $[H]$.

The accuracy of Algorithm 11.7 is reflected in the sharpness of the computed bounds for the solution of the linear programming problem P. The tightness of these bounds only depends on the algorithm LinSolve in Step 2 of Algorithm 11.1 for calculating the bounds of the linear systems (11.4).

11.3 Implementation and Examples

11.3.1 PASCAL–XSC Program Code

Here we list the program code to solve a linear programming problem determined by the triple $P = (A, b, c)$. These procedures return enclosures for the optimal value z_{opt}, the set of optimal basic index sets $\mathcal{V}_{opt}$, and the set of optimal basic feasible solutions $\mathcal{X}_{opt}$ satisfying conditions 1, 2, and 3 in Section 11.1.2.

The algorithm described above is implemented in separate modules for clarity. The module *lop_ari* supplies a linearly linked list to keep the list of basic index sets that are candidates for being optimal basic index sets. Module *rev_simp* supplies an implementation of the revised simplex algorithm and is used to compute an initial optimal basic index set for the given linear programming problem. Module *lop* uses the abstract data type of module *lop_ari* and uses Algorithms 11.1 through 11.7 to compute the enclosures mentioned above.

11.3.1.1 Module lop_ari

Module *lop_ari* handles the index sets and lists of index sets. This module offers a linearly linked list as a very simple data structure to handle the dynamically growing number of index sets and solution vectors of intervals.

```
{----------------------------------------------------------------------------}
{ Purpose: Definition of a linearly linked list as an abstract data type for }
{    the representation of a list of integer sets.                           }
{ Global constants, types, operators, functions, and procedures:             }
{    constant maxDim      : maximum number of indices in a set (restricted   }
{                           by the PASCAL-XSC compiler)                      }
{    type IndexSet        : representation of a set of indices               }
{    type BaseList        : representation of a linearly linked list of      }
{                           index sets                                       }
{    operator in          : returns TRUE, if index set is in list            }
{    operator :=          : assigns index set to a vector                    }
{    function NextIndex   : returns value of next index in an index set      }
{    function GetIndex    : returns value of i-th index in an index set      }
{    function size        : returns number of indices in an index set        }
{    function length      : returns number of elements in a list             }
{    function empty       : returns TRUE if list is empty                    }
{    function select      : selects first element of a list                  }
{    function extract     : extracts submatrix/subvector depending on the    }
{                           actual index set                                 }
{    procedure insert     : inserts index set at the head of a list          }
{    procedure delete     : deletes index set from list                      }
{    procedure append     : appends 2nd list to end of 1st list              }
{    procedure remove     : removes elements of 2nd list from 1st list       }
{    procedure FreeAll    : frees complete list                              }
{    procedure write      : prints data types IndexSet and BaseList          }
{----------------------------------------------------------------------------}
module lop_ari;

global const   maxDim = 255; { Maximum number of indices }

{----------------------------------------------------------------------------}
{ Global type definitions                                                    }
{----------------------------------------------------------------------------}
```

```
global type     IndexSet = set of 1..maxDim;
                BaseList = ↑BaseListElement;
                BaseListElement =
                          record
                            Member : IndexSet;
                            next   : ↑BaseListElement;
                          end;

{----------------------------------------------------------------------------}
{ Local variable storing list of free elements (automatic garbage recycling) }
{----------------------------------------------------------------------------}
var
  FreeList : BaseList;

{----------------------------------------------------------------------------}
{ Procedures for generating and freeing of list elements                    }
{----------------------------------------------------------------------------}
procedure NewPP (var P: BaseList);    { 'NewPP' generates a new list element }
begin                                 { or gets one from 'FreeList'.         }
  if FreeList = nil then              {--------------------------------------}
    begin
      new(P);  P↑.next:= nil;
    end
  else
    begin
      P:= FreeList;  FreeList:= FreeList↑.next;  P↑.next:= nil;
    end;
end;

procedure Free (var P: BaseList);            { 'Free' enters one element of a }
begin                                        { list in the 'FreeList'.        }
  if P <> nil then                           {--------------------------------}
  begin
     P↑.next:= FreeList;  FreeList := P;  P:= nil;
  end;
end;

global procedure FreeAll (var List: BaseList);{ 'FreeAll' enters all elements}
var  H : BaseList;                            { of 'List' in the 'FreeList'. }
begin                                         {------------------------------}
  if List <> nil then
  begin
    H:= List;
    while H↑.next <> nil do  H:= H↑.next;
    H↑.next:= FreeList;  FreeList:= List;  List:= nil;
  end;
end;

{----------------------------------------------------------------------------}
{ Global functions and procedures for index sets                            }
{----------------------------------------------------------------------------}
global function NextIndex(Index : integer;           { Determine next index }
                              B : IndexSet) : integer;{ in B following Index }
begin                                                 {----------------------}
  Index := Index + 1;
  while not (Index in B) and (Index < maxDim) do Index := Index + 1;
  NextIndex := Index;
end;

global function GetIndex(i : integer;             { Determine i-th index in B }
                         B : IndexSet) : integer;{---------------------------}
var Index, Counter : integer;
```

```
begin
  Index := 0; Counter := 0;
  repeat
    Index := NextIndex(Index, B);
    Counter := Counter+1;
  until (Counter=i);
  GetIndex := Index;
end;

global function size(B : IndexSet) : integer; { Determine number of indices  }
var Index, k : integer;                       { in B                         }
begin                                         {------------------------------}
  k := 0;
  for Index:=1 to maxDim do
    if (Index in B) then
      k := k+1;
  size := k;
end;

global function extract(A : rmatrix; B : IndexSet) : { Extract submatrix of  }
                    rmatrix[1..ub(A,1),1..size(B)]; { columns A[*,Ind] with }
var     Index, k : integer;                          { (Index in B)          }
begin                                                {-----------------------}
  k := 0;
  for Index:=1 to ub(A,2) do
    if (Index in B) then
      begin
        k := k+1;
        extract[*,k] := A[*,Index];
      end;
end;

global function extract(x : rvector; B : IndexSet) : { Extract subvector     }
                               rvector[1..size(B)]; { x[Index] with (Index  }
var     Index, k : integer;                          { in B)                 }
begin                                                {-----------------------}
  k := 0;
  for Index:=1 to ub(x) do
    if (Index in B) then
      begin
        k := k+1;
        extract[k] := x[Index];
      end;
end;

{----------------------------------------------------------------------------}
{ Global functions and procedures for lists of index sets                    }
{----------------------------------------------------------------------------}
global function empty(L : BaseList) : boolean;        { Check if L is empty }
begin                                                 {---------------------}
  empty := (L = nil);
end;

global function length(L : BaseList) : integer;       { Return length of L  }
var counter : integer;                                {---------------------}
begin
  counter := 0;
  while (L <> nil) do
    begin
      counter := counter + 1;
      L := L↑.next;
    end;
 length := counter;
end;
```

```
global operator in (B : IndexSet; L : BaseList)          { Check if B in L }
                             inlist : boolean;           {-----------------}
begin
  inlist := false;
  while (L <> nil) and (not inlist) do
    if (L↑.Member = B) then inlist := true
                             else L := L↑.next;
end;

global function select(L : BaseList) : IndexSet;    { Select first index set }
begin                                                { from L                 }
  select := L↑.Member;                               {------------------------}
end;

global procedure insert (var L : BaseList;  { Insert index set B to the head }
                             B : IndexSet); { of the list of index sets L    }
var P : BaseList;                           {--------------------------------}
begin
  if not (B in L) then
    begin
      P := L;
      NewPP(L);
      L↑.Member := B;
      L↑.next := P;
    end;
end;

global procedure delete (var L : BaseList;  { Delete index set B from list   }
                             B : IndexSet); { of index sets L                }
var P, Del : BaseList;                      {--------------------------------}
begin
  if not empty(L) then
    if (L↑.Member = B) then { B is 1st element }
      begin
        Del := L; L := L↑.next; Free(Del);
      end
    else { B is not 1st element }
      begin
        P := L;
        while (P↑.next <> nil) do
          if (P↑.next↑.Member = B) then
            begin
              Del := P↑.next; P↑.next := P↑.next↑.next; Free(Del);
            end
          else P := P↑.next;
      end;
end;

global procedure append (var Res : BaseList;  { Append list of index sets Add}
                         var Add : BaseList); { to list Res                  }
var P : BaseList;                             {------------------------------}
begin
  if empty(Res) then
    Res := Add
  else
    begin
      P := Res;
      while (P↑.next <> nil) do P := P↑.next;
      P↑.next := Add;
    end;
  Add := nil;
end;
```

```
global procedure remove (var Res : BaseList;  { Remove list of index sets Sub}
                             Sub : BaseList); { from Res                     }
var B : IndexSet;                             {------------------------------}
begin
  while (Sub <> nil) do
    begin
      B := Sub↑.Member;
      delete(Res,B);
      Sub := Sub↑.next;
    end;
end;

global procedure write(var out : text; B : IndexSet);       { Write index set }
var Index : integer;                                        {-----------------}
begin
  for Index:=1 to maxDim do
    if (Index in B) then write(out,Index,' ');
end;

global procedure write(var out : text;               { Write list of index sets }
                       BaseSet : BaseList);          {-------------------------}
begin
  while (BaseSet <> nil) do
    begin
      writeln(out,BaseSet↑.Member);
      BaseSet := BaseSet↑.next;
    end;
end;

global operator := (var x : rvector; B : IndexSet);     { Assign index set to }
var i, Index : integer;                                 { real vector         }
begin                                                   {---------------------}
  Index := 0;
  for i:=1 to ub(x) do
    begin
      Index := NextIndex(Index,B);
      x[i] := Index;
    end;
end;

{----------------------------------------------------------------------------}
{ Module initialization                                                      }
{----------------------------------------------------------------------------}
begin
  FreeList := nil;          { List of freed elements which can be used again  }
end.                        {-------------------------------------------------}
```

11.3.1.2 Module rev_simp

We provide in module *rev_simp* a procedure *RevSimplex* based on the *revised simplex method.* It solves a linear programming problem in standard form (11.1). The implementation closely follows the one published in Syslo, Deo, and Kowalic [84, p. 14 ff]. Beside the correction of an index calculation error (marked with (!)), the source code has just slightly been changed for our special needs. This variant returns an optimal value z_{opt}, an optimal basic feasible solution x_{opt}, and an optimal basic index set $v_{\text{opt}} = \mathcal{B}_{\text{start}}$ as required to start Algorithm 11.7.

```
{----------------------------------------------------------------------------}
{ Purpose: Determine  an optimal value 'z', an optimal basic index set 'v',  }
{    and an optimal solution vector 'x' for a linear programming problem     }
{    P = (A,b,c) given in the standard form:                                 }
{                   ( z = c↑t * x = max! )                                   }
{         (LP)      (     A * x = b      )                                   }
{                   (       x >= 0       ).                                  }
{ Method: An initial basic index set is determined using the 2-phases method.}
{    Then the revised simplex method described by Syslo, Deo, and Kowalic in }
{    'Discrete Optimization Algorithms with Pascal' [p.14], Prentice Hall,   }
{    New Jersey (1983) is used.                                              }
{ Global procedure:                                                          }
{    procedure RevSimplex(...)   : determines the values z, v, x for the     }
{                                  linear programming problem P=(A,b,c)      }
{----------------------------------------------------------------------------}
module rev_simp;
use mv_ari; { Matrix/vector arithmetic }

{----------------------------------------------------------------------------}
{ Purpose: Determination of the solutions of a linear optimization problem.  }
{ Parameters:                                                                }
{   In:     'A'        : matrix of constraints                               }
{           'b'        : right-hand side of constraints                      }
{           'c'        : objective function by its coefficients              }
{   Out:    'x'        : optimal solution vector                             }
{           'v'        : optimal basic index set                             }
{           'z'        : optimal value                                       }
{           'Err'      : error flag                                          }
{ Description: For a detailed description of the revised simplex method see  }
{   Syslo, Deo, and Kowalic.                                                 }
{----------------------------------------------------------------------------}
global procedure RevSimplex(A : rmatrix; b, c : rvector;
                            var x, v : rvector; var z : real;
                            var Err : integer               );

const eps = 1E-30;                 { if (|a| < eps), it is handled as (a = 0) }
      NoError           = 0;
      WrongDimension    = 1;
      NoOptimalSolution = 8;
      NoInitialVertex   = 9;

var
   min, s             : real;
   T                  : rmatrix[1..ub(b)+2,1..ub(c)+2];
   U                  : rmatrix[1..ub(b)+2,1..ub(b)+2];
   w, y               : rvector[1..ub(c)+2];
   ex, SolutionFound  : boolean;
   i, j, k, l, p, q,
   m, n, phase        : integer;

begin

  m := ub(b); n := ub(c);
  if (n <= m) or (m < 1) or (n < 1) then                { Check dimensions }
    Err := WrongDimension                               {------------------}
  else                                              { Err <> WrongDimension }
    begin                                           {-----------------------}
      Err := NoError; phase := 1;                        { Initialization }
      p := m+2; q := m+2; k := m+1;                      {----------------}
      SolutionFound := false;
      c := -c;

      for i:=1 to m do                                  { Compute tableau }
        for j:=1 to n do T[i,j] := A[i,j];              {-----------------}
```

```
for j:=1 to n do
  begin
    T[k,j] := c[j];
    s      := 0;
    for i:= 1 to m do s := s - T[i,j];
    T[p,j] := s;
  end;
U := id(U);

s := 0; w := 0;                       { Determine initial basic index set v, }
for i:=1 to m do                      { initial optimal value s, and initial }
  begin                               { optimal solution x                   }
    v[i] := n - m + i; {(!)}          {--------------------------------------}
    w[i] := b[i];
    s    := s - b[i];
  end;
w[k] := 0; w[p] := s;

repeat

  if (w[p] >= -eps*10) and (phase = 1) then            { Only phase 1 }
    begin                                     { Solution for phase 1 found }
      phase := 2; q := m+1;                   {----------------------------}
    end;

  min := 0;                                        { phase 1 and phase 2 }
  for j:= 1 to n do                        { Determine index  k that minimizes }
    begin                                  { U[q,*] * T[*,j] (j = 1,..,n)      }
      s := U[q,*] * T[*,j];                {-----------------------------------}
      if (s < min) then
        begin
          min := s; k := j;
        end;
    end;

  if (min > -eps) then                      { vector of reduced costs vanishes }
    if (phase = 1) then                     {----------------------------------}
      Err := NoInitialVertex
    else                                                    { phase = 2 }
      begin                                     { Optimal solution found }
        Solutionfound := true;                  {------------------------}
        z := w[q];
      end

  else
    begin                                       { Determine candidate l for }
      for i:=1 to q do                          { exchange of indices       }
        y[i] := U[i,*] * T[*,k];                {---------------------------}
      ex := true;
      for i:= 1 to m do
        if (y[i] >= eps) then
          begin
            s := w[i]/y[i];
            if ex or (s < min) then
              begin
                min := s; l := i;
              end;
            ex := false;
          end;

      if ex then                                      { No candidate found }
        Err := NoOptimalSolution                      {--------------------}
```

```
          else
            begin                              { Determine new basic index set }
              v[l] := k;                       { and compute new tableau       }
              s    := 1 / y[l];                {-------------------------------}
              for j:= 1 to m do U[l,j] := U[l,j] * s;
              if (l = 1) then i := 2
                         else i := 1;
              repeat
                s := y[i];
                w[i] := w[i] - min * s;
                for j:=1 to m do U[i,j] := U[i,j] - U[l,j] * s;
                if (i = l-1) then i := i + 2
                             else i := i + 1;
              until (i > q);
              w[l] := min;
            end;                           { Determine new basic index set ... }

        end;                               { Determine candidate l ...         }

      until SolutionFound or (Err <> NoError);

      if SolutionFound then                { Return optimal value z, optimal }
        begin                              { basic index set v, and optimal  }
          x := 0;                          { solution vector x               }
          for i:= 1 to m do                {---------------------------------}
            x[trunc(v[i])] := w[i];
        end;
    end;                                        { Err <> WrongDimension ... }
end; { procedure RevSimplex }

{----------------------------------------------------------------------------}
{ Module initialization part                                                 }
{----------------------------------------------------------------------------}
begin
  { Nothing to initialize }
end.
```

11.3.1.3 Module lop

The module *lop* contains routines for each of the Algorithms 11.1 trough 11.7 that are necessary to compute the enclosures mentioned above. It makes use of the arithmetic modules for intervals and interval vectors and matrices as well as of the abstract data type for lists of index sets. The interface to a calling main program consists of the two routines *LinOpt* and *LinOptErrMsg*. The procedure *LinOpt* needs as input data the linear optimization problem $P = (A, b, c)$ and an initial optimal basic index set $\mathcal{B}_{start}$. It returns the enclosures of $[z]$, $\mathcal{V}_{opt}$, $\mathcal{X}_{opt}$, the number of optimal solutions, and an error code. The function *LinOptErrMsg* returns a string containing an error message.

```
{----------------------------------------------------------------------------}
{ Purpose: Determine enclosures for the optimal value 'z_opt', for the set   }
{    of optimal basic index sets 'V_opt', and for the set of solution        }
{    vectors 'X_opt' for a linear programming problem P = (A,b,c) given in   }
{    the standard form:                                                      }
{               ( z = c↑t * x = max! )                                       }
{         (LP)  (     A * x = b      )                                       }
{               (       x >= 0       )                                       }
{    with an initial optimal basic index set.                                }
{ Method: Starting from an initial basic index set, all neighboring index    }
```

```
{    sets are examined to determine all optimal solutions. To validate the   }
{    optimality of a solution, all linear systems of equations are solved    }
{    using a verifying linear system solver.                                  }
{ Global procedures and functions:                                            }
{   procedure LinOpt(...)      : determines the enclosures of z_opt, V_opt,    }
{                                and X_opt for LP P = (A,b,c)                 }
{   function LinOptErrMsg(...) : delivers an error message text               }
{-----------------------------------------------------------------------------}
module lop; { Linear optimization module }
            { -       --                 }

use     i_ari,           { Interval arithmetic                 }
        mv_ari,          { Matrix/vector arithmetic            }
        mvi_ari,         { Matrix/vector interval arithmetic }
        linsys,          { Verifying linear system solver     }
        x_real,          { Extended real arithmetic            }
        lop_ari;         { Linear linked list for index sets }

const  kmax = 100;       { Maximum number of iterations       }
var    Rplus   : interval;
{-----------------------------------------------------------------------------}
{ Error codes used in this module.                                            }
{-----------------------------------------------------------------------------}
const
  NoError                  = 0;  { No error occurred                         }
  WrongDimension           = 1;  { Wrong dimension of problem                }
  SolutionSetIsEmpty       = 2;  { Set of feasible solutions is empty        }
  FunctionUnbounded        = 3;  { Objective function is unbounded           }
  SubmatrixSingular        = 4;  { Submatrix A_B is singular                 }
  StartIndexSetNotOptimal = 5;  { Initial basic index set not optimal       }
  IterationError           = 6;  { Maximum number of iteration exceeded      }
  SolutionMatrixTooSmall   = 7;  { Matrices for storage of solutions too small}
  NoOptimalSolution        = 8;  { No initial optimal solution found         }
  NoInitialVertex          = 9;  { No initial vertex found                   }

{-----------------------------------------------------------------------------}
{ Error messages depending on the error code.                                 }
{-----------------------------------------------------------------------------}
global function LinOptErrMsg(Err : integer) : string;
var
  Msg : string;
begin
  case Err of
    NoError            : Msg := '';
    WrongDimension     : Msg := 'Wrong dimension of problem (e.g. m >= n)';
    SolutionSetIsEmpty: Msg := 'Set of feasible solutions is probably empty';
    FunctionUnbounded : Msg := 'Objective function is probably unbounded';
    SubmatrixSingular : Msg := 'Submatrix A_B in [A_B] is probably singular';
    StartIndexSetNotOptimal
                       : Msg := 'Initial basic index set not optimal '
                                + '(i.e. B_start not in V_opt)';
    IterationError     : Msg := 'Maximum number of iterations (='
                                + image(kmax,0) + ') exceeded';
    SolutionMatrixTooSmall
                       : Msg := 'Matrices for storage of solutions is'
                                + ' too small. Increase index ranges of V_opt'
                                + ' and X_opt';
    NoOptimalSolution : Msg := 'No initial optimal solution found';
    NoInitialVertex    : Msg := 'No initial vertex found';
    else               : Msg := 'Code not defined';
  end;
  if (Err <> NoError) then Msg := 'Error: ' + Msg + '!';
  LinOptErrMsg := Msg;
end;
```

```
                                                  { Determine solution vector }
function Solution(TT              : imatrix;      { corresponding to Base     }
                  Base, NonBase : IndexSet)       {---------------------------}
                  : ivector[1..size(Base)+size(NonBase)];
var Index, i : integer;
begin
  Index := 0;
  Solution := 0;
  for i:=1 to size(Base) do
    begin
      Index := NextIndex(Index,Base);
      Solution[Index] := TT[i+1,1] ** Rplus;
    end;
end;


{-----------------------------------------------------------------------------}
{ Compute the interval tableau                                                }
{                             ([z]    | [d]↑t     )                           }
{                       [T] := (------+----------)                            }
{                             ([x]_B | [H]       ).                           }
{-----------------------------------------------------------------------------}
procedure ComputeTableau(        A             : rmatrix;
                                 b, c          : rvector;
                                 Base, NonBase : IndexSet;
                             var TT            : imatrix;
                             var Err           : integer);
var
  A_B                  : rmatrix[1..ub(A,1), 1..size(Base)];
  A_N                  : rmatrix[1..ub(A,1), 1..size(NonBase)];
  c_B                  : rvector[1..size(Base)];
  c_N                  : rvector[1..size(NonBase)];
  xx_B, yy             : ivector[1..size(Base)];
  dd                   : ivector[1..size(NonBase)];
  HH                   : imatrix[1..size(Base), 1..size(NonBase)];
  zz                   : interval;
  Index, local_Error : integer;
  i, j, m, n           : integer;

begin { procedure ComputeTableau }

  m := ub(A,1);
  n := ub(A,2);

  { Determine submatrices and subvectors according to index sets Base and    }
  { NonBase                                                                   }
  A_B := extract(A, Base); A_N := extract(A, NonBase);{ Determine submatrices}
  c_B := extract(c, Base); c_N := extract(c, NonBase);{ and subvectors accor-}
                                                      { ding to B and N      }
  { Solve linear systems of equations }               {----------------------}
  linsolve(A_B, b, xx_B, local_Error);

  if (local_Error = NoError) then
    linsolve(transp(A_B), c_B, yy, local_Error);

  Index := 0;
  if (local_Error = NoError) then
    for i:=1 to (n-m) do
      begin
        Index := NextIndex(Index,NonBase);
        if (local_Error=NoError) then
        linsolve(A_B, A[*,Index], HH[*,i], local_Error);
      end;
```

```
  if (local_Error = NoError) then
    begin
      dd      := transp(A_N) * yy - c_N;    { Compute components of interval }
      zz      := (c_B * xx_B) ** (b * yy);  { tableau                        }
      TT[1,1] := zz;                        {--------------------------------}
      for j:=2 to (n-m+1) do
        TT[1,j] := dd[j-1];
      for i:=2 to (m+1) do
        TT[i,1] := xx_B[i-1];
      for i:=2 to (m+1) do
        for j:=2 to (n-m+1) do
          TT[i,j] := HH[i-1,j-1];
    end
  else Err := SubmatrixSingular;

end; { procedure ComputeTableau }

{----------------------------------------------------------------------------}
{ Determine whether the interval tableau [T] represents a basis stable       }
{ solution, i.e. ([x]_B.inf > 0) and ([d].inf > 0).                          }
{----------------------------------------------------------------------------}
function BasisStable(TT : imatrix) : boolean;
var IsStable : boolean;
    i,j      : integer;
begin
  IsStable := true; i := 1; j := 1;
  while ( (i < ub(TT,1)) and IsStable ) do
    begin
      i := i+1; { [x]_B = [T_i,1] (i = 2 .. m+1) }
      if (TT[i,1].inf <= 0) then IsStable := false;
    end;
  while ( (j < ub(TT,2)) and IsStable ) do
    begin
      j := j+1; { [d] = [T_1,j] (j = 2 .. n-m+1) }
      if (TT[1,j].inf <= 0) then IsStable := false;
    end;
  BasisStable := IsStable;
end;

{----------------------------------------------------------------------------}
{ Determine whether the interval tableau [T] represents a possibly optimal   }
{ solution, i.e ([x]_B.sup >= 0) and ([d].sup >= 0)                          }
{----------------------------------------------------------------------------}
function PossiblyOptimalSolution(TT : imatrix) : boolean;
var IsOptimal : boolean;
    i,j       : integer;
begin
  IsOptimal := true; i := 1; j := 1;
  while ( (i < ub(TT,1)) and IsOptimal ) do
    begin
      i := i+1; { [x]_B = [T_i,1] (i = 2 .. m+1) }
      if not( TT[i,1].sup >= 0 ) then IsOptimal := false;
    end;
  while ( (j < ub(TT,2)) and IsOptimal ) do
    begin
      j := j+1; { [d] = [T_1,j] (j = 2 .. n-m+1) }
      if not( TT[1,j].sup >= 0 ) then IsOptimal := false;
    end;
  PossiblyOptimalSolution := IsOptimal;
end;
```

```
{----------------------------------------------------------------------------}
{ Check whether the set of feasible solutions is empty, i.e. not for all(beta}
{ in B) with (0 in [x]_beta) there exists a (nu in N) with ([H]_beta,nu < 0) }
{----------------------------------------------------------------------------}
function EmptySolutionSet(TT : imatrix) : boolean;
var IsEmpty : boolean;
    i,j     : integer;
begin
  IsEmpty := false; i := 1;
  while (not IsEmpty) and (i < ub(TT,1)) do
    begin
      i := i+1;
      if (0 in TT[i,1]) then
        begin
          IsEmpty := true;
          j := 1;
          while IsEmpty and (j < ub(TT,2)) do
            begin
              j := j+1;
              if (TT[i,j].sup < 0) then IsEmpty := false;
            end;
        end;
    end;
  EmptySolutionSet := IsEmpty;
end;

{----------------------------------------------------------------------------}
{ Check whether the objective function is unbounded, i.e. not for all (nu    }
{ in N) with (0 in [d]_nu) there exists a (beta in B) with ([H]_beta,nu > 0) }
{----------------------------------------------------------------------------}
function Unbounded(TT : imatrix) : boolean;
var IsUnbounded : boolean;
    i,j         : integer;
begin
  IsUnbounded := false; j := 1;
  while (not IsUnbounded) and (j < ub(TT,2)) do
    begin
      j := j+1;
      if (0 in TT[1,j]) then
        begin
          IsUnbounded := true;
          i := 1;
          while IsUnbounded and (i < ub(TT,1)) do
            begin
              i := i+1;
              if (TT[i,j].inf > 0) then IsUnbounded := false;
            end;
        end;
    end;
  Unbounded := IsUnbounded;
end;

{----------------------------------------------------------------------------}
{ Determine list of neighboring basic index sets L for index set Base that   }
{ are good candidates for being optimal basic index sets.                    }
{----------------------------------------------------------------------------}
function NeighboringList(TT : imatrix; Base, NonBase : IndexSet) : BaseList;
var
  L                  : BaseList;
  NewBase            : IndexSet;
  i, j, m, n         : integer;
  beta, nu, Counter  : integer;
  colmin, rowmax     : real;
  xx, dd, HH         : interval;
```

```
begin
  m := size(Base); n := m + size(NonBase);                    { Initialization }
  L := nil;                                                   {---------------}

  colmin := maxReal;                               { Search for candidates (nu   }
  for j := 1 to (n-m) do { for all (nu in N) } { in N) and (beta in B) by       }
    begin                                          { determining of the minimum  }
      dd := TT[1,j+1];                             { of the quotients            }
      if (0 in dd) then                            { [x]_beta / [H]_beta,nu      }
        begin                                      {-----------------------------}
          for i:=1 to m do { for all (beta in B) } { Determine minimum of   }
            begin                                      {   [x]_beta / [H]_beta,nu }
              xx := TT[i+1,1];                         { for column nu of [T]   }
              HH := TT[i+1,j+1];                       {------------------------}
              if ((HH.Inf > 0) and (xx.Sup/<HH.Inf < colmin)) then
                colmin := xx.Sup/<HH.Inf;
            end; { (beta in B) }
          for i:=1 to m do { for all (beta in B) }{ Determine candidate (beta}
            begin                                     { in B) for exchange of    }
              xx := TT[i+1,1];                        { indices                  }
              HH := TT[i+1,j+1];                      {--------------------------}
              if (HH.Sup > 0) and (xx.Inf/<HH.Sup <= colmin) then
                begin                                { Determine new index set and }
                  beta := GetIndex(i,Base);          { add it to L                 }
                  nu   := GetIndex(j,NonBase);       {-----------------------------}
                  NewBase := Base - [beta] + [nu];
                  insert(L, NewBase);
                end;
            end; { (beta in B) }
        end;
    end; { (nu in N) }

  rowmax := -maxReal;                              { Search for candidates (beta }
  for i := 1 to m do { for all (beta in B) }       { in B) and (nu in N) by      }
    begin                                          { determining of the maximum  }
      xx := TT[i+1,1];                             { of the quotients            }
      if (0 in xx) then                            { [d]_nu / [H]_beta,nu        }
        begin                                      {-----------------------------}
          for j:=1 to (n-m) do { for all (nu in N) } { Determine maximum of  }
            begin                                        { [d]_nu / [H]_beta,nu  }
              dd := TT[1,j+1];                           { for row beta of [T]   }
              HH := TT[i+1,j+1];                         {-----------------------}
              if ((HH.Sup < 0) and (dd.Inf/>HH.Sup > rowmax)) then
                rowmax := dd.Sup/>HH.Sup;
            end; { (nu in N) }
          for j:=1 to (n-m) do { for all (nu in N) } { Determine candidate   }
            begin                                        { (nu in N) for exchange}
              dd := TT[1,j+1];                           { of indices            }
              HH := TT[i+1,j+1];                         {-----------------------}
              if (HH.Inf < 0) and (dd.Inf/>HH.Inf >= rowmax) then
                begin                                { Determine new index set and }
                  beta := GetIndex(i,Base);          { add it to L                 }
                  nu := GetIndex(j,NonBase);         {-----------------------------}
                  NewBase := Base - [beta] + [nu];
                  insert(L, NewBase);
                end;
            end; { (nu in N) }
        end;
    end; { (beta in B) }

  NeighboringList := L;

end;  { function NeighboringList }
```

```
{-----------------------------------------------------------------------------}
{ Purpose: Determination of enclosures for the solutions of a linear opti-    }
{    mization problem:      ( z = c↑t * x = max! )                            }
{                      (LP) (     A * x = b      )                            }
{                           (       x >= 0       )                            }
{   with an initial optimal basic index set.                                  }
{ Parameters:                                                                 }
{    In: 'A'     : matrix of constraints                                      }
{        'b'     : right-hand side of constraints                             }
{        'c'     : represents the objective function by its coefficients      }
{        'B_st.': initial basic index set                                     }
{    Out:'z_opt': enclosure of the optimal value                              }
{        'V_opt': superset of all optimal basic index sets                    }
{        'X_opt': superset of all optimal solution vectors                    }
{ Description: Starting from an initial basic index set 'B_start', a true     }
{    superset of all neighboring index sets 'L' that are candidates for being}
{    optimal index sets are determined and stored in a list of candidates     }
{    'CList'. As long as there are candidates left, they are examined using   }
{    a verifying linear system solver. The linear system solver returns       }
{    enclosures for the solutions of the square linear systems A_B [x] = b,   }
{    where A_B denotes a submatrix of the constraint matrix A according to a }
{    basic index set B.                                                       }
{-----------------------------------------------------------------------------}
global procedure LinOpt (      A                       : rmatrix;
                               b, c, B_start_Vector    : rvector;
                           var z_opt                   : interval;
                           var V_opt                   : rmatrix;
                           var X_opt                   : imatrix;
                           var No, Err                 : integer    );

var
  B_start              : IndexSet;
  Base, NonBase        : IndexSet;
  L, CList, E          : BaseList;
  TT                   : imatrix[1..1+(ub(b)), 1..1+(ub(c)-ub(b))];
  k, m, n, i, maxNo    : integer;

begin { global procedure LinOpt }

  Err := NoError;

  m := ub(A,1); n := ub(A,2);                    { Determine dimension of problem }
                                                 {--------------------------------}
  if (n <= m) or (m < 1) or (n < 1) or (n > maxDim)       { Check dimensions }
             or (ub(V_opt,1) <> ub(X_opt,1)) then         {------------------}
    Err := WrongDimension
  else { Err <> WrongDimension }
    begin
      B_start := [];                                      { Initialization }
      for i:=1 to m do  B_start := B_start + [trunc(B_start_Vector[i])];
      k := 0; V_opt := 0; X_opt := 0; No := 0; maxNo := ub(V_opt,1);
      E := nil; CList := nil; insert(CList, B_start);

      repeat { until empty(CList) }

        Base := select(CList); NonBase := [1..n] - Base; { Select Base out of}
        delete(CList, Base); insert(E, Base);            { CList and update  }
                                                         { examined list E   }
        ComputeTableau(A, b, c, Base, NonBase, TT, Err); {-------------------}

        if (Err = NoError) then
```

```
        if BasisStable(TT) then
          begin
            No          := No + 1;                        { Store unique     }
            z_opt       := TT[1,1];                       { optimal solution }
            X_opt[No,*] := Solution(TT, Base, NonBase);  {------------------}
            V_opt[No,*] := Base;
          end

        else { not basis stable }
          begin
            if EmptySolutionSet(TT) then Err := SolutionSetIsEmpty
            else if Unbounded(TT)   then Err := FunctionUnbounded

            else if PossiblyOptimalSolution(TT) and (Err = NoError) then
              begin
                No          := No + 1;                      { Store optimal }
                if (No = 1) then z_opt:= TT[1,1]            { solution      }
                            else z_opt:= TT[1,1] +* z_opt; {---------------}
                X_opt[No,*] := Solution(TT, Base, NonBase);
                V_opt[No,*] := Base;

                L := NeighboringList(TT,Base,NonBase); { Determine list of }
                remove(L, E); remove(L, CList);        { neighboring basic }
                                                       { index sets        }
                                                       {-------------------}
                append(CList, L);          { Compute new list of candidates }
              end;                         {--------------------------------}
          end;

      k := k+1;

    until empty(CList) or (Err <> NoError) or (k = kmax) or (No = maxNo);

    if (Err = NoError) then                           { Determine error code }
      if (not empty(CList)) and (k = kmax) then       {----------------------}
                          Err := IterationError
      else if (No = 0) then Err := StartIndexSetNotOptimal
      else if (No = maxNo) and (not empty(CList)) then
                          Err := SolutionMatrixTooSmall;

    FreeAll(CList); FreeAll(E);

  end; { Err <> WrongDimension }

end;  { global procedure LinOpt }

{----------------------------------------------------------------------------}
{ Module initialization part                                                 }
{----------------------------------------------------------------------------}
begin
  Rplus := intval(0,maxReal);
end.
```

11.3.2 Examples

Example 11.1 We consider the following linear programming problem:

maximize $50x_1 + 9x_2$ subject to the linear constraints

$$\begin{array}{rcl} x_1 & \leq & 50 \\ x_2 & \leq & 200 \\ 100\,x_1 + 18\,x_2 & \leq & 5000 \\ x_1, x_2 & \geq & 0. \end{array} \tag{11.10}$$

This two-dimensional optimization problem is illustrated by Figure 11.1, where X (shaded area) denotes the set of feasible solutions.

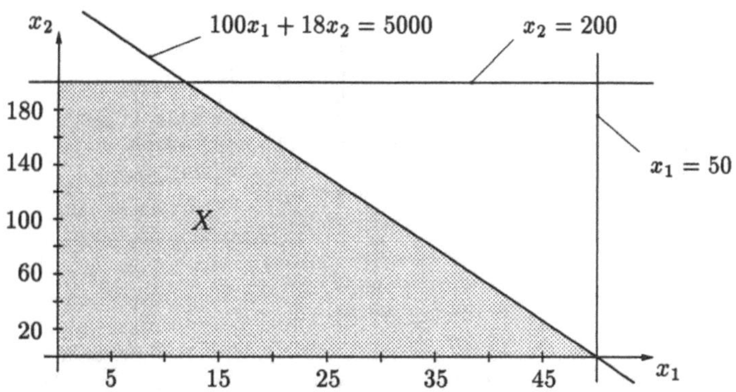

Figure 11.1: Two-dimensional optimization problem

We introduce the slack variables x_3, x_4, and x_5 and transform LP (11.10) into standard form (11.1).

maximize $c^T \cdot x$ subject to the linear constraints

$$\begin{array}{rcl} A \cdot x & = & b \\ x & \geq & 0, \end{array}$$

where

$$A = \begin{pmatrix} 1 & 0 & 1 & 0 & 0 \\ 0 & 1 & 0 & 1 & 0 \\ 100 & 18 & 0 & 0 & 1 \end{pmatrix}, \quad b = \begin{pmatrix} 50 \\ 200 \\ 5000 \end{pmatrix}, \quad c = \begin{pmatrix} 50 \\ 9 \\ 0 \\ 0 \\ 0 \end{pmatrix}.$$

The following sample main program determines enclosures for the optimal value $[z]$, the set of optimal basic index sets $\mathcal{V}_{opt}$, and the set of optimal basic feasible solutions $\mathcal{X}_{opt}$ for Example 11.1. The necessary work arrays are allocated depending on the dimension $n \times m$ of the optimization problem. Procedure *RevSimplex* computes an approximate solution and returns an initial basic index set. Procedure *LinOpt* determines the enclosures mentioned above and the results are returned by the main program.

```
program lop_example(input, output);

use     i_ari,                          { Interval Arithmetic               }
        mv_ari,                         { Matrix/Vector Arithmetic          }
        mvi_ari,                        { Matrix/Vector Interval Arithmetic }
        lop,                            { Linear Optimization Module        }
        rev_simp;                       { Revised Simplex Algorithm Module  }

const   maxSolNo = 50;
var     m,n : integer;

procedure main(m,n : integer);
var
        A                          : rmatrix[1..m,1..n];
        c, x_start                 : rvector[1..n];
        b, B_start                 : rvector[1..m];
        V_opt                      : rmatrix[1..maxSolNo,1..m];
        X_opt                      : imatrix[1..maxSolNo,1..n];
        z_start                    : real;
        z                          : interval;
        Error, i, NoOfSolutions    : integer;

begin {main}

  Error := 0;

  writeln;                                               { Read optimization   }
  writeln('Enter objective function (c[1]..c[n]): ');    { problem P = (A,b,c) }
  read(c); writeln;                                      {---------------------}
  writeln('Enter matrix of constraints (A[1,1]..A[m,n]): ');
  read(A); writeln;
  writeln('Enter right-hand side (b[1]..b[m]): ');
  read(b); writeln;
                                                           { Call revised      }
  RevSimplex(A, b, c, x_start, B_start, z_start, Error);   { simplex procedure }
                                                           {-------------------}
  if (Error <> 0) then writeln('Error occurred during approximation')
  else
    begin                                                  { Display results of }
      writeln('Results of approximation:');                { calculation        }
      writeln('Optimal value :  ',z_start); writeln;       {--------------------}
      write('Initial basic index set : ');
      writeln(B_start);
      write('Initial optimal solution : ');
      writeln(x_start);
    end;

  if (Error = 0) then                                   { Call verifying linear }
    begin                                               { optimization solver   }
      LinOpt(A, b, c, B_start, z, V_opt, X_opt, NoOfSolutions, Error);

      writeln('Results of verification:');
      if (Error = 0) or (Error = 6) or (Error = 7) then
        begin                                                 { Display results }
          writeln('Optimal value interval :  ',z); writeln;{ of calculation  }
          writeln('List of optimal basic index sets : ');   {-----------------}
          for i:=1 to NoOfSolutions do
            begin
              write('B',i:1,' =');  write(V_opt[i]);
            end;
          writeln;
          writeln('List of optimal basic feasible solutions : ');
          for i:=1 to NoOfSolutions do
            begin
```

```
            write('x of B',i:1,' ='); write(X_opt[i]);
          end;
        writeln;
      end;
  end;

  if (Error <> 0) then
    writeln(LinOptErrMsg(Error));

end; { procedure main }

begin

  writeln('Enter dimensions of linear ',                { Read dimension of    }
          'optimization problem (m and n): ');          { optimization problem }
  read(m); read(n);

  main(m,n);

end.
```

The procedure *LinOpt* calculates three optimal basic index sets and the corresponding three optimal solution vectors as listed below.

```
Enter dimensions of linear optimization problem (m and n):
3 5

Enter objective function (c[1]..c[n]):
50 9 0 0 0

Enter matrix of constraints (A[1,1]..A[m,n]):
  1  0 1 0 0
  0  1 0 1 0
100 18 0 0 1

Enter right-hand side (b[1]..b[m]):
50 200 5000

Results of approximation:
Optimal value :    2.500000000000000E+003

Initial basic index set :
 1.000000000000000E+000
 3.000000000000000E+000
 2.000000000000000E+000

Initial optimal solution :
 1.400000000000000E+001
 2.000000000000000E+002
 3.600000000000000E+001
 0.000000000000000E+000
 0.000000000000000E+000

Results of verification:
Optimal value interval :  [  2.500000000000000E+003,  2.500000000000000E+003 ]

List of optimal basic index sets :
B1 =
 1.000000000000000E+000
 2.000000000000000E+000
 3.000000000000000E+000
B2 =
 1.000000000000000E+000
```

```
 2.000000000000000E+000
 4.000000000000000E+000
B3 =
 1.000000000000000E+000
 3.000000000000000E+000
 4.000000000000000E+000

List of optimal basic feasible solutions :
x of B1 =
[  1.400000000000000E+001,  1.400000000000000E+001 ]
[  2.000000000000000E+002,  2.000000000000000E+002 ]
[  3.600000000000000E+001,  3.600000000000000E+001 ]
[  0.000000000000000E+000,  0.000000000000000E+000 ]
[  0.000000000000000E+000,  0.000000000000000E+000 ]
x of B2 =
[  5.000000000000000E+001,  5.000000000000000E+001 ]
[  0.000000000000000E+000,  0.000000000000000E+000 ]
[  0.000000000000000E+000,  0.000000000000000E+000 ]
[  2.000000000000000E+002,  2.000000000000000E+002 ]
[  0.000000000000000E+000,  0.000000000000000E+000 ]
x of B3 =
[  5.000000000000000E+001,  5.000000000000000E+001 ]
[  0.000000000000000E+000,  0.000000000000000E+000 ]
[  0.000000000000000E+000,  0.000000000000000E+000 ]
[  2.000000000000000E+002,  2.000000000000000E+002 ]
[  0.000000000000000E+000,  0.000000000000000E+000 ]
```

The standard simplex algorithm is neither able to recognize whether there exist more than one optimal solution nor if the computed basic index set and the corresponding solution are truly optimal. Example 11.1 has three optimal basic index sets representing two different vertices of the set of feasible solutions X.

Example 11.2 If we modify the objective function of Example 11.1 to

$$\text{maximize } 50x_1 + 9x_2 + 10^{-40}x_3$$

subject to the linear constraints (11.10), the standard algorithm still returns an optimal solution even though the objective function is unbounded on the set of feasible solutions:

```
Enter dimensions of linear optimization problem (m and n):
3 6

Enter objective function (c[1]..c[n]):
50 9 10E-40 0 0 0

Enter matrix of constraints (A[1,1]..A[m,n]):
  1  0 0 1 0 0
  0  1 0 0 1 0
100 18 0 0 0 1

Enter right-hand side (b[1]..b[m]):
50 200 5000

Results of approximation:
Optimal value :   2.500000000000000E+003

Initial basic index set :
 1.000000000000000E+000
```

```
 4.000000000000000E+000
 2.000000000000000E+000

Initial optimal solution :
 1.400000000000000E+001
 2.000000000000000E+002
 0.000000000000000E+000
 3.600000000000000E+001
 0.000000000000000E+000
 0.000000000000000E+000
```

11.3.3 Restrictions and Hints

Problem Size

This implementation does not take advantage of any special characteristic of the constraint matrix A in (11.1). It is always assumed to be dense so we can use module *linsys* from Chapter 10. If the problem turns out to be *basis stable* (cf. Theorem 11.1) (the most common case), or if it has only a few optimal basic index sets, the size of the problem is bounded by the memory of the host computer necessary to allocate the constraint matrix A and not by the calculation time.

Starting Problem

The determination of the initial basic index set $\mathcal{B}_{start}$ using the implementation of the revised simplex algorithm published by Syslo, Deo, and Kowalic [84] may be replaced by any other implementation. We require only that the approximate algorithm returns a set of basic indices of an optimal solution of the linear programming problem P.

Input Data

The procedure *LinOpt* assumes that all input data $P = (A, b, c)$ are exact, i.e. they are not afflicted with tolerances, and they are representable on the machine without conversion errors (see Section 3.7). Only in this case is a verification of the calculated results possible. Exercise 11.2 suggests a way around these difficulties.

11.4 Exercises

Exercise 11.1 Determine the optimal value, the set of optimal solutions, and the set of optimal basic index sets for the linear programming problem

$$\begin{array}{rcrcl} x_1 & + & 2\ x_2 & \leq & 80 \\ & & x_2 & \leq & 30 \\ 2\ x_1 & + & x_2 & \leq & 100 \\ & & x_1, x_2 & \geq & 0, \end{array}$$

using the sample program *lop_ex*.

In practice the input data of a linear programming problem often are afflicted with tolerances (e.g. measurement errors) and even more often the coefficients of the problem are not exactly representable on the machine. We can perform sensitivity and error analysis as well as achieve correct results if we specify the input data by a triple

$$[P] = ([A], [b], [c])$$

where $[A] \in I\!\!R^{m\times n}$, $[b] \in I\!\!R^m$, and $[c] \in I\!\!R^n$. The exact values for all input data are not necessarily known, but their lower and upper bounds are given.

The triple $[P]$ defines a whole set of linear programming problems $P \in [P]$ with real input data. This can be viewed as a linear parametric optimization problem, where each coefficient varies independently between the given lower and upper interval bounds. This approach overcomes the problem of conversion errors when reading decimal input data using a binary machine arithmetic (see 3.7).

Exercise 11.2 Write a *full interval version* of Algorithm 11.7 by changing the type definitions of the input data A, b, and c as well as of the submatrices and subvectors A_B, A_N, c_B, and c_N from `rmatrix` (resp. `rvector`) to `imatrix` (resp. `ivector`) (for details see Section 3.6). Also replace the module *linsys* for solving linear systems of equations (cf. Chapter 10) by the module *ilss* for solving interval systems of equations. The module *ilss* supplies a procedure *lss* to determine the solution of a whole set of linear systems of equations (e.g. $[A][x] = [b]$). Since all the theory of this chapter holds for the interval case (cf. [34]), the *full interval algorithm* determines all optimal solutions for the entire set of linear programming problems $P \in [P]$.

11.5 References and Further Reading

The standard simplex algorithm was first introduced by Dantzig [13]. A first verifying algorithm for linear programming problems was published by Jansson [34]. There are recent developments inspired by the work of Karmarkar [36] in the field of nonlinear methods for solving linear programming problems. Polynomial-time algorithms have been implemented and tested, but self-validating algorithms based on the projective and ellipsoid approaches of Karmarkar [36] and Khachiyan [42] have not yet appeared.

Chapter 12

Automatic Differentiation for Gradients, Jacobians, and Hessians

In Chapter 5, we considered automatic differentiation in one variable, but there are also many applications of numerical and scientific computing where it is necessary to compute derivatives of multi-dimensional functions. In this chapter, we extend the concept of automatic differentiation to the multi-dimensional case as given by Rall [66] and many others. We apply well-known differentiation rules for gradients, Jacobians, or Hessians with the computation of numerical values, combining the advantages of symbolic and numerical differentiation. Only the algorithm or formula for the function is required. No explicit formulas for the gradient, Jacobian, or Hessian have to be derived and coded.

In this chapter, we deal with automatic differentiation based on interval operations to get guaranteed enclosures for the function value and the values of the gradients, Jacobians, or Hessians. We will use gradients, Jacobians and Hessians generated by automatic differentiation in our algorithms for solving nonlinear systems (Chapter 13) and for global multi-dimensional optimization (Chapter 14). The techniques of automatic differentiation are also frequently applied in solving ordinary differential equations, sensitivity analysis, bifurcation studies, and continuation methods. Rall [66] and Griewank and Corliss [21] contain many applications of automatic differentiation.

12.1 Theoretical Background

Let $f : \mathbb{R}^n \to \mathbb{R}$ be a scalar-valued and twice continuously differentiable function. We are interested in computing the gradient

$$\nabla f(x) = \begin{pmatrix} \frac{\partial f}{\partial x_1}(x) \\ \frac{\partial f}{\partial x_2}(x) \\ \vdots \\ \frac{\partial f}{\partial x_n}(x) \end{pmatrix}$$

or the Hessian matrix

$$\nabla^2 f(x) = \begin{pmatrix} \frac{\partial^2 f}{\partial x_1^2}(x) & \frac{\partial^2 f}{\partial x_1 \partial x_2}(x) & \cdots & \frac{\partial^2 f}{\partial x_1 \partial x_n}(x) \\ \frac{\partial^2 f}{\partial x_2 \partial x_1}(x) & \frac{\partial^2 f}{\partial x_2^2}(x) & \cdots & \frac{\partial^2 f}{\partial x_2 \partial x_n}(x) \\ \vdots & \vdots & \vdots & \vdots \\ \frac{\partial^2 f}{\partial x_n \partial x_1}(x) & \frac{\partial^2 f}{\partial x_n \partial x_2}(x) & \cdots & \frac{\partial^2 f}{\partial x_n^2}(x) \end{pmatrix}.$$

In the case of functions of a single variable described in Chapter 5, we parsed the function to be differentiated into a code list (see [66]) consisting of a finite sequence of arithmetic operations $+$, $-$, $\cdot$, and $/$ and elementary functions sqrt, exp, sin, etc. Then the code list is interpreted in a differentiation arithmetic. We do exactly the same for multivariable functions in this chapter. We consider only first- and second-order derivative objects, although the same techniques have been used to generate much higher order partial derivatives [9].

Gradient and Hessian arithmetic is an arithmetic for ordered triples of the form

$$U = (u_f, u_g, u_h), \quad \text{with } u_f \in I\!R,\ u_g \in I\!R^n,\ u_h \in I\!R^{n\times n},$$

where the scalar u_f denotes the function value $u(x)$ of a twice differentiable function $u : I\!R^n \to I\!R$. The vector u_g and the matrix u_h denote the value of the gradient $\nabla u(x)$ and the value of the Hessian matrix $\nabla^2 u(x)$, respectively, each at a fixed point $x \in I\!R^n$. Our treatment is very similar to that of Rall (see [67]).

For the constant function $u(x) = c$, we set $U = (u_f, u_g, u_h) = (c, 0, 0)$. For the function $u(x) = x_k$ with $k \in \{1, 2, \ldots, n\}$, we define $U = (u_f, u_g, u_h) = (x_k, e^{(k)}, 0)$, where $e^{(k)} \in I\!R^n$ denotes the k-th unit vector, and 0 denotes the zero vector or zero matrix, respectively. The rules for the arithmetic are

$$W = U + V \Rightarrow \begin{cases} w_f = u_f + v_f, \\ w_g = u_g + v_g, \\ w_h = u_h + v_h, \end{cases} \tag{12.1}$$

$$W = U - V \Rightarrow \begin{cases} w_f = u_f - v_f, \\ w_g = u_g - v_g, \\ w_h = u_h - v_h, \end{cases} \tag{12.2}$$

$$W = U \cdot V \Rightarrow \begin{cases} w_f = u_f \cdot v_f, \\ w_g = u_f \cdot v_g + v_f \cdot u_g, \\ w_h = v_f \cdot u_h + u_g \cdot v_g^{\mathrm{T}} + v_g \cdot u_g^{\mathrm{T}} + u_f \cdot v_h, \end{cases} \tag{12.3}$$

$$W = U \,/\, V \Rightarrow \begin{cases} w_f = u_f / v_f, \\ w_g = (u_g - w_f \cdot v_g)/v_f, \\ w_h = (u_h - w_g \cdot v_g^{\mathrm{T}} - v_g \cdot w_g^{\mathrm{T}} - w_f \cdot v_h)/v_f, \end{cases} \tag{12.4}$$

where the familiar rules of calculus have to be used in the second and third components, and $v_f \neq 0$ is assumed in the case of division. The operations for w_f, w_g, and w_h in these definitions are operations on real numbers, vectors, and matrices. If the independent variables x_i of a formula for a function $f : I\!R^n \to I\!R$ and $x \mapsto f(x)$

are replaced by $X_i = (x_i, e^{(k)}, 0)$, and if all constants c are replaced by their $(c, 0, 0)$ representation, then evaluation of f using the rules of differentiation arithmetic gives

$$f(X) = f\left(\begin{pmatrix} X_1 \\ X_2 \\ \vdots \\ X_n \end{pmatrix}\right) = f\left(\begin{pmatrix} (x_1, e^{(1)}, 0) \\ (x_2, e^{(2)}, 0) \\ \vdots \\ (x_n, e^{(n)}, 0) \end{pmatrix}\right) = (f(x), \nabla f(x), \nabla^2 f(x)).$$

Example 12.1 We want to compute the function, gradient, and Hessian values of the function $f(x) = x_1 \cdot (4 + x_2)$ at the point $x = (1, 2)^T$. Differentiation arithmetic gives

$$\begin{aligned} f(X) = (f_f, f_g, f_h) &= (x_1, \begin{pmatrix} 1 \\ 0 \end{pmatrix}, \begin{pmatrix} 0 & 0 \\ 0 & 0 \end{pmatrix}) \cdot ((4,0,0) + (x_2, \begin{pmatrix} 0 \\ 1 \end{pmatrix}, \begin{pmatrix} 0 & 0 \\ 0 & 0 \end{pmatrix})) \\ &= (1, \begin{pmatrix} 1 \\ 0 \end{pmatrix}, \begin{pmatrix} 0 & 0 \\ 0 & 0 \end{pmatrix}) \cdot ((4,0,0) + (2, \begin{pmatrix} 0 \\ 1 \end{pmatrix}, \begin{pmatrix} 0 & 0 \\ 0 & 0 \end{pmatrix})) \\ &= (1, \begin{pmatrix} 1 \\ 0 \end{pmatrix}, \begin{pmatrix} 0 & 0 \\ 0 & 0 \end{pmatrix}) \cdot ((6, \begin{pmatrix} 0 \\ 1 \end{pmatrix}, \begin{pmatrix} 0 & 0 \\ 0 & 0 \end{pmatrix})) \\ &= (6, \begin{pmatrix} 6 \\ 1 \end{pmatrix}, \begin{pmatrix} 0 & 1 \\ 1 & 0 \end{pmatrix}). \end{aligned}$$

The result is $f(x) = 6$, $\nabla f(x) = \begin{pmatrix} 6 \\ 1 \end{pmatrix}$, and $\nabla^2 f(x) = \begin{pmatrix} 0 & 1 \\ 1 & 0 \end{pmatrix}$ for $x = \begin{pmatrix} 1 \\ 2 \end{pmatrix}$.

For an elementary function $s : I\!R \to I\!R$ and $U = (u_f, u_g, u_h)$, we define

$$W = s(U) \Rightarrow \begin{cases} w_f = s(u_f), \\ w_g = s'(u_f) \cdot u_g, \\ w_h = s''(u_f) \cdot u_g \cdot u_g^T + s'(u_f) \cdot u_h, \end{cases} \tag{12.5}$$

Here $s' : I\!R \to I\!R$ and $s'' : I\!R \to I\!R$ are the first and second derivatives of s, assuming they exist.

Rules (12.1)–(12.5) have assumed exact arithmetic. To get enclosures for the true values of the function, its gradient, and its Hessian, we use a differentiation arithmetic based on interval arithmetic. That is, the components u_f, u_g, and u_h are replaced by the corresponding interval values, and all arithmetic operations and function evaluations are executed in an interval arithmetic. The evaluation of a function $f : I\!R^n \to I\!R$ for an argument $[x] \in I I\!R^n$ using interval differentiation arithmetic delivers

$$f(X) = ([f_f], [f_g], [f_h])$$

satisfying

$$f([x]) \subseteq [f_f], \quad \nabla f([x]) \subseteq [f_g], \quad \text{and} \quad \nabla^2 f([x]) \subseteq [f_h].$$

Alternately, let $f : \mathbb{R}^n \to \mathbb{R}^n$ be a vector-valued and differentiable function, and let us compute the Jacobian matrix

$$J_f(x) = \begin{pmatrix} \frac{\partial f_1}{\partial x_1}(x) & \frac{\partial f_1}{\partial x_2}(x) & \cdots & \frac{\partial f_1}{\partial x_n}(x) \\ \frac{\partial f_2}{\partial x_1}(x) & \frac{\partial f_2}{\partial x_2}(x) & \cdots & \frac{\partial f_2}{\partial x_n}(x) \\ \vdots & \vdots & \vdots & \vdots \\ \frac{\partial f_n}{\partial x_1}(x) & \frac{\partial f_n}{\partial x_2}(x) & \cdots & \frac{\partial f_n}{\partial x_n}(x) \end{pmatrix}.$$

We can do this using our gradient arithmetic by computing the gradient for each component f_i with $i = 1, 2, \ldots, n$. In this case, we need not compute the Hessian components in our Rules (12.1)–(12.5).

12.2 Algorithmic Description

We now describe the algorithms for the elementary operators $+$, $-$, $\cdot$, and $/$, and for an elementary function $s \in \{$sqr, sqrt, power, exp, ln, sin, cos, tan, cot, arcsin, arccos, arctan, arccot, sinh, cosh, tanh, coth, arsinh, arcosh, artanh, arcoth$\}$ of a gradient and Hessian arithmetic for a twice continuously differentiable function $f : \mathbb{R}^n \to \mathbb{R}$.

In order to keep the notation of our implementation close to the discussion in this section, we use a special matrix representation for operands, arguments, or results $U := ([u_f], [u_g], [u_h])$ with $[u_f] \in I\mathbb{R}$, $[u_g] \in I\mathbb{R}^n$, and $[u_h] \in I\mathbb{R}^{n\times n}$. We first define the set of $(n+1)$-dimensional interval matrices with index ranges $0, 1, \ldots, n$ by

$$I\mathbb{R}^{\hat{n}\times\hat{n}} := \{[A] \in I\mathbb{R}^{(n+1)\times(n+1)} \mid [A] = ([a]_{ij})_{\substack{i=0,1,\ldots,n \\ j=0,1,\ldots,n}}\}. \tag{12.6}$$

Then following our algorithmic description, we fix the following identities for U and $[U] \in I\mathbb{R}^{\hat{n}\times\hat{n}}$:

$$[u_f] = [u]_{00}, \tag{12.7}$$

$$[u_g] = ([u]_{01}, [u]_{02}, \ldots, [u]_{0n})^{\mathrm{T}}, \tag{12.8}$$

$$[u_h] = \begin{pmatrix} [u]_{11} & [u]_{12} & \cdots & [u]_{1n} \\ [u]_{21} & [u]_{22} & \cdots & [u]_{2n} \\ \vdots & \vdots & \vdots & \vdots \\ [u]_{n1} & [u]_{n2} & \cdots & [u]_{nn} \end{pmatrix}. \tag{12.9}$$

That is, $[U] \in I\mathbb{R}^{\hat{n}\times\hat{n}}$ is partitioned into

$$[U] = \left(\begin{array}{c|c} [u_f] & [u_g]^{\mathrm{T}} \\ \hline & [u_h] \end{array}\right).$$

According to the symmetry of the Hessian matrix we really have to compute only the components $[u]_{ij}$ with $i = 0, \ldots, n$ and $j = 1, \ldots, i$.

Algorithm 12.1: $+ \ ([U],[V])$ {Operator}

1. $[w]_{00} := [u]_{00} + [v]_{00}$; {Function value}
2. **for** $i := 1$ **to** n **do**
   (a) $[w]_{0i} := [u]_{0i} + [v]_{0i}$; {Gradient components}
   (b) **for** $j := 1$ **to** i **do** {Hessian components}
   $[w]_{ij} := [u]_{ij} + [v]_{ij}$;
3. **return** $[W]$;

Algorithm 12.2: $- \ ([U],[V])$ {Operator}

1. $[w]_{00} := [u]_{00} - [v]_{00}$; {Function value}
2. **for** $i := 1$ **to** n **do**
   (a) $[w]_{0i} := [u]_{0i} - [v]_{0i}$; {Gradient components}
   (b) **for** $j := 1$ **to** i **do** {Hessian components}
   $[w]_{ij} := [u]_{ij} - [v]_{ij}$;
3. **return** $[W]$;

Algorithm 12.3: $\cdot \ ([U],[V])$ {Operator}

1. $[w]_{00} := [u]_{00} \cdot [v]_{00}$; {Function value}
2. **for** $i := 1$ **to** n **do**
   (a) $[w]_{0i} := [v]_{00} \cdot [u]_{0i} + [u]_{00} \cdot [v]_{0i}$; {Gradient components}
   (b) **for** $j := 1$ **to** i **do** {Hessian components}
   $[w]_{ij} := [v]_{00} \cdot [u]_{ij} + [u]_{0i} \cdot [v]_{0j} + [v]_{0i} \cdot [u]_{0j} + [u]_{00} \cdot [v]_{ij}$;
3. **return** $[W]$;

In Algorithm 12.4, we do not take care of the case $0 \in [v]_{00}$, because it does not make sense to go any further in computations when this case occurs. In an implementation, the standard error handling (runtime error) should be invoked if a division by zero occurs while computing the function value. We chose a special form of the rule for the differentiation of a quotient to be close to our implementation, where this form can save some computation time.

Algorithm 12.4: $/ \ ([U],[V])$ {Operator}

1. $[w]_{00} := [u]_{00}/[v]_{00}$; {Function value}
2. **for** $i := 1$ **to** n **do**
   (a) $[w]_{0i} := ([u]_{0i} - [w]_{00} \cdot [v]_{0i})/[v]_{00}$; {Gradient components}

for $j := 1$ **to** i **do** {Hessian components}
$[w]_{ij} := ([u]_{ij} - [w]_{0i} \cdot [v]_{0j} - [v]_{0i} \cdot [w]_{0j} - [w]_{00} \cdot [v]_{ij}) / [v]_{00}$;

3. **return** $[W]$;

In Algorithm 12.5, the rules from calculus for derivatives of the elementary functions are applied to compute the temporary values. These functions are listed in Table 5.1 in Chapter 5. As in the case of division by zero, we do not have to treat specially the case when the interval argument $[u]_{00}$ does not lie in the domain specified for the interval functions s (or s' for sqrt). In an implementation of Algorithm 12.5, the standard error handling of interval arithmetic should be invoked for domain violations in steps 1 and 2. See also the discussion of Table 5.1 in Chapter 5.

Algorithm 12.5: s $([U])$ {Function}

1. $[w]_{00} := s([u]_{00})$; {Function value}
2. $[h_1] := s'([u]_{00})$; $[h_2] := s''([u]_{00})$; {Temporary values}
3. **for** $i := 1$ **to** n **do**
   (a) $[w]_{0i} := [h_1] \cdot [u]_{0i}$; {Gradient components}
   (b) **for** $j := 1$ **to** i **do** {Hessian components}
   $[w]_{ij} := [h_2] \cdot [u]_{0i} \cdot [u]_{0j} + [h_1] \cdot [u]_{ij}$;
4. **return** s $:= [W]$;

12.3 Implementation and Examples

12.3.1 PASCAL–XSC Program Code

In the following sections, we present two modules for multi-dimensional automatic differentiation. We give a module supplying differentiation arithmetic for gradients *and* Hessians for scalar-valued functions in Section 12.3.1.1. The module can also be used to compute Jacobians of vector-valued functions. We give a special gradient arithmetic *grad_ari* for computing only gradients for scalar-valued functions and Jacobians for vector-valued functions without the storage overhead for the Hessian components in Section 12.3.1.2.

12.3.1.1 Module hess_ari

Module *hess_ari* supplies the type definition, operators, and elementary functions for an interval differentiation arithmetic for gradients and Hessians. The local variable *HessOrder* is used to select the highest order of derivative desired. This enables the user to save computation time in computing only the function value or the gradient and no Hessian, although storage for Hessian elements is still allocated. The default value of *HessOrder* is 2, so normally the gradient and the Hessian are computed.

The procedures *fEvalH*, *fgEvalH*, *fghEvalH*, *fEvalJ*, and *fJEvalJ* simplify the mechanism of function evaluating and automate the setting and resetting of the *HessOrder* variable. The $*$*EvalH* procedures can be applied to scalar-valued functions (result type *HessType*), whereas the $*$*EvalJ* procedures can be applied to vector-valued functions (result type *HTvector*).

For a scalar-valued function of type *HessType*, *fEvalH* computes and returns only the function value by setting *HessOrder* to 0 before the evaluation is done. If the gradient also is desired, *fgEvalH* can be used to set *HessOrder* to 1 before the evaluation is done. The procedure *fghEvalH* uses the default value of *HessOrder*, computes, and returns the values of $f(x)$, $\nabla f(x)$, and $\nabla^2 f(x)$. For a vector-valued function of type *HTvector*, *fEvalJ* computes and returns only the function value by setting *HessOrder* to 0 before the evaluation is done. If the Jacobian matrix also is desired, *fJEvalJ* can be used to set *HessOrder* to 1 before the evaluation is done.

All operators and functions in module *hess_ari* are implemented using the "modified call by reference" of PASCAL–XSC (see [45, Section 2.7.9] for details) to avoid the very inefficient allocation of local memory for the copies of the actual parameters.

```
{----------------------------------------------------------------------------}
{ Purpose: Definition of a multi-dimensional interval differentiation        }
{    arithmetic which allows function evaluation with automatic differen-    }
{    tiation up to second order (i.e. Hessian matrix).                       }
{ Method: Overloading of operators and elementary functions for operations   }
{    of data types 'HessType' and 'HTvector'. Note that all operators and    }
{    functions are implemented using the "modified call by reference" of     }
{    PASCAL-XSC (see Language Reference for details) to avoid the very       }
{    inefficient allocation of local memory for the copies of the actual     }
{    parameters.                                                             }
{ Global types, operators, functions, and procedures:                        }
{    type       HessType, HTvector: data types of differentiation arithmetic }
{    operators  +, -, *, /        : operators of differentiation arithmetic  }
{    operator   :=                : to define 'HessType' constants           }
{    function   HessVar           : to define 'HessType' variables           }
{    functions  fValue, gradValue,                                           }
{               JacValue,                                                    }
{               hessValue         : to get function and derivative values    }
{    functions  sqr, sqrt, power,                                            }
{               exp, sin, cos,... : elementary functions of diff. arithmetic }
{    procedures fEvalH, fEvalJ    : to compute function value only           }
{    procedures fgEvalH, fJEvalJ  : to compute function and first derivative }
{                                   value (gradient or Jacobian)             }
{    procedure  fghEvalH          : to compute function value, gradient, and }
{                                   Hessian matrix value                     }
{----------------------------------------------------------------------------}
module hess_ari;

use i_ari, i_util;        { interval arithmetic, interval utility functions }

{----------------------------------------------------------------------------}
{ Global type definition and local variable                                  }
{----------------------------------------------------------------------------}
global type
  HessType       = dynamic array [*,*] of interval;
                     { The index range must be 0..n,0..n. The component [0,0]  }
                     { contains the function value. The components of the row  }
                     { 0, i.e. [0,1],...,[0,n], contain the value of the       }
                     { gradient. The components [1,1],...,[n,n] contain the    }
                     { value of the Hessian matrix. In fact, only the lower    }
                     { triangle is used (and computed).                        }
```

```
  HTvector      = global dynamic array [*] of HessType;
                   { The index range must be 1..n (i.e.: 1..n,0..n,0..n).     }

var                    { The local variable 'HessOrder' is used to select the  }
  HessOrder : 0..2;    { highest order of derivative which is computed. Its    }
                       { default value is 2, and normally the gradient and the }
                       { Hessian matrix are computed.                          }
{-----------------------------------------------------------------------------}
{ Transfer operators and functions for constants and variables                }
{-----------------------------------------------------------------------------}
global operator := (var u: HessType; x: interval);        { Generate constant }
var i, j: integer;                                        {-------------------}
begin
  u[0,0]:= x;
  for j:=1 to ub(u) do
  begin
    u[0,j]:= 0;
    for i:=1 to ub(u) do u[i,j]:= 0;
  end;
end;

global operator := (var u: HessType; r: real);            { Generate constant }
begin                                                     {-------------------}
  u := intval(r);
end;

global function HessVar (x: ivector)
       : HTvector[1..ub(x)-lb(x)+1, 0..ub(x)-lb(x)+1, 0..ub(x)-lb(x)+1];
var                                                       { Generate variable }
  i, j, k, ubd, d : integer;                              {-------------------}
begin
  ubd := ub(x)-lb(x)+1;   d := 1-lb(x);
  for i:=1 to ubd do
  begin
    HessVar[i,0,0] := x[i-d];  { HessVar[*,*,0]:=x; }
    for k:=1 to ubd do
    begin
      if i=k then HessVar[i,0,k]:= 1 else HessVar[i,0,k]:= 0;
      for j:=1 to ubd do HessVar[i,j,k]:= 0;
    end;
  end;
end;

global function HessVar (v: rvector)
       : HTvector[1..ub(v)-lb(v)+1, 0..ub(v)-lb(v)+1, 0..ub(v)-lb(v)+1];
var                                                       { Generate variable }
  u : ivector[lb(v)..ub(v)];                              {-------------------}
  i : integer;
begin
  for i:=lb(v) to ub(v) do u[i]:= v[i];
  HessVar:= HessVar (u);
end;

{-----------------------------------------------------------------------------}
{ Access functions for the function value, the gradient, or the Hessian       }
{-----------------------------------------------------------------------------}
global function fValue (var f: HessType) : interval;    { Get function value }
begin                                                   {--------------------}
  fValue:= f[0,0];
end;

global function gradValue (var f: HessType) : ivector[1..ub(f)];
var i: integer;                                         { Get gradient value }
begin                                                   {--------------------}
```

```
  for i:=1 to ub(f) do gradValue[i]:= f[0,i];
end;

global function hessValue (var f: HessType) : imatrix[1..ub(f),1..ub(f)];
var i,j: integer;                                    { Get Hessian value  }
begin                                                {--------------------}
  for i:=1 to ub(f) do
  begin
    hessValue[i,i]:= f[i,i];
    for j:=1 to i-1 do
    begin
      hessValue[i,j]:= f[i,j];  hessValue[j,i]:= f[i,j];
    end;
  end;
end;

{----------------------------------------------------------------------------}
{ Access functions for vector-valued functions (function value or Jacobian)  }
{----------------------------------------------------------------------------}
global function fValue (var f: HTvector) : ivector[1..ub(f,1)];
begin                                             { Get function value of }
  fValue:= f[*,0,0];                              { n-dimensional vector- }
end;                                              { valued function       }
                                                  {-----------------------}

global function JacValue (var f: HTvector) : imatrix[1..ub(f,1),1..ub(f,2)];
var i, j: integer;                                { Get Jacobian value of }
begin                                             { n-dimensional scalar- }
  for i:=1 to ub(f,1) do                          { valued function       }
    for j:=1 to ub(f,2) do                        {-----------------------}
      JacValue[i,j]:= f[i,0,j];
end;

{----------------------------------------------------------------------------}
{ Monadic operators + and - for HessType operands                            }
{----------------------------------------------------------------------------}
global operator + (var u : HessType ) uplus : HessType[0..ub(u),0..ub(u)];
begin
  uplus:= u;
end;

global operator - (var u : HessType ) umin : HessType[0..ub(u),0..ub(u)];
var
  i, j: integer;
begin
  umin[0,0]:= -u[0,0];
  if (HessOrder > 0) then
    for i:=1 to ub(u) do
    begin
      umin[0,i]:= -u[0,i];
      if (HessOrder > 1) then
        for j:=1 to i do umin[i,j]:= -u[i,j];
    end;
end;

{----------------------------------------------------------------------------}
{ Operators +, -, *, and / for two HessType operands                         }
{----------------------------------------------------------------------------}
global operator + (var u,v: HessType ) add : HessType[0..ub(u),0..ub(u)];
var
  i, j: integer;
begin
  add[0,0]:= u[0,0]+v[0,0];
  if (HessOrder > 0) then
```

```
    for i:=1 to ub(u) do
    begin
      add[0,i]:= u[0,i]+v[0,i];
      if (HessOrder > 1) then
        for j:=1 to i do add[i,j]:= u[i,j]+v[i,j];
    end;
end;

global operator - (var u,v: HessType ) sub : HessType[0..ub(u),0..ub(u)];
var
  i, j: integer;
begin
  sub[0,0]:= u[0,0]-v[0,0];
  if (HessOrder > 0) then
    for i:=1 to ub(u) do
    begin
      sub[0,i]:= u[0,i]-v[0,i];
      if (HessOrder > 1) then
        for j:=1 to i do sub[i,j]:= u[i,j]-v[i,j];
    end;
end;

global operator * (var u,v: HessType ) mul : HessType[0..ub(u),0..ub(u)];
var
  i, j: integer;
begin
  mul[0,0]:= u[0,0]*v[0,0];
  if (HessOrder > 0) then
    for i:=1 to ub(u) do
    begin
      mul[0,i]:= v[0,0]*u[0,i]+u[0,0]*v[0,i];
      if (HessOrder > 1) then
        for j:=1 to i do
          mul[i,j]:= v[0,0]*u[i,j]+u[0,i]*v[0,j]+v[0,i]*u[0,j]+u[0,0]*v[i,j];
    end;
end;

global operator / (var u,v: HessType ) divis: HessType[0..ub(u),0..ub(u)];
var
  h    : HessType[0..ub(u),0..ub(u)];
  i, j: integer;
begin
  h[0,0]:= u[0,0]/v[0,0];  { Can propagate 'division by zero' error }
  if (HessOrder > 0) then
    for i:=1 to ub(u) do
    begin
      h[0,i]:= (u[0,i]-h[0,0]*v[0,i])/v[0,0];
      if (HessOrder > 1) then
        for j:=1 to i do
          h[i,j]:= (u[i,j]-h[0,i]*v[0,j]-v[0,i]*h[0,j]-h[0,0]*v[i,j])/v[0,0];
    end;
  divis:=h;
end;

{-----------------------------------------------------------------------------}
{ Operators +, -, *, and / for one interval and one HessType operand          }
{-----------------------------------------------------------------------------}
global operator + (var u: HessType;
                       b: interval) add : HessType[0..ub(u),0..ub(u)];
begin
  add := u;  add[0,0]:= u[0,0]+b;
end;

global operator - (var u: HessType;
```

```
                         b: interval) sub : HessType[0..ub(u),0..ub(u)];
begin
  sub := u;  sub[0,0]:= u[0,0]-b;
end;

global operator * (var u: HessType;
                         b: interval) mul : HessType[0..ub(u),0..ub(u)];
var
  i, j: integer;
begin
  mul[0,0]:= u[0,0]*b;
  if (HessOrder > 0) then
    for i:=1 to ub(u) do
    begin
      mul[0,i]:= b*u[0,i];
      if (HessOrder > 1) then
        for j:=1 to i do mul[i,j]:= b*u[i,j];
    end;
end;

global operator / (var u: HessType;
                         b: interval) divis : HessType[0..ub(u),0..ub(u)];
var
  i, j: integer;
begin
  divis[0,0]:= u[0,0]/b;    { Can propagate 'division by zero' error }
  if (HessOrder > 0) then
    for i:=1 to ub(u) do
    begin
      divis[0,i]:= u[0,i]/b;
      if (HessOrder > 1) then
        for j:=1 to i do divis[i,j]:= u[i,j]/b;
    end;
end;

global operator + (     a: interval;
                    var v: HessType) add : HessType[0..ub(v),0..ub(v)];
begin
  add := v;  add[0,0]:= a+v[0,0];
end;

global operator - (     a: interval;
                    var v: HessType) sub : HessType[0..ub(v),0..ub(v)];
var
  i, j: integer;
begin
  sub[0,0]:= a-v[0,0];
  if (HessOrder > 0) then
    for i:=1 to ub(v) do
    begin
      sub[0,i]:= -v[0,i];
      if (HessOrder > 1) then
        for j:=1 to i do sub[i,j]:= -v[i,j];
    end;
end;

global operator * (     a: interval;
                    var v: HessType) mul : HessType[0..ub(v),0..ub(v)];
var
  i, j: integer;
begin
  mul[0,0]:= a*v[0,0];
  if (HessOrder > 0) then
    for i:=1 to ub(v) do
```

```
    begin
      mul[0,i]:= a*v[0,i];
      if (HessOrder > 1) then
        for j:=1 to i do mul[i,j]:= a*v[i,j];
    end;
end;

global operator / (     a: interval;
                    var v: HessType) divis : HessType[0..ub(v),0..ub(v)];
var
  h    : HessType[0..ub(v),0..ub(v)];
  i, j: integer;
  p, q: interval;
begin
  h[0,0]:= a/v[0,0];       { Can propagate 'division by zero' error }
  if (HessOrder > 0) then
  begin
    p:= -h[0,0]/v[0,0];  q:= -2*p/v[0,0];
    for i:=1 to ub(v) do
    begin
      h[0,i]:= p*v[0,i];
      if (HessOrder > 1) then
        for j:=1 to i do h[i,j]:= p*v[i,j]+q*v[0,i]*v[0,j];
    end;
  end;
  divis:=h;
end;

{----------------------------------------------------------------------------}
{ Operators +, -, *, and / for one real and one HessType operand             }
{----------------------------------------------------------------------------}
global operator + (var u : HessType;
                       b : real     ) add : HessType[0..ub(u),0..ub(u)];
begin
  add:= u + intval(b);
end;

global operator - (var u: HessType;
                       b: real     ) sub : HessType[0..ub(u),0..ub(u)];
begin
  sub:= u - intval(b);
end;

global operator * (var u: HessType;
                       b: real     ) mul : HessType[0..ub(u),0..ub(u)];
begin
  mul:= u * intval(b);
end;

global operator / (var u: HessType;
                       b: real     ) divs : HessType[0..ub(u),0..ub(u)];
begin
  divs:= u / intval(b);  { Can propagate 'division by zero' error }
end;

global operator + (     a: real;
                    var v: HessType) add : HessType[0..ub(v),0..ub(v)];
begin
  add:= intval(a) + v;
end;

global operator - (     a: real;
                    var v: HessType) sub : HessType[0..ub(v),0..ub(v)];
begin
```

```
  sub:= intval(a) - v;
end;

global operator * (     a: real;
                    var v: HessType) mul : HessType[0..ub(v),0..ub(v)];
begin
  mul:= intval(a) * v;
end;

global operator / (     a: real;
                    var v: HessType) divs: HessType[0..ub(v),0..ub(v)];
begin
   divs:= intval(a) / v;  { Can propagate 'division by zero' error }
end;

{----------------------------------------------------------------------------}
{ Elementary functions for HessType arguments                                }
{----------------------------------------------------------------------------}
global function sqr (var u: HessType) : HessType[0..ub(u),0..ub(u)];
var
  i, j: integer;
  h1  : interval;
begin
  sqr[0,0]:= sqr(u[0,0]);
  if (HessOrder > 0) then
  begin
    h1:= 2*u[0,0];
    for i:=1 to ub(u) do
    begin
      sqr[0,i]:= h1*u[0,i];
      if (HessOrder > 1) then
        for j:=1 to i do sqr[i,j]:= h1*u[i,j]+2*u[0,i]*u[0,j];
    end;
  end;
end;

global function power (var u: HessType;
                           k: integer) : HessType[0..ub(u),0..ub(u)];
var
  h1   : interval;
  i, j : integer;
begin
  if (k = 0) then
    power:= 1
  else if (k = 1) then
    power:= u
  else if (k = 2) then
    power:= sqr(u)
  else
    begin
      power[0,0]:= power(u[0,0], k);
      if (HessOrder > 0) then
      begin
        h1:= k * power(u[0,0], k-1);
        for i:=1 to ub(u) do
        begin
          power[0,i]:= h1 * u[0,i];
          if (HessOrder > 1) then
            for j:=1 to i do
              power[i,j]:= h1*u[i,j] + k*(k-1)*power(u[0,0],k-2)*u[0,i]*u[0,j]
        end;
      end;
    end;
end;
```

```
global function sqrt (var u: HessType) : HessType[0..ub(u),0..ub(u)];
var
  i, j       : integer;
  h0, h1, h2 : interval;
begin
  h0:= sqrt(u[0,0]);  { Can propagate domain error }
  sqrt[0,0]:= h0;
  if (HessOrder > 0) then
  begin
    h1:= 0.5/h0;  h2:= -0.5*h1/u[0,0];
    for i:=1 to ub(u) do
    begin
      sqrt[0,i]:= h1*u[0,i];
      if (HessOrder > 1) then
        for j:=1 to i do sqrt[i,j]:= h1*u[i,j]+h2*u[0,i]*u[0,j];
    end;
  end;
end;

global function exp (var u: HessType) : HessType[0..ub(u),0..ub(u)];
var
  h0  : interval;
  i, j: integer;
begin
  h0:= exp(u[0,0]);  exp[0,0]:= h0;
  if (HessOrder > 0) then
    for i:=1 to ub(u) do
    begin
      exp[0,i]:= h0*u[0,i];
      if (HessOrder > 1) then
        for j:=1 to i do exp[i,j]:= h0*(u[i,j]+u[0,i]*u[0,j]);
    end;
end;

global function ln (var u: HessType) : HessType[0..ub(u),0..ub(u)];
var
  i, j   : integer;
  h1, h2 : interval;
begin
  ln[0,0]:= ln(u[0,0]);     { Can propagate domain error }
  if (HessOrder > 0) then
  begin
    h1:= 1/u[0,0];  h2:= -sqr(h1);
    for i:=1 to ub(u) do
    begin
      ln[0,i]:= h1*u[0,i];
      if (HessOrder > 1) then
        for j:=1 to i do ln[i,j]:= h1*u[i,j]+h2*u[0,i]*u[0,j];
    end;
  end;
end;

global function sin (var u: HessType) : HessType[0..ub(u),0..ub(u)];
var
  i, j       : integer;
  h0, h1, h2 : interval;
begin
  h0:= sin(u[0,0]);  sin[0,0]:= h0;
  if (HessOrder > 0) then
  begin
    h1:= cos(u[0,0]);  h2:= -h0;
    for i:=1 to ub(u) do
    begin
```

```
      sin[0,i]:= h1*u[0,i];
      if (HessOrder > 1) then
        for j:=1 to i do sin[i,j]:= h1*u[i,j]+h2*u[0,i]*u[0,j];
    end;
  end;
end;

global function cos (var u: HessType) : HessType[0..ub(u),0..ub(u)];
var
 i, j       : integer;
 h0, h1, h2 : interval;
begin
  h0:= cos(u[0,0]);  cos[0,0]:= h0;
  if (HessOrder > 0) then
  begin
    h1:= -sin(u[0,0]);  h2:= -h0;
    for i:=1 to ub(u) do
    begin
      cos[0,i]:= h1*u[0,i];
      if (HessOrder > 1) then
        for j:=1 to i do cos[i,j]:= h1*u[i,j]+h2*u[0,i]*u[0,j];
    end;
  end;
end;

global function tan (var u: HessType) : HessType[0..ub(u),0..ub(u)];
var
 i, j       : integer;
 h0, h1, h2 : interval;
begin
  h0:= tan(u[0,0]);  { Can propagate domain error }
  tan[0,0]:= h0;
  if (HessOrder > 0) then
  begin                                 { The subdistributive law implies  }
    h1:= sqr(h0)+1;  h2:= 2*h0*h1;      {   h0 * (h0↑2 + 1) <= h0↑3 + h0   }
    for i:=1 to ub(u) do                { So, we use the first form.       }
    begin
      tan[0,i] := h1*u[0,i];
      if (HessOrder > 1) then
        for j:=1 to i do tan[i,j]:= h1*u[i,j]+h2*u[0,i]*u[0,j];
    end;
  end;
end;

global function cot (var u: HessType) : HessType[0..ub(u),0..ub(u)];
var
 i, j       : integer;
 h0, h1, h2 : interval;
begin
  h0:= cot(u[0,0]);  { Can propagate domain error }
  cot[0,0]:= h0;
  if (HessOrder > 0) then
  begin                                 { The subdistributive law implies  }
    h1:= -(sqr(h0)+1);  h2:= -2*h0*h1;  {   h0 * (h0↑2 + 1) <= h0↑3 + h0   }
    for i:=1 to ub(u) do                { So, we use the first form.       }
    begin
      cot[0,i] := h1*u[0,i];
      if (HessOrder > 1) then
        for j:=1 to i do cot[i,j]:= h1*u[i,j]+h2*u[0,i]*u[0,j];
    end;
  end;
end;

global function arcsin (var u: HessType) : HessType[0..ub(u),0..ub(u)];
```

```
var
 i, j      : integer;
 h, h1, h2 : interval;
begin
  arcsin[0,0]:= arcsin(u[0,0]);  { Can propagate domain error }
  if (HessOrder > 0) then
  begin
    h := 1-sqr(u[0,0]);  h1:= 1/sqrt(h);  h2:= u[0,0]*h1/h;
    for i:=1 to ub(u) do
    begin
      arcsin[0,i]:= h1*u[0,i];
      if (HessOrder > 1) then
      for j:=1 to i do arcsin[i,j]:= h1*u[i,j]+h2*u[0,i]*u[0,j];
    end;
  end;
end;

global function arccos (var u: HessType) : HessType[0..ub(u),0..ub(u)];
var
 i, j      : integer;
 h, h1, h2 : interval;
begin
  arccos[0,0]:= arccos(u[0,0]);  { Can propagate domain error }
  if (HessOrder > 0) then
  begin
    h := 1-sqr(u[0,0]);  h1:= -1/sqrt(h);  h2:= u[0,0]*h1/h;
    for i:=1 to ub(u) do
    begin
      arccos[0,i]:= h1*u[0,i];
      if (HessOrder > 1) then
        for j:=1 to i do arccos[i,j]:= h1*u[i,j]+h2*u[0,i]*u[0,j];
    end;
  end;
end;

global function arctan (var u: HessType) : HessType[0..ub(u),0..ub(u)];
var
 i, j   : integer;
 h1, h2 : interval;
begin
  arctan[0,0]:= arctan(u[0,0]);  { Can propagate domain error }
  if (HessOrder > 0) then
  begin
    h1:= 1/(1+sqr(u[0,0]));  h2:= -2*u[0,0]*sqr(h1);
    for i:=1 to ub(u) do
    begin
      arctan[0,i]:= h1*u[0,i];
      if (HessOrder > 1) then
        for j:=1 to i do arctan[i,j]:= h1*u[i,j]+h2*u[0,i]*u[0,j];
    end;
  end;
end;

global function arccot (var u: HessType) : HessType[0..ub(u),0..ub(u)];
var
 i, j   : integer;
 h1, h2 : interval;
begin
  arccot[0,0]:= arccot(u[0,0]);  { Can propagate domain error }
  if (HessOrder > 0) then
  begin
    h1:= -1/(1+sqr(u[0,0]));  h2:= 2*u[0,0]*sqr(h1);
    for i:=1 to ub(u) do
    begin
```

```
      arccot[0,i]:= h1*u[0,i];
      if (HessOrder > 1) then
        for j:=1 to i do arccot[i,j]:= h1*u[i,j]+h2*u[0,i]*u[0,j];
    end;
  end;
end;

global function sinh (var u: HessType) : HessType[0..ub(u),0..ub(u)];
var
  i, j       : integer;
  h0, h1, h2 : interval;
begin
  h0:= sinh(u[0,0]);  sinh[0,0]:= h0;
  if (HessOrder > 0) then
  begin
    h1:= cosh(u[0,0]);  h2:= h0;
    for i:=1 to ub(u) do
    begin
      sinh[0,i]:= h1*u[0,i];
      if (HessOrder > 1) then
        for j:=1 to i do sinh[i,j]:= h1*u[i,j]+h2*u[0,i]*u[0,j];
    end;
  end;
end;

global function cosh (var u: HessType) : HessType[0..ub(u),0..ub(u)];
var
  i, j       : integer;
  h0, h1, h2 : interval;
begin
  h0:= cosh(u[0,0]);  cosh[0,0]:= h0;
  if (HessOrder > 0) then
  begin
    h1:= sinh(u[0,0]);  h2:= h0;
    for i:=1 to ub(u) do
    begin
      cosh[0,i] := h1*u[0,i];
      if (HessOrder > 1) then
        for j:=1 to i do cosh[i,j]:= h1*u[i,j]+h2*u[0,i]*u[0,j];
    end;
  end;
end;

global function tanh (var u: HessType) : HessType[0..ub(u),0..ub(u)];
var
  i, j       : integer;
  h0, h1, h2 : interval;
begin
  h0:= tanh(u[0,0]);  tanh[0,0]:= h0;
  if (HessOrder > 0) then
  begin
    h1:= 1-sqr(h0);  h2:= -2*h0*h1;        { The subdistributive law implies  }
    for i:=1 to ub(u) do                   {   h0 * (h0↑2 - 1) <= h0↑3 - h0   }
    begin                                  { So, we use the first form.       }
      tanh[0,i] := h1*u[0,i];
      if (HessOrder > 1) then
        for j:=1 to i do tanh[i,j]:= h1*u[i,j]+h2*u[0,i]*u[0,j];
    end;
  end;
end;

global function coth (var u: HessType) : HessType[0..ub(u),0..ub(u)];
var
  i, j       : integer;
```

```
 h0, h1, h2 : interval;
begin
  h0:= coth(u[0,0]);  { Can propagate domain error }
  coth[0,0]:= h0;
  if (HessOrder > 0) then
  begin                                   { The subdistributive law implies  }
    h1:= 1-sqr(h0);  h2:= -2*h0*h1;       {   h0 * (h0↑2 - 1) <= h0↑3 - h0   }
    for i:=1 to ub(u) do                  { So, we use the first form.       }
    begin
      coth[0,i] := h1*u[0,i];
      if (HessOrder > 1) then
        for j:=1 to i do coth[i,j]:= h1*u[i,j]+h2*u[0,i]*u[0,j];
    end;
  end;
end;

global function arsinh (var u: HessType) : HessType[0..ub(u),0..ub(u)];
var
 i, j       : integer;
 h, h1, h2 : interval;
begin
  arsinh[0,0]:= arsinh(u[0,0]);  { Can propagate domain error }
  if (HessOrder > 0) then
  begin
    h:= 1+sqr(u[0,0]);  h1:= 1/sqrt(h);  h2:= -u[0,0]*h1/h;
    for i:=1 to ub(u) do
    begin
      arsinh[0,i]:= h1*u[0,i];
      if (HessOrder > 1) then
        for j:=1 to i do arsinh[i,j]:= h1*u[i,j]+h2*u[0,i]*u[0,j];
    end;
  end;
end;

global function arcosh (var u: HessType) : HessType[0..ub(u),0..ub(u)];
var
  i, j       : integer;
  h, h1, h2 : interval;
begin
  arcosh[0,0]:= arcosh(u[0,0]);  { Can propagate domain error }
  if (HessOrder > 0) then
  begin
    h:= sqr(u[0,0])-1;  h1:= 1/sqrt(h);  h2:= -u[0,0]*h1/h;
    for i:=1 to ub(u) do
    begin
      arcosh[0,i]:= h1*u[0,i];
      if (HessOrder > 1) then
        for j:=1 to i do arcosh[i,j]:= h1*u[i,j]+h2*u[0,i]*u[0,j];
    end;
  end;
end;

global function artanh (var u: HessType) : HessType[0..ub(u),0..ub(u)];
var
 i, j    : integer;
 h1, h2 : interval;
begin
  artanh[0,0]:= artanh(u[0,0]);  { Can propagate domain error }
  if (HessOrder > 0) then
  begin
    h1:= 1/(1-sqr(u[0,0]));  h2:= 2*u[0,0]*sqr(h1);
    for i:=1 to ub(u) do
    begin
      artanh[0,i]:= h1*u[0,i];
```

```
      if (HessOrder > 1) then
        for j:=1 to i do artanh[i,j]:= h1*u[i,j]+h2*u[0,i]*u[0,j];
    end;
  end;
end;

global function arcoth (var u: HessType) : HessType[0..ub(u),0..ub(u)];
var
 i, j   : integer;
 h1, h2 : interval;
begin
  arcoth[0,0]:= arcoth(u[0,0]);  { Can propagate domain error }
  if (HessOrder > 0) then
  begin
    h1:= 1/(1-sqr(u[0,0]));  h2:= 2*u[0,0]*sqr(h1);
    for i:=1 to ub(u) do
    begin
      arcoth[0,i]:= h1*u[0,i];
      if (HessOrder > 1) then
        for j:=1 to i do arcoth[i,j]:= h1*u[i,j]+h2*u[0,i]*u[0,j];
    end;
  end;
end;


{----------------------------------------------------------------------------}
{ Predefined routines for evaluation of HessType-functions                  }
{----------------------------------------------------------------------------}
{ Purpose: Evaluation of function 'f' for argument 'x' in differentiation   }
{    arithmetic computing only the function value.                          }
{ Parameters:                                                                }
{    In      : 'f'       : function of 'HessType'.                          }
{              'x'       : argument for evaluation of 'f'.                  }
{    Out     : 'fx'      : returns the function value 'f(x)'.               }
{ Description: This procedure sets 'HessOrder' to 0, evaluates 'f(x)' in    }
{    differentiation arithmetic, and returns the function value only.       }
{----------------------------------------------------------------------------}
global procedure fEvalH (function f(y:HTvector) : HessType[lb(y,2)..ub(y,2),
                                                          lb(y,3)..ub(y,3)];
                                                 x : ivector;
                                  var           fx : interval);
begin
  HessOrder:= 0;  fx:= fValue(f(HessVar(x)));  HessOrder:= 2;
end;

{----------------------------------------------------------------------------}
{ Purpose: Evaluation of function 'f' for argument 'x' in differentiation   }
{    arithmetic computing the function value and the gradient value.        }
{ Parameters:                                                                }
{    In      : 'f'       : function of 'HessType'.                          }
{              'x'       : argument for evaluation of 'f'.                  }
{    Out     : 'fx'      : returns the function value 'f(x)'.               }
{              'gx'      : returns the gradient value 'grad f(x)'.          }
{ Description: This procedure sets 'HessOrder' to 1, evaluates 'f(x)' in    }
{    differentiation arithmetic, and returns the function value and the     }
{    value of the gradient.                                                 }
{----------------------------------------------------------------------------}
global procedure fgEvalH (function f(y:HTvector) : HessType[lb(y,2)..ub(y,2),
                                                           lb(y,3)..ub(y,3)];
                                                  x : ivector;
                                   var           fx : interval;
                                   var           gx : ivector);
var
  fxH : HessType[0..ub(x),0..ub(x)];
```

```
begin
  HessOrder:= 1;
  fxH:= f(HessVar(x));  fx:= fValue(fxH);  gx:= gradValue(fxH);
  HessOrder := 2;
end;

{----------------------------------------------------------------------------}
{ Purpose: Evaluation of function 'f' for argument 'x' in differentiation   }
{    arithmetic computing the function value, the gradient value, and the   }
{    Hessian matrix value.                                                   }
{ Parameters:                                                                }
{    In     : 'f'        : function of 'HessType'.                           }
{             'x'        : argument for evaluation of 'f'.                   }
{    Out    : 'fx'       : returns the function value 'f(x)'.                }
{             'gx'       : returns the gradient value 'grad f(x)'.           }
{             'hx'       : returns the Hessian matrix value 'hess f(x)'.     }
{ Description: This procedure keeps 'HessOrder' = 2, evaluates 'f(x)' in    }
{    differentiation arithmetic, and returns the function value, the value   }
{    of the gradient, and the value of the Hessian matrix.                   }
{----------------------------------------------------------------------------}
global procedure fghEvalH (function f(y:HTvector): HessType[lb(y,2)..ub(y,2),
                                                        lb(y,3)..ub(y,3)];
                                                   x : ivector;
                                  var             fx : interval;
                                  var             gx : ivector;
                                  var             hx : imatrix);
var
  fxH : HessType[0..ub(x),0..ub(x)];
begin
  fxH:= f(HessVar(x));
  fx:= fValue(fxH);  gx:= gradValue(fxH);  hx:= hessValue(fxH);
end;

{----------------------------------------------------------------------------}
{ Predefined routines for evaluation of HTvector-functions (with Jacobians)  }
{----------------------------------------------------------------------------}
{ Purpose: Evaluation of function 'f' for argument 'x' in differentiation   }
{    arithmetic computing only the function value.                           }
{ Parameters:                                                                }
{    In     : 'f'        : function of type 'HTvector'.                      }
{             'x'        : argument for evaluation of 'f'.                   }
{    Out    : 'fx'       : returns the vector function value 'f(x)'.         }
{ Description: This procedure sets 'HessOrder' to 0, evaluates 'f(x)' in    }
{    differentiation arithmetic, and returns the function value only.        }
{----------------------------------------------------------------------------}
global procedure fEvalJ (function f(y:HTvector) : HTvector[lb(y,1)..ub(y,1),
                                                       lb(y,2)..ub(y,2),
                                                       lb(y,3)..ub(y,3)];
                                                  x : ivector;
                                var              fx : ivector);
begin
  HessOrder:= 0;  fx:= fValue(f(HessVar(x)));  HessOrder:= 2;
end;

{----------------------------------------------------------------------------}
{ Purpose: Evaluation of function 'f' for argument 'x' in differentiation   }
{    arithmetic computing the function value and the Jacobian matrix value.  }
{ Parameters:                                                                }
{    In     : 'f'        : function of type 'HTvector'.                      }
{             'x'        : argument for evaluation of 'f'.                   }
{    Out    : 'fx'       : returns the function value 'f(x)'.                }
{             'Jx'       : returns the Jacobian value 'Jf(x)'.               }
{ Description: This procedure sets 'HessOrder' to 1, evaluates 'f(x)' in    }
{    differentiation arithmetic, and returns the function value and the      }
```

```
{    value of the Jacobian matrix.                                          }
{---------------------------------------------------------------------------}
global procedure fJEvalJ (function f(y:HTvector) : HTvector[lb(y,1)..ub(y,1),
                                                            lb(y,2)..ub(y,2),
                                                            lb(y,3)..ub(y,3)];
                                               x : ivector;
                              var             fx : ivector;
                              var             Jx : imatrix);
var
  fxGTv : HTvector[1..ub(x),0..ub(x),0..ub(x)];
begin
  HessOrder:= 1;
  fxGTv:= f(HessVar(x));  fx:= fValue(fxGTv);  Jx:= JacValue(fxGTv);
  HessOrder:= 2;
end;

{---------------------------------------------------------------------------}
{ Module initialization part                                                }
{---------------------------------------------------------------------------}
begin
  HessOrder := 2;
end.
```

12.3.1.2 Module grad_ari

If we only want to compute gradients or Jacobians, the module presented in Section 12.3.1.1 has the disadvantage of alllocating storage that is not used for the Hessian matrix. We now present a module for gradients, which supplies type definition, operators, and elementary functions for an interval differentiation arithmetic for gradients. The module can also be used to compute Jacobians of vector-valued functions.

The local variable *GradOrder* is used to select the highest order of derivative desired. This enables the user to save computation time in computing only the function value and no gradient. The default value of *GradOrder* is 1, so normally the gradient is computed.

The procedures *fEvalG*, *fgEvalG*, *fEvalJ*, and *fJEvalJ* simplify the mechanism of function evaluating and automate the setting and resetting of the *GradOrder* variable. The **EvalG* procedures can be applied to scalar-valued functions (result type *GradType*), whereas the **EvalJ* procedures can be applied to vector-valued functions (result type *GTvector*).

For a scalar-valued function of type *GradType*, *fEvalG* computes and returns only the function value by setting *GradOrder* to 0 before the evaluation is done. The procedure *fgEvalG* uses the default value of *GradOrder*, computes, and returns the values of $f(x)$ and $\nabla f(x)$. For a vector-valued function of type *GTvector*, *fEvalJ* computes and returns only the function value by setting *GradOrder* to 0 before the evaluation is done. If the Jacobian matrix also is desired, *fJEvalJ* can be used.

As in module *hess_ari*, all operators and functions of module *grad_ari* are implemented using the "modified call by reference" of PASCAL–XSC (see [45, Section 2.7.9] for details) to avoid the very inefficient allocation of local memory for the copies of the actual parameters.

```
{----------------------------------------------------------------------------}
{ Purpose: Definition of a multi-dimensional interval differentiation        }
{    arithmetic which allows function evaluation with automatic differen-    }
{    tiation up to first order (i.e. gradient or Jacobian).                  }
{ Method: Overloading of operators and elementary functions for operations   }
{    of data type 'GradType' and 'GTvector'. Note that all operators and     }
{    functions are implemented using the "modified call by reference" of     }
{    PASCAL-XSC (see Language Reference for details) to avoid the very       }
{    inefficient allocation of local memory for the copies of the actual     }
{    parameters.                                                             }
{ Global types, operators, functions, and procedures:                        }
{    type       GradType, GTvector: data types of differentiation arithmetic }
{    operators  +, -, *, /        : operators of differentiation arithmetic  }
{    operator   :=                : to define 'GradType' constants           }
{    function   GradVar           : to define 'GradType' variables           }
{    functions  fValue, gradValue,                                           }
{               JacValue          : to get function and derivative values    }
{    functions  sqr, sqrt, power,                                            }
{               exp, sin, cos,... : elementary functions of diff. arithmetic }
{    procedures fEvalG, fEvalJ    : to compute function value only           }
{    procedures fgEvalG, fJEvalJ  : to compute function and first derivative }
{                                   value (gradient or Jacobian)             }
{----------------------------------------------------------------------------}
module grad_ari;

use i_ari, i_util;  { interval arithmetic, interval utility functions }

{----------------------------------------------------------------------------}
{ Global type definitions and variable                                       }
{----------------------------------------------------------------------------}
global type
  GradType     = dynamic array [*] of interval;
                   { The index range must be 0..n. The component [0]         }
                   { contains the function value, the components [1],...,[n] }
                   { contain the gradient.                                   }
  GTvector     = global dynamic array [*] of GradType;
                   { The index range must be 1..n (i.e.: 1..n,0..n).         }

var                   { The local variable 'GradOrder' is used to select the }
  GradOrder : 0..1;   { highest order of derivative which is computed. Its   }
                      { default value is 1, and normally the gradient or the }
                      { Jacobian matrix are computed.                        }
{----------------------------------------------------------------------------}
{ Transfer operators and functions for constants and variables               }
{----------------------------------------------------------------------------}
global operator := (var f: GradType; u: interval);        { Generate constant }
var i : integer;                                          {-------------------}
begin
  f[0]:= u;
  for i:=1 to ub(f) do f[i]:= 0;
end;

global operator := (var f: GradType; r: real);            { Generate constant }
begin                                                     {-------------------}
  f:= intval(r);
end;

global function GradVar (x : ivector)
                          : GTvector[1..ub(x)-lb(x)+1, 0..ub(x)-lb(x)+1];
var                                                       { Generate variable }
  i, k, ubd, d : integer;                                 {-------------------}
begin
  ubd:= ub(x)-lb(x)+1;   d:= 1-lb(x);
  for i:=1 to ubd do
```

```
  begin
    GradVar[i,0]:= x[i-d];  { GradVar[*,0]:=x; }
    for k:=1 to ubd do
      if i=k then GradVar[i,k]:= 1 else GradVar[i,k]:= 0;
  end;
end;

global function GradVar (v : rvector)
                         : GTvector[1..ub(v)-lb(v)+1, 0..ub(v)-lb(v)+1];
var                                                 { Generate variable }
  u : ivector[lb(v)..ub(v)];                        {-------------------}
  i : integer;
begin
  for i:=lb(v) to ub(v) do u[i]:= v[i];
  GradVar:= GradVar(u);
end;

{-----------------------------------------------------------------------------}
{ Access functions for the function value, the gradient, or the Jacobian     }
{-----------------------------------------------------------------------------}
global function fValue (var f: GradType) : interval; { Get function value of }
begin                                                { n-dimensional scalar- }
  fValue:= f[0];                                     { valued function       }
end;                                                 {-----------------------}

global function gradValue (var f: GradType) : ivector[1..ub(f)];
var i: integer;                                      { Get gradient value of }
begin                                                { n-dimensional scalar- }
  for i:=1 to ub(f) do gradValue[i]:= f[i];          { valued function       }
end;                                                 {-----------------------}

global function fValue (var f: GTvector) : ivector[1..ub(f,1)];
begin                                                { Get function value of }
  fValue:= f[*,0];                                   { n-dimensional vector- }
end;                                                 { valued function       }
                                                     {-----------------------}

global function JacValue (var f: GTvector) : imatrix[1..ub(f,1),1..ub(f,2)];
var i, j: integer;                                   { Get Jacobian value of }
begin                                                { n-dimensional scalar- }
  for i:=1 to ub(f,1) do                             { valued function       }
    for j:=1 to ub(f,2) do                           {-----------------------}
      JacValue[i,j]:= f[i,j];
end;

{-----------------------------------------------------------------------------}
{ Monadic operators + and - for GradType operands                             }
{-----------------------------------------------------------------------------}
global operator + (var u : GradType ) uplus : GradType[0..ub(u)];
begin
  uplus:= u;
end;

global operator - (var u : GradType ) umin : GradType[0..ub(u)];
var
  i : integer;
begin
  umin[0]:= -u[0];
  if (GradOrder > 0) then
    for i:=1 to ub(u) do
      umin[i]:= -u[i];
end;

{-----------------------------------------------------------------------------}
```

```
{ Operators +, -, *, and / for two GradType operands                          }
{-----------------------------------------------------------------------------}
global operator + (var u,v : GradType ) add : GradType[0..ub(u)];
var
  i: integer;
begin
  add[0]:= u[0]+v[0];
  if (GradOrder > 0) then
    for i:=1 to ub(u) do add[i]:= u[i]+v[i];
end;

global operator - (var u,v : GradType ) sub : GradType[0..ub(u)];
var
  i: integer;
begin
  sub[0]:= u[0]-v[0];
  if (GradOrder > 0) then
    for i:=1 to ub(u) do sub[i]:= u[i]-v[i];
end;

global operator * (var u,v : GradType ) mul : GradType[0..ub(u)];
var
  i: integer;
begin
  mul[0]:= u[0]*v[0];
  if (GradOrder > 0) then
    for i:=1 to ub(u) do mul[i]:= v[0]*u[i]+u[0]*v[i];
end;

global operator / (var u,v: GradType ) divis : GradType[0..ub(u)];
var
  h: GradType[0..ub(u)];
  i: integer;
begin
  h[0]:= u[0]/v[0];  { Can propagate 'division by zero' error }
  if (GradOrder > 0) then
    for i:=1 to ub(u) do h[i]:= (u[i]-h[0]*v[i])/v[0];
  divis:=h;
end;

{-----------------------------------------------------------------------------}
{ Operators +, -, *, and / for one interval and one GradType operand          }
{-----------------------------------------------------------------------------}
global operator + (var u: GradType; b: interval) add : GradType[0..ub(u)];
begin
  add:= u;  add[0]:= u[0]+b;
end;

global operator - (var u: GradType; b: interval) sub : GradType[0..ub(u)];
begin
  sub:= u;  sub[0]:= u[0]-b;
end;

global operator * (var u: GradType; b: interval) mul : GradType[0..ub(u)];
var
  i: integer;
begin
  mul[0]:= u[0]*b;
  if (GradOrder > 0) then
    for i:=1 to ub(u) do mul[i]:= b*u[i];
end;

global operator / (var u: GradType; b: interval) divis : GradType[0..ub(u)];
var
```

```
  i: integer;
begin
  divis[0]:= u[0]/b;  { Can propagate 'division by zero' error }
  if (GradOrder > 0) then
    for i:=1 to ub(u) do divis[i]:= u[i]/b;
end;

global operator + (a: interval; var v: GradType) add : GradType[0..ub(v)];
begin
  add:= v;  add[0]:= a+v[0];
end;

global operator - (a: interval; var v: GradType) sub : GradType[0..ub(v)];
var
  i: integer;
begin
  sub[0]:= a-v[0];
  if (GradOrder > 0) then
    for i:=1 to ub(v) do sub[i]:= -v[i];
end;

global operator * (a: interval; var v: GradType) mul : GradType[0..ub(v)];
var
  i: integer;
begin
  mul[0]:= a*v[0];
  if (GradOrder > 0) then
    for i:=1 to ub(v) do mul[i]:= a*v[i];
end;

global operator / (a: interval; var v: GradType) divis : GradType[0..ub(v)];
var
  h: GradType[0..ub(v)];
  i: integer;
  p: interval;
begin
  h[0]:= a/v[0];  { Can propagate 'division by zero' error }
  if (GradOrder > 0) then
  begin
    p:= -h[0]/v[0];
    for i:=1 to ub(v) do h[i]:= p*v[i];
  end;
  divis:=h;
end;

{-----------------------------------------------------------------------}
{ Operators +, -, *, and / for one real and one GradType operand        }
{-----------------------------------------------------------------------}
global operator + (var u: GradType; b: real) add : GradType[0..ub(u)];
begin
  add:= u + intval(b);
end;

global operator - (var u: GradType; b: real) sub : GradType[0..ub(u)];
begin
  sub:= u - intval(b);
end;

global operator * (var u: GradType; b: real) mul : GradType[0..ub(u)];
begin
  mul:= u * intval(b);
end;

global operator / (var u: GradType; b: real) divs : GradType[0..ub(u)];
```

```
begin
  divs:= u / intval(b);  { Can propagate 'division by zero' error }
end;

global operator + (a: real; var v: GradType) add : GradType[0..ub(v)];
begin
  add:= intval(a) + v;
end;

global operator - (a: real; var v: GradType) sub : GradType[0..ub(v)];
begin
  sub:= intval(a) - v;
end;

global operator * (a: real; var v: GradType) mul : GradType[0..ub(v)];
begin
  mul:= intval(a) * v;
end;

global operator / (a: real; var v: GradType) divs: GradType[0..ub(v)];
begin
  divs:= intval(a) / v;  { Can propagate 'division by zero' error }
end;

{----------------------------------------------------------------------------}
{ Elementary functions for GradType arguments                                }
{----------------------------------------------------------------------------}
global function sqr (var u: GradType) : GradType[0..ub(u)];
var
  i: integer;
  h: interval;
begin
  sqr[0]:= sqr(u[0]);
  if (GradOrder > 0) then
  begin
    h:= 2*u[0];
    for i:=1 to ub(u) do sqr[i]:= h*u[i];
  end;
end;

global function power (var u: GradType; k: integer) : GradType[0..ub(u)];
var
  h : interval;
  i : integer;
begin
  if (k = 0) then
    power:= 1
  else if (k = 1) then
    power:= u
  else if (k = 2) then
    power:= sqr(u)
  else
    begin
      power[0]:= power(u[0], k);
      if (GradOrder > 0) then
      begin
        h:= k * power(u[0], k-1);
        for i:=1 to ub(u) do power[i]:= h * u[i];
      end;
    end;
end;

global function sqrt (var u: GradType) : GradType[0..ub(u)];
var
```

```
  i: integer;
  h: interval;
begin
  h:= sqrt(u[0]);  { Can propagate domain error }
  sqrt[0]:= h;
  if (GradOrder > 0) then
  begin
    h:= 0.5/h;
    for i:=1 to ub(u) do sqrt[i]:= h*u[i];
  end;
end;

global function exp (var u: GradType) : GradType[0..ub(u)];
var
  h: interval;
  i: integer;
begin
  h:= exp(u[0]);  exp[0]:= h;
  if (GradOrder > 0) then
    for i:=1 to ub(u) do exp[i]:= h*u[i];
end;

global function ln (var u: GradType) : GradType[0..ub(u)];
var
  i: integer;
  h: interval;
begin
  ln[0]:= ln(u[0]);  { Can propagate domain error }
  if (GradOrder > 0) then
  begin
    h:= 1/u[0];
    for i:=1 to ub(u) do ln[i]:= h*u[i];
  end;
end;

global function sin (var u: GradType) : GradType[0..ub(u)];
var
  i: integer;
  h: interval;
begin
  sin[0]:= sin(u[0]);
  if (GradOrder > 0) then
  begin
    h:= cos(u[0]);
    for i:=1 to ub(u) do sin[i]:= h*u[i];
  end;
end;

global function cos (var u: GradType) : GradType[0..ub(u)];
var
 i: integer;
 h: interval;
begin
  cos[0]:= cos(u[0]);
  if (GradOrder > 0) then
  begin
    h:= -sin(u[0]);
    for i:=1 to ub(u) do cos[i]:= h*u[i];
  end;
end;

global function tan (var u: GradType) : GradType[0..ub(u)];
var
 i: integer;
```

```
  h: interval;
begin
  h:= tan(u[0]);  { Can propagate domain error }
  tan[0]:= h;
  if (GradOrder > 0) then
  begin                                      { The subdistributive law implies  }
    h:= sqr(h)+1;                            {   h0 * (h0↑2 + 1) <= h0↑3 + h0   }
    for i:=1 to ub(u) do tan[i]:= h*u[i];    { So, we use the first form.       }
  end;
end;

global function cot (var u: GradType) : GradType[0..ub(u)];
var
  i: integer;
  h: interval;
begin
  h:= cot(u[0]);  { Can propagate domain error }
  cot[0]:= h;
  if (GradOrder > 0) then
  begin                                      { The subdistributive law implies  }
    h:= -(sqr(h)+1);                         {   h0 * (h0↑2 + 1) <= h0↑3 + h0   }
    for i:=1 to ub(u) do cot[i]:= h*u[i];    { So, we use the first form.       }
  end;
end;

global function arcsin (var u: GradType) : GradType[0..ub(u)];
var
  i: integer;
  h: interval;
begin
  arcsin[0]:= arcsin(u[0]);  { Can propagate domain error }
  if (GradOrder > 0) then
  begin
    h:= 1/sqrt(1-sqr(u[0]));
    for i:=1 to ub(u) do arcsin[i]:= h*u[i];
  end;
end;

global function arccos (var u: GradType) : GradType[0..ub(u)];
var
  i: integer;
  h: interval;
begin
  arccos[0]:= arccos(u[0]);  { Can propagate domain error }
  if (GradOrder > 0) then
  begin
    h:= -1/sqrt(1-sqr(u[0]));
    for i:=1 to ub(u) do arccos[i]:= h*u[i];
  end;
end;

global function arctan (var u: GradType) : GradType[0..ub(u)];
var
  i: integer;
  h: interval;
begin
  arctan[0]:= arctan(u[0]);  { Can propagate domain error }
  if (GradOrder > 0) then
  begin
    h:= 1/(1+sqr(u[0]));
    for i:=1 to ub(u) do arctan[i]:= h*u[i];
  end;
end;
```

```
global function arccot (var u: GradType) : GradType[0..ub(u)];
var
 i: integer;
 h: interval;
begin
  arccot[0]:= arccot(u[0]);  { Can propagate domain error }
  if (GradOrder > 0) then
  begin
    h:= -1/(1+sqr(u[0]));
    for i:=1 to ub(u) do arccot[i]:= h*u[i];
  end;
end;

global function sinh (var u: GradType) : GradType[0..ub(u)];
var
  i: integer;
  h: interval;
begin
  sinh[0]:= sinh(u[0]);
  if (GradOrder > 0) then
  begin
    h:= cosh(u[0]);
    for i:=1 to ub(u) do sinh[i]:= h*u[i];
  end;
end;

global function cosh (var u: GradType) : GradType[0..ub(u)];
var
 i: integer;
 h: interval;
begin
  cosh[0]:= cosh(u[0]);
  if (GradOrder > 0) then
  begin
    h:= sinh(u[0]);
    for i:=1 to ub(u) do cosh[i]:= h*u[i];
  end;
end;

global function tanh (var u: GradType) : GradType[0..ub(u)];
var
 i: integer;
 h: interval;
begin
  h:= tanh(u[0]);  tanh[0]:= h;
  if (GradOrder > 0) then
  begin                                { The subdistributive law implies  }
    h:= 1-sqr(h);                      {   h0 * (h0↑2 - 1) <= h0↑3 - h0   }
    for i:=1 to ub(u) do tanh[i]:= h*u[i];{ So, we use the first form.    }
  end;
end;

global function coth (var u: GradType) : GradType[0..ub(u)];
var
 i: integer;
 h: interval;
begin
  h:= coth(u[0]);  { Can propagate domain error }
  coth[0]:= h;
  if (GradOrder > 0) then
  begin                                { The subdistributive law implies  }
    h:= 1-sqr(h);                      {   h0 * (h0↑2 - 1) <= h0↑3 - h0   }
    for i:=1 to ub(u) do coth[i]:= h*u[i];{ So, we use the first form.    }
  end;
```

```
end;

global function arsinh (var u: GradType) : GradType[0..ub(u)];
var
 i: integer;
 h: interval;
begin
  arsinh[0]:= arsinh(u[0]);  { Can propagate domain error }
  if (GradOrder > 0) then
  begin
    h:= 1/sqrt(1+sqr(u[0]));
    for i:=1 to ub(u) do arsinh[i]:= h*u[i];
  end;
end;

global function arcosh (var u: GradType) : GradType[0..ub(u)];
var
  i: integer;
  h: interval;
begin
  arcosh[0]:= arcosh(u[0]);  { Can propagate domain error }
  if (GradOrder > 0) then
  begin
    h:= 1/sqrt(sqr(u[0])-1);
    for i:=1 to ub(u) do arcosh[i]:= h*u[i];
  end;
end;

global function artanh (var u: GradType) : GradType[0..ub(u)];
var
 i: integer;
 h: interval;
begin
  artanh[0]:= artanh(u[0]);  { Can propagate domain error }
  if (GradOrder > 0) then
  begin
    h:= 1/(1-sqr(u[0]));
    for i:=1 to ub(u) do artanh[i]:= h*u[i];
  end;
end;

global function arcoth (var u: GradType) : GradType[0..ub(u)];
var
 i: integer;
 h: interval;
begin
  arcoth[0]:= arcoth(u[0]);  { Can propagate domain error }
  if (GradOrder > 0) then
  begin
    h:= 1/(1-sqr(u[0]));
    for i:=1 to ub(u) do arcoth[i]:= h*u[i];
  end;
end;

{----------------------------------------------------------------------------}
{ Predefined routines for evaluation of GradType-functions                   }
{----------------------------------------------------------------------------}
{ Purpose: Evaluation of function 'f' for argument 'x' in differentiation   }
{    arithmetic computing only the function value.                           }
{ Parameters:                                                                }
{    In     : 'f'      : function of 'GradType'.                             }
{             'x'      : argument for evaluation of 'f'.                     }
{    Out    : 'fx'     : returns the function value 'f(x)'.                  }
{ Description: This procedure sets 'GradOrder' to 0, evaluates 'f(x)' in   }
```

```
{    differentiation arithmetic, and returns the function value only.          }
{----------------------------------------------------------------------------}
global procedure fEvalG (function f(y:GTvector) : GradType[lb(y,2)..ub(y,2)];
                                                       x : ivector;
                                  var                 fx : interval);
begin
  GradOrder:= 0;  fx:= fValue(f(GradVar(x)));  GradOrder:= 1;
end;

{----------------------------------------------------------------------------}
{ Purpose: Evaluation of function 'f' for argument 'x' in differentiation    }
{    arithmetic computing the function value and the gradient value.         }
{ Parameters:                                                                }
{    In      : 'f'        : function of 'GradType'.                          }
{              'x'        : argument for evaluation of 'f'.                  }
{    Out     : 'fx'       : returns the function value 'f(x)'.               }
{              'gx'       : returns the gradient value 'grad f(x)'.          }
{ Description: This procedure keeps 'GradOrder' = 1, evaluates 'f(x)' in    }
{    differentiation arithmetic, and returns the function value and the      }
{    value of the gradient.                                                  }
{----------------------------------------------------------------------------}
global procedure fgEvalG (function f(y:GTvector) : GradType[lb(y,2)..ub(y,2)];
                                                       x : ivector;
                                  var                 fx : interval;
                                  var                 gx : ivector);
var
  fxG : GradType[0..ub(x)];
begin
  fxG:= f(GradVar(x));  fx:= fValue(fxG);  gx:= gradValue(fxG);
end;

{----------------------------------------------------------------------------}
{ Predefined routines for evaluation of GTvector-functions (with Jacobians)  }
{----------------------------------------------------------------------------}
{ Purpose: Evaluation of function 'f' for argument 'x' in differentiation    }
{    arithmetic computing only the function value.                           }
{ Parameters:                                                                }
{    In      : 'f'        : function of type 'GTvector'.                     }
{              'x'        : argument for evaluation of 'f'.                  }
{    Out     : 'fx'       : returns the vector function value 'f(x)'.        }
{ Description: This procedure sets 'GradOrder' to 0, evaluates 'f(x)' in    }
{    differentiation arithmetic, and returns the function value only.        }
{----------------------------------------------------------------------------}
global procedure fEvalJ (function f(y:GTvector) : GTvector[lb(y,1)..ub(y,1),
                                                     lb(y,2)..ub(y,2)];
                                                       x : ivector;
                                  var                 fx : ivector);
begin
  GradOrder:= 0;  fx:= fValue(f(GradVar(x)));  GradOrder:= 1;
end;

{----------------------------------------------------------------------------}
{ Purpose: Evaluation of function 'f' for argument 'x' in differentiation    }
{    arithmetic computing the function value and the Jacobian matrix value.  }
{ Parameters:                                                                }
{    In      : 'f'        : function of type 'GTvector'.                     }
{              'x'        : argument for evaluation of 'f'.                  }
{    Out     : 'fx'       : returns the function value 'f(x)'.               }
{              'Jx'       : returns the Jacobian value 'Jf(x)'.              }
{ Description: This procedure keeps 'GradOrder' = 1, evaluates 'f(x)' in    }
{    differentiation arithmetic, and returns the function value and the      }
{    value of the Jacobian matrix.                                           }
{----------------------------------------------------------------------------}
global procedure fJEvalJ (function f(y:GTvector) : GTvector[lb(y,1)..ub(y,1),
```

```
                                                      lb(y,2)..ub(y,2)];
                                          x : ivector;
                             var         fx : ivector;
                             var         Jx : imatrix);
var
  fxGTv : GTvector[1..ub(x),0..ub(x)];
begin
  fxGTv:= f(GradVar(x));  fx:= fValue(fxGTv);  Jx:= JacValue(fxGTv);
end;

{---------------------------------------------------------------------------}
{ Module initialization part                                                }
{---------------------------------------------------------------------------}
begin
  GradOrder:= 1;
end.
```

12.3.2 Examples

In this section, we illustrate applications of automatic differentiation. The well-known approximate Newton's method applied to optimization and root-finding problems makes effective use of automatic differentiation, but we do not treat sophisticated methods in that area in this section. A detailed discussion of methods with automatic result verification using Newton's method is given in Chapters 13 and 14.

We first illustrate the use of our differentiation arithmetic in a program to solve Example 12.1 on page 227.

Example 12.2 We must define a PASCAL–XSC function for the function $f(x) = x_1 \cdot (4 + x_2)$:

```
function f (x: HTvector) : HessType[0..2,0..2];
begin
  f := x[1] * (4 + x[2]);
end;
```

With the declarations

```
var x          : HTvector[1..2,0..2,0..2];
    fx         : HessType[0..2,0..2];
    y, grady   : ivector[1..2];
    fy         : interval;
    hessy      : imatrix[1..2,1..2];
```

we can compute function values and derivative values directly by the sequence

```
    y[1]  := 123;  y[2]  := 456;
       x  := HessVar(y);
      fx  := f(x);
      fy  := fValue(fx);          { Function value f(y) }
   grady  := gradValue(fx);       { Gradient g(y)       }
   hessy  := hessValue(fx);       { Hessian H(y)        }
```

Alternatively, we can use the evaluating procedures:

```
   y[1] := 123;  y[2] := 456;
   fghEvalH(f,y,fy,grady,hessy);
```

Example 12.3 The well known method of Newton for computing an approximation of a zero of a system of nonlinear equations (see [83] for example) can also be used for computing a stationary point of a twice continuously differentiable multi-dimensional function. Newton's method is applied to the gradient of the function, i.e. to the equation $\nabla f(x) = 0$, starting from an approximation $x^{(0)} \in I\!R^n$. The method can be stated as

$$\left.\begin{array}{rcl} R^{(k)} & := & (\nabla^2 f(x^{(k)}))^{-1} \\ x^{(k+1)} & := & x^{(k)} - R^{(k)} \cdot \nabla f(x^{(k)}) \end{array}\right\} \; k = 0,1,2,\ldots$$

In our sample program *hess_ex* for this method, all real operations are replaced by interval operations, because we make use of the interval differentiation arithmetic. That is, we iterate

$$\left.\begin{array}{rcl} R^{(k)} & := & (m(\nabla^2 f([x]^{(k)})))^{-1} \\ [x]^{(k+1)} & := & [x]^{(k)} - R^{(k)} \cdot \nabla f([x]^{(k)}) \end{array}\right\} \; k = 0,1,2,\ldots$$

The influence of rounding errors is directly demonstrated by the increasing interval diameters. The approximate inverse $R^{(k)}$ is computed by applying the procedure *MatInv* from Chapter 10 to the midpoint matrix of the interval Hessian evaluation.

We terminate the iteration if an interval evaluation of ∇f over the current interval iterate $[x]^{(k)}$ contains zero, or if the number of iterations is greater than or equal to 100 to prevent endless iterations. Although interval operations are applied, the result is an enclosure of the sequence of values $x^{(k+1)}$ that would be computed in exact real arithmetic. The result is *not* a validated enclosure of the zero of the gradient because this approach does *not* enclose the truncation errors in the Newton iteration. For a more sophisticated interval method for finding *all* global minimizers of a nonlinear function with guarantee and high accuracy, we refer to Chapter 14.

We apply our iteration to the problem of finding a stationary point of Rosenbrock's function [75] $f(x) = 100 \cdot (x_2 - x_1^2)^2 + (x_1 - 1)^2$, which has become accepted as a difficult test problem for optimization methods.

```
{----------------------------------------------------------------------------}
{ This is an implementation of Newton's Method for computing a stationary    }
{ point of a twice continuously differentiable multi-dimensional function.   }
{                                                                            }
{    given:        the function f(x) and the starting value x0               }
{    iteration:    x[n+1] := x[n] - InvHf(x[n])*Gf(x[n]) , n = 0,1,...       }
{                                                                            }
{ where InvHf(x) denotes the inverse of the Hessian matrix of f(x).          }
{ All real operations are replaced by interval operations, all function and }
{ derivative evaluations are calculated by differentiation arithmetic.       }
{----------------------------------------------------------------------------}

program hess_ex;

use
  i_ari,       { Interval arithmetic                }
  mv_ari,      { Real matrix/vector arithmetic      }
  mvi_ari,     { Interval matrix/vector arithmetic  }
  matinv,      { Matrix inversion                   }
  hess_ari;    { Hessian differentiation arithmetic }
```

```
const
  fDim = 2;
  nmax = 100;
var
  fx     : interval;
  x, Gfx : ivector[1..fDim];
  Hfx    : imatrix[1..fDim,1..fDim];
  InvHfx : rmatrix[1..fDim,1..fDim];
  n, Err : integer;

function f (x: HTvector) : HessType[0..fDim,0..fDim];
begin
  f:= 100*sqr(x[2] - sqr(x[1])) + sqr(x[1] - 1);
end;

begin
  writeln('Newton''s method for finding a stationary point of Rosenbrock''s');
  writeln('function:  f(x) = 100*sqr(x[2] - sqr(x[1])) + sqr(x[1] - 1)');
  writeln;
  write('Starting vector x = ');  read(x);
  writeln;
  writeln('Iteration:');
  fghEvalH(f, x, fx, Gfx, Hfx);
  n := 0;
  repeat
    n := n + 1;
    writeln('x: ', x, 'Gf(x): ', Gfx);
    MatInv(mid(Hfx), InvHfx, Err);
    if Err = 0 then
    begin
      x:= x - InvHfx * Gfx;
      fghEvalH(f, x, fx, Gfx, Hfx);
    end;
  until ((0 in Gfx[1]) and (0 in Gfx[2])) or (n >= nmax) or (Err <> 0);
  if Err = 0 then
    begin
      writeln;
      writeln('Stationary point: ', x, 'Gradient value: ', Gfx);
      writeln('Expected solution: ');
      writeln(' 1.0 '); writeln(' 1.0 ');
    end
  else
    writeln(MatInvErrMsg(Err));
end.
```

If we run our sample program for Newton's method we get the following runtime output.

```
Newton's method for finding a stationary point of Rosenbrock's
function:  f(x) = 100*sqr(x[2] - sqr(x[1])) + sqr(x[1] - 1)

Starting vector x = 3
                    3

Iteration:
x:
[  3.000000000000000E+000,  3.000000000000000E+000 ]
[  3.000000000000000E+000,  3.000000000000000E+000 ]
Gf(x):
[  7.204000000000000E+003,  7.204000000000000E+003 ]
[ -1.200000000000000E+003, -1.200000000000000E+003 ]
```

```
x:
[ 2.998334721065778E+000, 2.998334721065779E+000 ]
[ 8.990008326394672E+000, 8.990008326394675E+000 ]
Gf(x):
[       3.99999537960E+000,      3.99999537962E+000 ]
[         -5.54630786E-004,        -5.54630784E-004 ]

x:
[       1.00110772357E+000,      1.00110772359E+000 ]
[       -2.9866990054E+000,      -2.9866990052E+000 ]
Gf(x):
[        1.5973359336E+003,       1.5973359337E+003 ]
[       -7.9778313591E+002,      -7.9778313589E+002 ]

x:
[        1.0011063367E+000,       1.0011063369E+000 ]
[         1.002213897E+000,        1.002213898E+000 ]
Gf(x):
[              2.2126E-003,             2.2128E-003 ]
[               -3.2E-008,                3.1E-008 ]

x:
[           9.999999E-001,          1.000001E+000 ]
[           9.999986E-001,          9.999989E-001 ]
Gf(x):
[                3.9E-004,               5.9E-004 ]
[               -3.0E-004,              -1.9E-004 ]


Stationary point:
[               9.999E-001,              1.001E+000 ]
[               9.998E-001,              1.001E+000 ]
Gradient value:
[                -1.6E-001,                1.6E-001 ]
[                -8.0E-002,                8.0E-002 ]

Expected solution:
 1.0
 1.0
```

This approximate method using interval operations converges, but the intervals grow wider. This shows the importance of using true *interval* algorithms, as opposed to *point* algorithms in interval arithmetic. The algorithm presented in Chapter 14 uses a method to generate a sequence of intervals that is guaranteed to enclose the global minimizer, and can achieve high accuracy.

Example 12.4 In this example, we demonstrate the use of *grad_ari* to compute Jacobians. We again use the method of Newton, now applied to the vector-valued function itself, i.e. to the problem $f(x) = 0$, starting from an approximation $x^{(0)} \in I\!R^n$. Newton's method is

$$\left.\begin{array}{rcl} R^{(k)} & := & (J_f(x^{(k)}))^{-1} \\ x^{(k+1)} & := & x^{(k)} - R^{(k)} \cdot f(x^{(k)}) \end{array}\right\} \; k = 0, 1, 2, \ldots$$

In our sample program *jac_ex* for this method, all real operations are replaced by interval operations, because we make use of the interval differentiation arithmetic.

That is, we start with $[x]_0$ and iterate

$$\left.\begin{array}{rcl} R^{(k)} & := & (m(J_f([x]^{(k)})))^{-1} \\ [x]^{(k+1)} & := & [x]^{(k)} - R^{(k)} \cdot f([x]^{(k)}) \end{array}\right\} \; k = 0,1,2,\ldots$$

The influence of rounding errors is directly demonstrated by the increasing interval diameters. The approximate inverse $R^{(k)}$ is computed by applying the procedure *MatInv* from Chapter 10 to the midpoint matrix of the interval Jacobian evaluation.

We terminate the iteration if an interval evaluation of f over the current interval iterate $[x]^{(k)}$ contains zero, or if the number of iterations is greater than or equal to 100. Although, this approximate method can *not* guarantee to find and enclose a zero of f in the last iterate. For a more sophisticated interval method for finding *all* zeros of a nonlinear system of equations with guarantee and high accuracy, we refer to Chapter 13.

We apply our iteration to the problem of finding a zero of Hansen's function [29]

$$f(x) = \begin{pmatrix} 6x_1^5 - 25.2x_1^3 + 24x_1 - 6x_2 \\ 12x_2 - 6x_1 \end{pmatrix}.$$

```
{-----------------------------------------------------------------------------}
{ This is an implementation of Newton's Method for computing a zero of        }
{ a continuously differentiable multi-dimensional function.                   }
{                                                                             }
{    given:       the function f(x) and the starting value x[0]               }
{    iteration:   x[n+1] := x[n] - InvJf(x[n])*f(x[n]) , n = 0,1,...          }
{                                                                             }
{ where InvJf(x) denotes the inverse of the Jacobian matrix of f(x).          }
{ All real operations are replaced by interval operations, all function and   }
{ derivative evaluations are calculated by differentiation arithmetic.        }
{-----------------------------------------------------------------------------}
program jac_ex;
use
  i_ari,        { Interval arithmetic                  }
  mv_ari,       { Real matrix/vector arithmetic        }
  mvi_ari,      { Interval matrix/vector arithmetic    }
  grad_ari,     { Gradient differentiation arithmetic  }
  matinv;       { Matrix inversion                     }

const
  fDim = 2;
  nmax = 100;

var
  fx, x  : ivector[1..fDim];
  Jfx    : imatrix[1..fDim,1..fDim];
  InvJfx : rmatrix[1..fDim,1..fDim];
  n, Err : integer;

function f (x: GTvector) : GTvector[1..fDim,0..fDim];
var
  x1sqr : GradType[0..fDim];
begin
  x1sqr:= sqr(x[1]);
  f[1]:= ((6*x1sqr - intval(252)/10)*x1sqr + 24)*x[1] - 6*x[2];
  f[2]:= 12*x[2] - 6*x[1];
end;
```

```
begin
  writeln('Newton'' s method for finding a zero of Hansen''s function:');
  writeln('f1(x) = 6 x1↑5 - 25.2 x1↑3 + 24 x1 - 6 x2');
  writeln('f2(x) = 12 x2 - 6 x1');
  writeln;
  write('Starting vector x: ');  read (x);
  writeln('Iteration');
  fJEvalJ(f, x, fx, Jfx);
  n := 0;
  repeat
    n := n + 1;
    writeln('x: ', x, 'f(x): ', fx);
    MatInv(mid(Jfx), InvJfx, Err);
    if Err = 0 then
    begin
      x:= x - InvJfx * fx;
      fJEvalJ(f, x, fx, Jfx);
    end;
  until ((0 in fx[1]) and (0 in fx[2])) or (n >= nmax) or (Err <> 0);
  writeln;
  if Err = 0 then
    begin
     writeln('Zero: ', x, 'Function value: ', fx);
     writeln('Expected zeros:');
     writeln('-1.7475... or -1.0705... or 0.0 or 1.0705... or 1.7475...');
     writeln('-0.8737...    -0.5352...    0.0    0.5352... or 0.8737...');
    end
  else
    writeln(MatInvErrMsg(Err));
end.
```

If we run our sample program for Newton's method, we get the report

```
Newton' s method for finding a zero of Hansen's function:
f1(x) = 6 x1^5 - 25.2 x1^3 + 24 x1 - 6 x2
f2(x) = 12 x2 - 6 x1

Starting vector x: 0.5
                   0.25

Iteration
x:
[  5.000000000000000E-001,  5.000000000000000E-001 ]
[  2.500000000000000E-001,  2.500000000000000E-001 ]
f(x):
[  7.537499999999999E+000,  7.537500000000002E+000 ]
[  0.000000000000000E+000,  0.000000000000000E+000 ]

x:
[ -1.396226415094340E+000, -1.396226415094339E+000 ]
[ -6.981132075471700E-001, -6.981132075471696E-001 ]
f(x):
[    7.4335163272550E+000,    7.4335163272552E+000 ]
[               -5.4E-015,                5.4E-015 ]

x:
[   -7.9519024745212E-001,   -7.9519024745211E-001 ]
[   -3.9759512372607E-001,   -3.9759512372605E-001 ]
f(x):
[   -5.9355981090667E+000,   -5.9355981090662E+000 ]
[               -6.0E-014,                6.0E-014 ]
```

```
x:
[   -1.1960053610903E+000,   -1.1960053610901E+000 ]
[   -5.9800268054511E-001,   -5.9800268054507E-001 ]
f(x):
[     3.312990631505E+000,     3.312990631509E+000 ]
[                -3.0E-013,                3.0E-013 ]

x:
[    -1.067378601460E+000,    -1.067378601459E+000 ]
[    -5.336893007299E-001,    -5.336893007296E-001 ]
f(x):
[      -8.293597506E-002,       -8.293597504E-002 ]
[                -1.4E-012,                1.4E-012 ]

x:
[    -1.070545193620E+000,    -1.070545193618E+000 ]
[    -5.352725968098E-001,    -5.352725968092E-001 ]
f(x):
[          7.613889E-005,           7.613895E-005 ]
[                -5.5E-012,                5.5E-012 ]


Zero:
[     -1.07054229183E+000,     -1.07054229182E+000 ]
[     -5.35271145915E-001,     -5.35271145911E-001 ]
Function value:
[                -2.5E-011,                1.5E-010 ]
[                -2.2E-011,                2.2E-011 ]

Expected zeros:
-1.7475... or -1.0705... or 0.0 or 1.0705... or 1.7475...
-0.8737...    -0.5352...    0.0    0.5352... or 0.8737...
```

This runtime output shows a very important fact in connection with the local convergence property of the classical Newton's method. Using the starting point $x^{(0)} = (0.5, 0.25)^T$, we might expect the method to converge to the zero $(0,0)^T$ or to the zero $(1.0705..., 0.5352...)^T$. However, the iteration stops at $(-1.0705..., -0.5352...)^T$. An approximation method cannot guarantee that we reach the "nearest" zero of $x^{(0)}$. We again refer to Chapter 13, where we introduce a method for finding *all* zeros, which generates a sequence of intervals that is guaranteed to enclose the roots, and can achieve high accuracy.

12.3.3 Restrictions and Hints

The implementations in modules *hess_ari* and *grad_ari* use the standard error handling of PASCAL–XSC if the interval argument of an elementary function does not lie in the domain specified for this interval function (see [65]). The same holds for an interval containing zero as the second operand of the division operator. These cases can also occur during a function evaluation using differentiation arithmetic because of the known overestimation effects of interval arithmetic (see Section 3.1).

Note that the rules for getting true enclosures in connection with conversion errors (see Section 3.7) also apply to interval differentiation arithmetic. To compute enclosures for the values of the gradients or Hessians of function f in Example 12.2 at point x with $x_1 = 0.1$, for example, you must assure that the machine interval argument $[x]_1$ used as argument for the interval function evaluation satisfies $x_1 \in [x]_1$. The same rule applies to real coefficients in the definition of f. That is why we expressed 25.2 as `intval(252)/10` in Example 12.4.

12.4 Exercises

Exercise 12.1 Implement a *real* version of module *hess_ari* by replacing all *interval* data types by the corresponding *real* data types. Then use the new module and apply it to Example 12.3.

Exercise 12.2 Implement a *real* version of module *grad_ari* by replacing all *interval* data types by the corresponding *real* data types. Then use the new module and apply it to Example 12.4.

12.5 References and Further Reading

The implementations in this chapter have followed those of Rall [67] in Pascal–SC. We use the *forward mode* of automatic differentiation in which the values of the derivative objects are propagated through the code list in the same order as intermediate values during the evaluation of f in ordinary arithmetic (see [66] or [68] for example).

It is possible to optimize the time complexity of computing gradients or Hessians by using the *backward or reverse mode* (also called fast automatic differentiation, see [16] for example). The reverse mode is faster than the forward mode for computing most large gradients, but it requires more storage space and is trickier to program.

Automatic differentiation methods can also be used to compute interval slopes (see [64]). Further applications and differentiation arithmetics can be found in [20], [21], and [69]. A large bibliography on automatic differentiation is given in [12].

Chapter 13

Nonlinear Systems of Equations

In Chapter 6, we considered the problem of finding zeros (or roots) of nonlinear functions of a single variable. Now, we consider its generalization, the problem of finding the solution vectors of a system of nonlinear equations. We give a method for finding *all* solutions of a nonlinear system of equations $f(x) = 0$ for a continuously differentiable function $f : \mathbb{R}^n \to \mathbb{R}^n$ in a given interval vector (box). Our method computes close bounds on the solution vectors, and it delivers information about existence and uniqueness of the computed solutions. The method we present is a variant of the *interval Gauss-Seidel method* based on the method of Hansen and Sengupta [3], [29], and a modification of Ratz [73]. Our method makes use of the extended interval operations defined in Section 3.3.

In classical numerical analysis, methods for solving nonlinear systems start from some approximate trial solution and iterate to (hopefully) improve the approximation until some convergence or termination criterion is satisfied. These non-interval methods have their difficulties finding *all* solutions of the given problem. It is a commonly held myth that it is impossible for a numerical method to find all solutions to a nonlinear system in a specified region. The myth is true if an algorithm only uses information from point evaluations of f. However, the algorithm of this chapter uses values of f and its derivatives evaluated at interval arguments to show that the myth is *false*.

13.1 Theoretical Background

Let $f : \mathbb{R}^n \to \mathbb{R}^n$ be a continuously differentiable function, and let $x \in [x]$ with $[x] \in I\mathbb{R}^n$. We address the problem of finding all solutions of the equation

$$f(x) = 0. \tag{13.1}$$

Let J_f denote the Jacobian matrix of f. We will compute J_f using automatic differentiation as described in Chapter 12. Interval Newton methods are used in general to sharpen bounds on the solutions of (13.1). They can be derived from the mean value form

$$f(m([x])) - f(x^*) = J_f(\xi)(m([x]) - x^*),$$

where $x^* \in [x]$, $\xi = (\xi_1, \ldots, \xi_n)$, and $\xi_i \in [x]$ for $i = 1, \ldots, n$. If we assume x^* to be a zero of f, we get

$$f(m([x])) = J_f(\xi) \cdot (m([x]) - x^*).$$

If we assume $J_f(\xi) \in I\!R^{n\times n}$ and all real matrices in $J_f([x]) \in I\,I\!R^{n\times n}$ to be nonsingular, we have

$$\begin{aligned} x^* &= m([x]) - (J_f(\xi))^{-1} \cdot f(m([x])) \\ &\in \underbrace{m([x]) - (J_f([x]))^{-1} \cdot f(m([x]))}_{=:\, N([x])}. \end{aligned}$$

$N([x])$ is the multi-dimensional interval Newton operator. Every zero of f in $[x]$ also lies in $N([x])$ and therefore in $N([x]) \cap [x]$. However, this method is only applicable if the interval matrix $J_f([x])$ is regular, i.e. if every real matrix $B \in J_f([x])$ is regular.

Thus, we are lead to another form of the interval Newton method that relaxes the requirement that $J_f([x])$ be nonsingular. We compute an enclosure of the set of solutions x^* of

$$f(m([x])) = J_f([x]) \cdot (m([x]) - x^*). \tag{13.2}$$

This method works better if we first *precondition* (13.2), i.e. if we multiply by a real matrix $R \in I\!R^{n\times n}$ to obtain

$$R \cdot f(m([x])) = R \cdot J_f([x]) \cdot (m([x]) - x^*).$$

Frequently, the inverse of the midpoint matrix $m(J_f([x]))$ is used as a preconditioner R. We will do so, too. Define

$$R := (m(J_f([x])))^{-1}, \quad b := R \cdot f(m([x])), \quad \text{and} \quad [A] := R \cdot J_f([x]), \tag{13.3}$$

and consider the interval linear equation

$$b = [A] \cdot (m([x]) - x^*). \tag{13.4}$$

The method we describe for solving equation (13.4) is based on the interval *Gauss-Seidel iteration*, which can also be applied to singular systems (cf. [64]).

13.1.1 Gauss-Seidel Iteration

We are interested in the solution set

$$S := \{x \in [x] \mid A \cdot (c - x) = b, \quad \text{for } A \in [A]\}$$

of the interval linear equation

$$[A] \cdot (c - x) = b$$

with fixed $c \approx m([x])$. In fact, we can use an arbitrary $c \in [x]$. Gauss-Seidel iteration is based on writing the linear system $A \cdot (c - x) = b$ explicitly in components as

$$\sum_{j=1}^{n} A_{ij} \cdot (c_j - x_j) = b_i, \quad i = 1, \ldots, n,$$

and solving the ith equation for the ith variable, assuming that $A_{ii} \neq 0$. Then we have

$$\begin{aligned} x_i &= c_i - \Big(b_i + \sum_{\substack{j=1\\ j\neq i}}^{n} A_{ij} \cdot (x_j - c_j) \Big) \Big/ A_{ii} \\ &\in \underbrace{c_i - \Big(b_i + \sum_{\substack{j=1\\ j\neq i}}^{n} [A]_{ij} \cdot ([x]_j - c_j) \Big) \Big/ [A]_{ii}}_{=:\,[z]_i} \end{aligned} \tag{13.5}$$

for $i = 1, \ldots, n$ if $0 \notin [A]_{ii}$ for all i. We obtain a new enclosure $[z]$ for x by computing the interval vector components $[z]_i$ according to (13.5) yielding

$$S \subseteq [z] \cap [x].$$

Moreover, it is possible to improve the enclosure $[z]$ using the fact that in the ith step improved enclosures $[z]_1, \ldots, [z]_{i-1}$ are already available. Thus, if we compute

$$\left.\begin{aligned} [y] &:= [x] \\ [y]_i &:= \Big(c_i - \Big(b_i + \sum_{\substack{j=1\\ j\neq i}}^{n} [A]_{ij} \cdot ([y]_j - c_j) \Big) \Big/ [A]_{ii} \Big) \cap [y]_i \\ & \qquad i = 1, \ldots, n \\ N_{\text{GS}}([x]) &:= [y] \end{aligned}\right\} \tag{13.6}$$

we have

$$S \subseteq N_{\text{GS}}([x]) \subseteq [z] \cap [x],$$

and every zero of f which lies in $[x]$ also lies in $N_{\text{GS}}([x])$. In (13.6), it is not necessary to compute the $[y]_i$ in fixed order $i = 1, \ldots, n$.

In summary, one interval Newton Gauss-Seidel step for the interval vector $[x]$ yields $N_{\text{GS}}([x])$ by first computing b and $[A]$ according to (13.3) and then computing $[y]$ according to (13.6). The interval Newton Gauss-Seidel iteration starts with an interval vector $[x]^{(0)}$ and iterates according to

$$[x]^{(k+1)} := N_{\text{GS}}([x]^{(k)}), \quad k = 0, 1, 2, \ldots \tag{13.7}$$

The intersections performed in (13.6) prevent the method from diverging. If an empty intersection occurs, we set $N_{\text{GS}}([x]^{(k)}) := [y]^{(k)} := \emptyset$, and we know that $[x]^{(k)}$ contains no zero of f.

If $0 \in [A]_{ii}$ for some i, the method can be used if extended interval arithmetic (see Chapters 3 and 6) is applied. In this case, a gap can be produced in the corresponding components $[y]_i^{(k)}$ of $[y]^{(k)}$, i.e. $[y]_i^{(k)}$ is given by one or two intervals resulting from the extended interval division and the succeeding intersection with the old value $[x]_i^{(k)}$. Therefore, $N_{\text{GS}}([x]^{(k)})$ is given by one or more interval vectors, and the next step of the interval Newton Gauss-Seidel iteration must be applied to each of these boxes which possibly contain solutions of (13.1). Thus, it is possible to compute *all* zeros of f in the starting interval vector $[x]^{(0)}$.

In the following theorem, we summarize the most important properties of the interval Newton Gauss-Seidel method.

Theorem 13.1 *Let $f : D \subseteq I\!R^n \to I\!R^n$ be a continuously differentiable function, and let $[x] \in I I\!R^n$ be an interval vector with $[x] \subseteq D$. Then $N_{\mathrm{GS}}([x])$ defined by (13.6) has the following properties:*

1. *Every zero $x^* \in [x]$ of f satisfies $x^* \in N_{\mathrm{GS}}([x])$.*
2. *If $N_{\mathrm{GS}}([x]) = \emptyset$, then there exists no zero of f in $[x]$.*
3. *If $N_{\mathrm{GS}}([x]) \mathring{\subset} [x]$, then there exists a unique zero of f in $[x]$ and hence in $N_{\mathrm{GS}}([x])$.*

For proofs see [28] or [64].

Remark: The conditions of Theorem 13.1 can be checked on a computer. For example, if $N_{\diamond\mathrm{GS}}([x])$ denotes the machine interval computation of $N_{\mathrm{GS}}([x])$ and if condition 3 is satisfied for $N_{\diamond\mathrm{GS}}$, then we have

$$N_{\mathrm{GS}}([x]) \subseteq N_{\diamond\mathrm{GS}}([x]) \mathring{\subset} [x].$$

Thus, the condition is fulfilled for $N_{\mathrm{GS}}([x])$, too. On the other hand, if we cannot fulfill conditions 2 and 3 of the Theorem, then $[x]$ *may* contain one or more zeros. This is especially the case if the problem has infinitely many solutions. Our interval method can find and bound an infinite set of solutions consisting of a continuum of points, but of course, none will be unique.

13.2 Algorithmic Description

The main algorithm AllNLSS (Algorithm 13.4) consists of two parts. The first part is the extended interval Newton iteration itself, including the interval Newton Gauss-Seidel step with intermediate checks for the uniqueness of a zero in a computed enclosure. The second part is an additional verification step which tries to verify the local uniqueness for enclosures that have not already been marked as enclosing a unique zero.

Algorithm 13.1 describes the interval Newton Gauss-Seidel step, where (13.6) is applied to the box $[y] \in I I\!R^n$ resulting in at most $n + 1$ boxes $[V]_i$. After some initializations, we use Algorithm 10.1 (MatInv) of Chapter 10 to compute an approximate inverse of the midpoint matrix of the interval Jacobian $J_f([y])$. In Step 4, we compute $[A]$, $[b]$, and $[y_c]$ followed by some initializations for the loop. Interval arithmetic must be used to compute $f(c)$ to bound all rounding errors (denoted by $f_\diamond(c)$ in Step 4).

We first perform the single component steps of the Gauss-Seidel step for all i with $0 \notin [A]_{ii}$ (Step 6) and then for the remaining indices with $0 \in [A]_{ii}$ (Step 7). Using this strategy, it is possible that the intervals $[y]_i$ become smaller by the intersections with the old values $[y]_i$ in Step 6(b) before the first splitting is produced in Step 7(b). If a gap is produced in this case, we store one part of the actual box $[y]$ in the pth row of the interval matrix $[V]$, and we continue the iteration for the other part of $[y]$. The flag *NoSolution* signals that an empty intersection has occurred resulting in $p := 0$. If the Gauss-Seidel step has only produced one box ($p = 1$) and if no improvement for $[y]$ is achieved, $[y]$ gets bisected in Step 10.

Algorithm 13.1: NewtonStep $(f, [y], [V], p)$ {Procedure}

1. $[y_{\text{in}}] := [y]$; $c := m([y])$; {Initializations}
2. **MatInv** $(m(J_f([y])), R, \mathit{InvErr})$; {Invert the midpoint matrix}
3. **if** ($\mathit{InvErr} \neq$ "No Error") **then** $R := I$;
4. $[A] := R \cdot J_f([y])$; $[b] := R \cdot f_{\Diamond}(c)$; $[y_c] := [y] - c$;
5. $p := 0$; $\mathit{NoSolution} := \mathit{false}$; {Initializations for loop}
6. **for** $i := 1$ **to** n **do** {Interval Gauss-Seidel step for $0 \notin [A]_{ii}$}
   (a) **if** $(0 \in [A]_{ii})$ **then next**$_i$;
   (b) $[y]_i := \left(c_i - \left([b]_i + \sum_{\substack{j=1 \\ j \neq i}}^{n} [A]_{ij} \cdot [y_c]_j \right) / [A]_{ii} \right) \cap [y]_i$;
   (c) **if** $[y]_i = \emptyset$ **then** $\mathit{NoSolution} := \mathit{true}$; **exit**$_{i\text{-loop}}$;
   (d) $[y_c]_i := [y]_i - c_i$;
7. **for** $i := 1$ **to** n **do** {Extended interval Gauss-Seidel step for $0 \in [A]_{ii}$}
   (a) **if** $(0 \notin [A]_{ii})$ **then next**$_i$;
   (b) $[z] := \left(c_i - \left([b]_i + \sum_{\substack{j=1 \\ j \neq i}}^{n} [A]_{ij} \cdot [y_c]_j \right) / [A]_{ii} \right) \cap [y]_i$; $\{[z] = [z]_1 \cup [z]_2\}$
   (c) **if** $([z] = \emptyset)$ **then** $\mathit{NoSolution} := \mathit{true}$; **exit**$_{i\text{-loop}}$;
   (d) $[y]_i := [z]_1$; $[y_c]_i := [y]_i - c_i$;
   (e) **if** $([z]_2 \neq \emptyset)$ **then**
   $p := p + 1$; $[V]_p := [y]$; $[V]_{pi} := [z]_2$; {Store part of $[y]$ in $[V]_p$}
8. **if** $\mathit{NoSolution}$ **then return** $p = 0$;
9. $p := p + 1$; $[V]_p := [y]$;
10. **if** $(p = 1)$ **and** $([y] = [y_{\text{in}}])$ **then** {Bisect the box}
    Find l with $d([y]_l) = d_\infty([y])$; {Component with maximum diameter}
    $p := 2$; $[V]_2 := [y]$; $[V]_{1l} := [\underline{V}_{1l}, c_l]$; $[V]_{2l} := [c_l, \overline{V}_{1l}]$; {Bisection}
11. **return** $[V], p$;

Algorithm 13.2 is a recursive procedure for the execution of the extended interval Newton Gauss-Seidel iteration for the function f. The input parameter $[y]$ specifies the actual box. The input parameter ε corresponds to the desired relative accuracy or tolerance for the resulting interval enclosures of the zeros. The calling procedure **AllNLSS** guarantees that ε is not chosen less than the relative machine accuracy (1 ulp). The input parameter *yUnique* signals whether we have already verified that the incoming interval vector $[y]$ contains a locally unique zero.

If $0 \notin f([y])$, then we know that no zero can lie within the interval vector $[y]$. Hence, a single interval function evaluation of f can guarantee that a complete range of real values cannot contain a zero of f. If $f([y])$ contains zero, then the extended interval Newton step given by (13.7) is applied to $[y]$ resulting in p non-empty boxes $[Y_p]_i$ with $i = 1, \ldots, p$ and $p \leq n+1$ (Step 2).

If the Newton step results in only one interval vector and if the local uniqueness of a zero in $[y]$ is not already proven, then the algorithm checks Condition 3 of Theorem 13.1 for the resulting box $[Y_p]_1$ and sets the flag *yUnique* in Step 3. In Step 4 the actual box $[Y_p]_i$ gets stored in the rectangular interval matrix $[Sol]$ if the tolerance criterion is satisfied. Otherwise, the procedure **XINewton** is recursively called (else-branch of Step 4). The corresponding information on the uniqueness of the zero is stored in the flag vector *Info*. N represents the number of enclosures of zeros stored in $[Sol]$.

Procedure **XINewton** terminates when no more recursive calls are necessary, that is if $0 \notin f([y])$ or if $(d_{\mathrm{rel},\infty}([y_p]_i) < \varepsilon)$ for $i = 1, \ldots, p$. The bisection in **NewtonStep** guarantees that this second condition is fulfilled at some stage of the recursion.

Algorithm 13.2: XINewton $(f, [y], \varepsilon, yUnique, [Sol], Info, N)$ {Proced.}

1. **if** $0 \notin f([y])$ **then return**;
2. NewtonStep $(f, [y], [Y_p], p)$;
3. **if** $p = 1$ **then**
   $yUnique := yUnique$ **or** $([y_p]_1 \overset{\circ}{\subset} [y])$; {Inner inclusion ⇒ uniqueness}
   else
   $yUnique := false$;
4. **for** $i := 1$ **to** p **do**
   if $(d_{\mathrm{rel},\infty}([Y_p]_i) < \varepsilon)$ **then** {Store enclosure of zero and uniqueness info}
   if $(0 \in f([Y_p]_i))$ **then**
   $N := N + 1$; $[Sol]_N := [Y_p]_i$; $Info_N := yUnique$;
   else {Recursive call of XINewton for $[Y_p]_i$}
   XINewton $(f, [Y_p]_i, \varepsilon, yUnique, [Sol], Info, N)$;
5. **return** $[Sol], Info, N$;

Algorithm 13.3 describes an additional verification step which checks the local uniqueness of the solution of the nonlinear system of equations enclosed in the interval vector $[y]$. The function f may have many zeros in the original box $[x]$. We attempt to find subboxes $[y]$, each of which contains a single zero of f. That is what we refer to as local uniqueness. Algorithm 13.3 can be used for boxes which have not yet been guaranteed to enclose a locally unique zero. This is done according to condition 3 of Theorem 13.1 by applying interval Newton steps including an epsilon inflation of the iterates $[y]$, which can help to verify zeros lying on the edge of $[y]$. We use $k_{\max} = 10$ as the maximum number of iterations, $\varepsilon = 0.25$ as the starting value for the epsilon inflation, and a factor of 8 to increase ε within the iterations. It turned out that these are good values for minimizing the effort if no verification is possible (see also [46]).

Algorithm 13.3: VerificationStep ($f, [y], yUnique$) {Procedure}

1. $k_{\max} := 10;\ k := 0;\ [y_{\text{in}}] := [y];\ \varepsilon := 0.25;\ yUnique := false$ {Initializations}
2. **while** (**not** $yUnique$) **and** ($k < k_{\max}$) **do** {Do $k_{\max}$ loops to achieve inclusion}
   (a) $k := k + 1;\ [y_{\text{old}}] := [y] \bowtie \varepsilon;$ {Epsilon inflation of $[y]$}
   (b) NewtonStep ($f, [y_{\text{old}}], [Y_p], p$);
   (c) **if** $p \neq 1$ **then exit**$_{\text{while-loop}}$; {No verification possible}
   (d) $[y] := [Y_p]_1;$
   (e) $yUnique := ([Y_p]_1 \overset{\circ}{\subset} [y_{\text{old}}]);$ {Inner inclusion $\Rightarrow$ uniqueness}
   (f) **if** $[y] = [y_{\text{old}}]$ **then** $\varepsilon := \varepsilon \cdot 8;$ {Increase ε}
3. **if not** $yUnique$ **then** $[y] := [y_{\text{in}}];$ {Reset $[y]$ to starting value}
4. **return** $[y], yUnique$;

Algorithm 13.4 now combines Algorithm 13.2 and Algorithm 13.3 to compute enclosures for all solutions of equation $f(x) = 0$ within the input interval vector $[x]$ and tries to prove the local uniqueness of a solution in each enclosure computed. The desired accuracy (relative diameter) of the interval enclosures is specified by the input parameter ε. 1 ulp accuracy is chosen if the specified value of ε is too small (for example 0). The enclosures for the solutions are returned row by row in the interval matrix $[Sol]$. The corresponding information on the local uniqueness of the solution is returned in the Boolean vector *Info*. The number of enclosures computed is returned in the integer variable N.

We use a function called CheckParameters as an abbreviation for the error checks for the parameters of AllNLSS which are necessary in an implementation. If no error occurs, AllNLSS delivers the N enclosures $[Sol]_i$, $i = 1, 2, \ldots, N$, satisfying

if $Info_i = true$, then $[Sol]_i$ encloses a locally unique zero of f,
if $Info_i = false$, then $[Sol]_i$ may enclose a zero of f.

If $N = 0$, then it is guaranteed that there is no zero of f in the starting box $[x]$.

Algorithm 13.4: AllNLSS ($f, [x], \varepsilon, [Sol], Info, N, Err$) {Procedure}

1. $Err :=$ CheckParameters;
2. **if** $Err \neq$ "No Error" **then return** Err;
3. $N := 0$;
4. Set ε to "1 ulp accuracy" if ε is too small;
5. XINewton ($f, [x], \varepsilon, false, [Sol], Info, N$);
6. **for** $i := 1$ **to** N **do**
   if $Info_i \neq true$ **then** VerificationStep ($f, [Sol]_i, Info_i$);
7. **return** $[Sol], Info, N, Err$;

Applicability of the Algorithm

To keep our algorithm and implementation as simple as possible, Algorithm 13.2 uses only the very simple stopping criteria

$$d_{\text{rel},\infty}([Y_p]_i) < \varepsilon.$$

Therefore, if the interval Newton steps do not improve the actual interval vector iterate, the method corresponds to a bisection method. The same applies to problems where the solution set is given by a continuum of points. For more sophisticated stopping criteria, see [28]. On the other hand, if the actual interval vector iterate $[y]$ causes no more extended interval divisions in **NewtonStep**, then the asymptotic rate of convergence to a zero of f in $[y]$ is quadratic.

The algorithm cannot verify the existence and the uniqueness of a multiple zero x^* of f in the enclosing interval vector. Nevertheless, the zero is bounded to the desired accuracy specified by ε. In this case, the corresponding component of the *Info*-vector is *false*. As a consequence of the splitting of the intervals, it may happen that a zero lying exactly on the splitting point is enclosed in several boxes. The method can be extended by a procedure to determine a single interval vector representing a set of abutting boxes (see [73] for example).

13.3 Implementation and Examples

13.3.1 PASCAL–XSC Program Code

Our implementation of Algorithm 13.4 uses the extended interval arithmetic module *xi_ari* (Section 3.3 and Chapter 6) and the automatic differentiation module *grad_ari* (Chapter 12).

13.3.1.1 Module nlss

Module *nlss* supplies the global routines *AllNLSS* (the implementation of Algorithm 13.4) and the corresponding function *AllNLSSErrMsg* to get an error message for the error code returned by *AllNLSS*. The procedures *NewtonStep*, *XINewton*, and *VerificationStep* are defined locally. All derivatives are evaluated in the differentiation arithmetic.

The procedure *AllNLSS* uses the *GTvector* function f and the starting interval vector *Start* as input parameters and stores all computed enclosures in the interval matrix *SoluVector* (row by row) which is also passed to and from the procedure *XINewton*. Before storing each interval vector in *SoluVector*, we check whether the interval matrix has free components left. If not, the corresponding error code is returned together with the complete *SoluVector* containing all solutions already computed. The user must then increase the upper index bound of the first index range of *SoluVector* to compute *all* zeros.

The same applies to the information about the uniqueness stored in the Boolean vector *InfoVector*. Therefore, the user shall declare both vectors with lower index bound equal to 1 and with upper index bounds which are equal. These conditions, as well as the condition *Epsilon* $\geq$ *MinEpsilon* are checked at the beginning of procedure *AllNLSS*. *Epsilon* in our program corresponds to the parameter ε in the algorithms, and *MinEpsilon* corresponds to 1 ulp accuracy.

Some of the routines in module *nlss* are implemented using the "modified call by reference" of PASCAL–XSC (see [45, Section 2.7.9] for details) to avoid the very inefficient allocation of local memory for the copies of the actual parameters.

```
{----------------------------------------------------------------------------}
{ Purpose: Computing enclosures for all solutions of systems of nonlinear    }
{    equations given by continuously differentiable functions.               }
{ Method: Extended interval Newton Gauss-Seidel method.                      }
{    Note that some routines are implemented using the "modified call by     }
{    reference" of PASCAL-XSC (see Language Reference for details) to avoid  }
{    the very inefficient allocation of local memory for the copies of the   }
{    actual parameters.                                                      }
{ Global procedures and functions:                                           }
{    procedure AllNLSS(...)       : computes enclosures for all solutions    }
{    function  AllNLSSErrMsg(...) : delivers an error message text           }
{----------------------------------------------------------------------------}
module nlss;

use
  i_ari,       { Interval arithmetic                   }
  xi_ari,      { Extended interval arithmetic          }
  grad_ari,    { Gradient differentiation arithmetic }
  mv_ari,      { Real matrix/vector arithmetic         }
  mvi_ari,     { Interval matrix/vector arithmetic     }
  b_util,      { Boolean utilities                     }
  i_util,      { Interval utilities                    }
  mvi_util,    { Interval matrix/vector utilities      }
  matinv;      { Matrix inversion                      }

{----------------------------------------------------------------------------}
{ Error codes used in this module.                                           }
{----------------------------------------------------------------------------}
const
  NoError        = 0;  { No error occurred.                                  }
```

```
  lbSoluVecNot1 = 1;  { Lower bound of variable SoluVector is not equal to 1.}
  lbInfoVecNot1 = 2;  { Lower bound of variable InfoVector is not equal to 1.}
  VecsDiffer    = 3;  { Bounds of SoluVector and InfoVector do not match.    }
  VecTooSmall   = 4;  { SoluVector too small. Not all zeros can be stored.   }

{----------------------------------------------------------------------------}
{ Error messages depending on the error code.                                }
{----------------------------------------------------------------------------}
global function AllNLSSErrMsg ( Err : integer ) : string;
var
  Msg : string;
begin
  case Err of
    NoError       : Msg := '';
    lbSoluVecNot1 : Msg := 'Lower bound of SoluVector is not equal to 1';
    lbInfoVecNot1 : Msg := 'Lower bound of InfoVector is not equal to 1';
    VecsDiffer    : Msg := 'Bounds of SoluVector and InfoVector do not match';
    VecTooSmall   : Msg := 'Not all zeros found. SoluVector is too small';
    else          : Msg := 'Code not defined';
  end;
  if (Err <> NoError) then Msg := 'Error: ' + Msg + '!';
  AllNLSSErrMsg := Msg;
end;

{----------------------------------------------------------------------------}
{ Check if every component of interval vector 'iv' contains the value 'n'    }
{----------------------------------------------------------------------------}
operator in (n: integer; var iv: ivector) ResIn : boolean;
var
  k : integer;
begin
  k:= lb(iv);
  while (k < ub(iv)) and (n in iv[k]) do k:= k+1;
  ResIn:= (n in iv[k]);
end;

{----------------------------------------------------------------------------}
{ Purpose: Execution of one single interval Newton Gauss-Seidel step for the }
{    interval vector 'Y' and the function 'f'.                               }
{ Parameters:                                                                }
{    In     : 'f'          : must be declared for the type 'GTvector' to     }
{                            enable the internal use of the differentiation  }
{                            arithmetic 'grad_ari'.                          }
{             'Y'          : starting interval.                              }
{             'JfY'        : Jacobian matrix of 'f(Y)', already computed     }
{                            outside of 'NewtonStep'.                        }
{    Out    : 'V'          : enclosures 'V[i]' for the splittings            }
{                            generated by the Newton step.                   }
{             'p'          : number of non-empty interval vectors 'V[i]'.    }
{ Description:                                                               }
{    'NewtonStep' executes the extended interval Newton Gauss-Seidel step    }
{    for 'Y' with result interval vector(s) 'V[i]' which can be empty.      }
{    'p' gives the number of non-empty interval vectors stored as vectors    }
{    'V[1]',...,'V[p]'.                                                      }
{----------------------------------------------------------------------------}
procedure NewtonStep (function f (x: GTvector) : GTvector[1..ub(x,1),
                                                          0..ub(x,2)];
                                             Y : ivector;
                      var                  JfY : imatrix;
                      var                    V : imatrix;
                      var                    p : integer);
var
  c                       : rvector[1..ub(Y)];
  fC, b, Yin, Y_minus_c : ivector[1..ub(Y)];
```

```
    R                        : rmatrix[1..ub(Y),1..ub(Y)];
    A                        : imatrix[1..ub(Y),1..ub(Y)];
    i, i0, n, InvErr, j      : integer;
    NoSolution               : boolean;
    h                        : interval;
    z                        : ivector[1..2];
  begin
    Yin:= Y;
    c:= mid(Y);  fEvalJ(f, intval(c), fC);      { Midpoint evaluation of 'f'      }
                                                {---------------------------------}
    MatInv(mid(JfY),R,InvErr);                  { Invert the midpoint matrix      }
    if InvErr <> 0 then R:= id(R);              {---------------------------------}

    A:= R * JfY;  b:= R * fC;               { Compute data for Gauss-Seidel step }
    Y_minus_c:= Y - c;                      {-------------------------------------}

    p:= 0; i:= 0; i0:= 0; n:= ub(Y); { Initializations, A[i0,i0] contains zero }
    NoSolution:= false;              {-----------------------------------------}
    while (i < n) and (not NoSolution) do       { Interval Gauss-Seidel step for }
    begin                                       { non-zero A[i,i] elements        }
      i:= succ(i);                              {---------------------------------}
      if not (0 in A[i,i]) then
        begin
          h:= ## (b[i] + (for j:=1 to i-1 sum (A[i,j] * Y_minus_c[j]) )
                       + (for j:=i+1 to n sum (A[i,j] * Y_minus_c[j]) ) );
          h:= c[i] - h / A[i,i];

          if (Y[i] >< h) then
            NoSolution:= true
          else
            begin
              Y[i]:= Y[i] ** h;  Y_minus_c[i]:= Y[i] - c[i];
            end;
        end
      else
        i0:= i;                                       { Largest i with 0 in A[i,i] }
    end;                                              {-----------------------------}

    i:= 0;
    while (not NoSolution) and (i < i0) do      { Interval Gauss-Seidel step for }
    begin                                       { zero A[i,i] elements            }
      i:= succ(i);                              {---------------------------------}
      if (0 in A[i,i]) then
      begin
        h:= ## ( b[i] + ( for j:=1 to i-1 sum (A[i,j] * Y_minus_c[j]) )
                      + ( for j:=i+1 to n sum (A[i,j] * Y_minus_c[j]) ) );
        z:= Y[i] ** (c[i] - h div A[i,i]);            { Extended interval division }
                                                      {-----------------------------}
        if (z[1] = EmptyIntval) then                  { z[1] = z[2] = EmptyIntval  }
          NoSolution:= true                           {-----------------------------}
        else
          begin                                                { Compute new 'Y' }
            Y[i]:= z[1];  Y_minus_c[i]:= Y[i] - c[i];          {-----------------}
            if z[2] <> EmptyIntval then
            begin                                        { Store further bisections }
              p:= p+1; V[p]:= Y; V[p][i]:= z[2];         {---------------------------}
            end
          end;
      end; { if 0 in A[i,i] ... }
    end; { while (not NoSolution) ... }

    if NoSolution then
      p := 0
    else
```

```
    begin
      p:= p+1;  V[p]:= Y;

      if (p = 1) and (Y = Yin) then                        { Bisect the box }
      begin                                                {----------------}
        i0:= 1;
        for i:=2 to ub(Y) do
          if diam(Y[i]) > diam(Y[i0]) then i0:= i;
        p:= 2;  V[2]:= Y;
        V[1][i0].sup:= c[i0]; V[2][i0].inf:= c[i0];            { Bisection }
      end;                                                     {-----------}
    end;
end; { NewtonStep }

{----------------------------------------------------------------------------}
{ Purpose: Recursive procedure for the execution of the extended interval    }
{    Newton Gauss_Seidel method for the function 'f'.                        }
{ Parameters:                                                                }
{    In     : 'f'          : must be declared for the type 'GTvector' to     }
{                            enable the internal use of the differentiation  }
{                            arithmetic 'grad_ari'.                          }
{             'y'          : starting interval.                              }
{             'Epsilon'    : desired relative accuracy                       }
{                            (interval diameter) of the result intervals.    }
{             'yUnique'    : signals whether it is already verified that the }
{                            actual interval 'y' contains a unique zero.     }
{    Out    : 'SoluVector' : enclosures for the zeros of 'f'.                }
{             'InfoVector' : corresponding information on the uniqueness of  }
{                            the zero in each of these enclosures.           }
{    In/Out : 'SoluNo'     : represents the number of the zero computed last }
{                            (in) and the total number of enclosures         }
{                            computed at the end of the recursion (out).     }
{ Description:                                                               }
{    The procedure 'XINewton' is recursively called whenever the extended    }
{    interval Newton step results in a splitting of the actual interval 'y'  }
{    in two or more intervals, and the tolerance condition is not fulfilled  }
{    yet. Otherwise, the enclosures for the zeros of 'f' are stored in the   }
{    interval matrix 'SoluVector' row by row, the corresponding information  }
{    on the uniqueness of the zero in each of these enclosures is stored in  }
{    the Boolean vector 'InfoVector'. The number of enclosures computed is   }
{    returned in the variable 'SoluNo'.                                      }
{----------------------------------------------------------------------------}
procedure XINewton (function f (x: GTvector) : GTvector[1..ub(x,1),
                                                        0..ub(x,2)];
                                          y : ivector;
                     var            Epsilon : real;
                                    yUnique : boolean;
                     var         SoluVector : imatrix;
                     var         InfoVector : bvector;
                     var             SoluNo : integer);
var
  fy        : ivector[1..ub(y)];
  Jfy       : imatrix[1..ub(y),1..ub(y)];
  yp        : imatrix[1..ub(y)+1,1..ub(y)];
  i, p      : integer;
begin
  fJEvalJ(f, y, fy, Jfy);                   { Compute f(y) and Jf(y)            }
  if 0 in fy then                           { Start if 0 in f(y), else do nothing }
  begin                                     {-----------------------------------}
    NewtonStep(f,y,Jfy,yp,p);               { Extended interval Newton step with }
                                            { results yp[i]                      }
    if (p = 1) then                         {-----------------------------------}
      yUnique := yUnique or (yp[1] in y)       { Inner inclusion ===> uniqueness }
    else                                       {--------------------------------}
```

```
      yUnique := false;

    for i:=1 to p do
    begin
      if (MaxRelDiam(yp[i]) < Epsilon) then
        begin { No more Newton steps }
          fEvalJ(f, yp[i], fy);                    { Compute f(yp[i])          }
          if (0 in fy) then                        { Store enclosure and info  }
          begin                                    {---------------------------}
            SoluNo := SoluNo + 1;
            if (SoluNo <= ub(SoluVector)) then
            begin
              SoluVector[SoluNo] := yp[i];       { Store enclosure of the zero }
              InfoVector[SoluNo] := yUnique;     { Store uniqueness info       }
            end;                                 {-----------------------------}
          end;
        end
      else { Recursive call of 'XINewton' for interval 'yp[i]' }
        XINewton(f,yp[i],Epsilon,yUnique,SoluVector,InfoVector,SoluNo);
    end;
  end;
end;

{----------------------------------------------------------------------------}
{ Purpose: Execution of a verification step including the use of an epsilon  }
{    inflation.                                                               }
{ Parameters:                                                                 }
{    In     : 'f'       : function of 'DerivType'.                            }
{    Out    : 'yUnique' : returns 'true' if the verification is successful.   }
{    In/Out : 'y'       : interval enclosure to be verified.                  }
{ Description: This procedure checks the uniqueness of the zero enclosed in   }
{    the variable 'y' by an additional verification step including the use    }
{    of an epsilon inflation of the iterates.                                 }
{----------------------------------------------------------------------------}
procedure VerificationStep (function f (x: GTvector) : GTvector[1..ub(x,1),
                                                                0..ub(x,2)];
                            var          y : ivector;
                            var    yUnique : boolean);
const
  kmax = 10;
var
  yIn, fY, yOld : ivector[1..ub(y)];
  JfY           : imatrix[1..ub(y),1..ub(y)];
  yp            : imatrix[1..ub(y)+1,1..ub(y)];
  k, p          : integer;
  eps           : real;
begin
  yIn := y;  k := 0;  eps:= 0.25;                           { Initializations }
  yUnique := false;                                         {-----------------}
  while (not yUnique) and (k < kmax) do { Do kmax loops to achieve inclusion }
  begin                                 {------------------------------------}
    yOld := blow(y,eps);                { Epsilon inflation of 'y'           }
                                        {------------------------------------}
    k := k+1;  fJEvalJ(f, yOld, fY, JfY);     { Perform interval Newton step }
    NewtonStep(f, yOld, JfY, yp, p);          {------------------------------}

    if (p <> 1) then                                { No verification possible }
      k := kmax                                     {--------------------------}
    else if (yp[1] = yOld) then                     { Increase 'eps'           }
      eps := eps * 8                                {--------------------------}
    else
      begin
        y:= yp[1];  yUnique:= y in yOld;   { Inner inclusion ===> uniqueness }
      end;                                 {---------------------------------}
```

```
    end;
    if not yUnique then y:= yIn;
  end;

{----------------------------------------------------------------------------}
{ Purpose: Computation of enclosures for all zeros of a continuously         }
{    differentiable multi-dimensional, vector-valued function.               }
{ Parameters:                                                                }
{    In     : 'f'            : objective function, must be declared for the }
{                              'GTvector' to enable the internal use of      }
{                              the differentiation arithmetic 'grad_ari'.    }
{             'Start',       : starting interval.                            }
{             'Epsilon'.     : desired relative accuracy                     }
{                              (interval diameter) of the result intervals. }
{    Out    : 'SoluVector'   : stores and returns the enclosures for the     }
{                              zeros of 'f'.                                 }
{             'InfoVector'   : corresponding information on the uniqueness  }
{                              of the zeros in each of these enclosures.     }
{             'NumberOfSolus' : number of enclosures computed.               }
{             'Err'          : error code.                                   }
{ Description:                                                               }
{    The enclosures for the zeros of 'f' are computed by calling procedure  }
{    'XINewton'. Then an additional verification step is applied to those   }
{    enclosures which have not been verified.                                }
{    If an error occurs, the value of 'Err' is different from 0.             }
{----------------------------------------------------------------------------}
global procedure AllNLSS (function f (x: GTvector) : GTvector[1..ub(x,1),
                                                              0..ub(x,2)];
                                                    Start : ivector;
                                                  Epsilon : real;
                                        var    SoluVector : imatrix;
                                        var    InfoVector : bvector;
                                        var NumberOfSolus : integer;
                                        var           Err : integer);

var
  i          : integer;
  MinEpsilon : real;
  Start1n    : ivector[1..ub(Start)-lb(Start)+1];
begin
  NumberOfSolus:= 0;
  if (lb(SoluVector) <> 1) then       { Check index bounds of result vectors }
    Err:= lbSoluVecNot1               {--------------------------------------}
  else if (lb(InfoVector) <> 1) then
    Err:= lbInfoVecNot1
  else if (ub(InfoVector) <> ub(SoluVector)) then
    Err:= VecsDiffer
  else
    begin                            { Start extended interval Newton method }
      Err:= NoError;                 {---------------------------------------}
      MinEpsilon:= succ(1.0) - 1.0;  { Relative machine accuracy (1 ulp)     }
      Start1n:= Start;               { Resize to standard bounds 1..n        }
                                     {---------------------------------------}
      if (Epsilon < MinEpsilon) then Epsilon := MinEpsilon;  { Set 'Epsilon' }
                                                             { to 1 ulp acc. }
      XINewton(f, Start1n, Epsilon, false,                   {---------------}
               SoluVector, InfoVector, NumberOfSolus);

                                           { Check if there are more zeros }
      if ub(SoluVector) < NumberOfSolus then { than storage space          }
      begin                                {-------------------------------}
        Err:= VecTooSmall;  NumberOfSolus:= ub(SoluVector);
      end;
```

```
      for i:=1 to NumberOfSolus do      { Verification step for the enclosures }
        if InfoVector[i] <> true then {---------------------------------------}
          VerificationStep(f,SoluVector[i],InfoVector[i]);
    end;
end;

{----------------------------------------------------------------------------}
{ Module initialization part                                                 }
{----------------------------------------------------------------------------}
begin
  { Nothing to initialize }
end.
```

13.3.2 Example

Our sample program uses *nlss* to compute all zeros of the function $f : I\!R^n \to I\!R^n$ with

$$f_i(x) = 0.6 \cdot x_i - 2 + 0.49 \cdot x_i \cdot \sum_{j=1}^{n} x_j^2, \quad i = 1, \ldots, n$$

in a specified starting interval vector. The function is taken from [29], where it is chosen because it is easily programmable for arbitrary dimension. Its zero $x = (x_i)$ is given by

$$x_i := \frac{4\sqrt{5}}{7\sqrt{n}} \cdot \sinh\left(\frac{1}{3} \cdot \operatorname{arsinh}\left(\frac{7\sqrt{5n}}{2}\right)\right), \quad i = 1, \ldots, n. \tag{13.8}$$

```
{----------------------------------------------------------------------------}
{ This program uses module 'nlss' to compute the solution of a system of     }
{ nonlinear equations given by Hansen. A starting interval vector and a      }
{ tolerance must be entered.                                                 }
{----------------------------------------------------------------------------}
program nlss_ex;

use
  i_ari,      { Interval arithmetic                    }
  mvi_ari,    { Interval matrix/vector arithmetic      }
  grad_ari,   { Gradient differentiation arithmetic    }
  b_util,     { Boolean utilities                      }
  nlss;       { Nonlinear System Solver                }

const
  n    = 20;  { Maximum number of solutions to be computed }
  name = 'Hansen''s Function';

var
  fDim : integer;

{----------------------------------------------------------------------------}
{ Function to prompt for the desired value of 'fDim'. It is called only once }
{ within the declaration part of the first dynamic variable. This enables    }
{ full dynamics within the main program.                                     }
{----------------------------------------------------------------------------}
function set_fDim : integer;
begin
  write('Desired dimension     : ');  read(fDim);
```

```
  set_fDim := fDim;
end;

var
  SearchInterval    : ivector[1..set_fDim];       { Declaration using a call }
  Tolerance         : real;                       { of function 'set_fDim'.  }
  Solu              : imatrix[1..n,1..fDim];
  Unique            : bvector[1..n];
  NumberOfSolus, i  : integer;
  ErrCode           : integer;

{----------------------------------------------------------------------------}
{ Definition of Hansen's function in 'fDim' variables.                       }
{----------------------------------------------------------------------------}
function f (x: GTvector) : GTvector[1..fDim,0..fDim];
var
  sqrsum : GradType[0..fDim];
  i      : integer;
begin
  sqrsum:= 0;
  for i:= 1 to fDim do
    sqrsum:= sqrsum + sqr(x[i]);
  for i:= 1 to fDim do
    f[i] := 6*x[i]/10 - 2 + 49*x[i]*sqrsum/100;
end;

{ Main program }
{--------------}
begin
  writeln;
  writeln('Computing all solutions for ', Name, ' in ', fDim, ' variables');
  writeln;
  write('Search interval      : ');  read(SearchInterval);
  write('Tolerance (relative) : ');  read(Tolerance);
  writeln;
  AllNLSS(f, SearchInterval, Tolerance, Solu, Unique, NumberOfSolus, ErrCode);
  for i:=1 to NumberOfSolus do
  begin
    write(Solu[i]);
    if unique[i] then
      writeln('encloses a locally unique zero!')
    else
      writeln('may contain a zero (not verified unique)!')
  end;
  if ErrCode <> 0 then writeln(AllNLSSErrMsg(ErrCode));
  writeln;  writeln(NumberOfSolus:1, ' interval enclosure(s)');  writeln;
  if (NumberOfSolus = 1) and (unique[1]) then
    writeln('We have validated that there is a globally unique zero!');
end.
```

If we execute this program for $n = 2$ and $n = 6$, we get the following runtime outputs:

```
Desired dimension    : 2

Computing all solutions for Hansen's Function in 2 variables

Search interval      : [-1,1] [-1,1]
Tolerance (relative) : 1e-10

0 interval enclosure(s)
```

```
Desired dimension      : 6

Computing all solutions for Hansen's Function in 6 variables

Search interval        : [-1,1] [-1,1] [-1,1] [-1,1] [-1,1] [-1,1]
Tolerance (relative) : 1e-10

[       8.02350910330E-001,       8.02350910335E-001 ]
[       8.02350910332E-001,       8.02350910334E-001 ]
[      8.023509103324E-001,      8.023509103329E-001 ]
[      8.023509103325E-001,      8.023509103328E-001 ]
[     8.0235091033261E-001,     8.0235091033268E-001 ]
[     8.0235091033264E-001,     8.0235091033265E-001 ]
encloses a locally unique zero!

1 interval enclosure(s)

We have validated that there is a globally unique zero!
```

For $n = 2$, we know from (13.8) that the solution x satisfies $x_i > 1$ for $i = 1,2$. Thus, our routine delivers no solution in the search interval and therefore confirms the fact that the solution given in [29] does not exist. For $n = 6$, our algorithm guarantees that there is only one zero of our sample function within the specified starting box.

13.3.3 Restrictions, Hints, and Improvements

The function f can only contain operators and elementary functions supported by module *grad_ari*.

The procedure *AllNLSS* stores all enclosures in the rows of the interval matrix *SoluVector*, which must be of sufficient length. If the first run of *AllNLSS* is not able to compute all zeros because of insufficient length, then the routine must be called again.

The method is not very fast if a very small value of ε (*Epsilon*) is used and the interval Newton step does not improve the actual iterates because of rounding and overestimation effects of the machine interval arithmetic. In this case, the method is equivalent with a bisection method. The method can be improved by incorporating a local iteration procedure, i.e. a non-interval Newton method which tries to find an approximation of a zero starting from the midpoint of the current interval vector iterate. The computed approximation can be used as the point of expansion in the interval Gauss-Seidel step (see [28] or [72] for details on this improvement).

In *XINewton*, the evaluation of the function with differentiation arithmetic can cause a runtime error if the interval argument of an elementary function does not lie in the domain specified for this interval function (see [65]) or if a division by an interval containing zero occurs. This is also due to the known overestimation effects of interval arithmetic (see Section 3.1). To get rid of these errors, the user may try to split the starting interval vector in several parts and call *AllNLSS* for each of these parts.

The rules for getting true enclosures in the presence of conversion errors (see Section 3.7) also apply here. That is why we expressed $0.49x_i$ as `49*x[i]/100` in our example in Section 13.3.2.

13.4 Exercises

Exercise 13.1 Use the procedure *AllNLSS* to compute the zeros of the function $f : I\!R^2 \to I\!R^2$ with

$$\begin{aligned} f_1(x) &= x_1^2 - 20x_1 + x_2^2 - 2x_2 + 100, \\ f_2(x) &= x_1^2 - 22x_1 + x_2^2 - 2x_2 + 121. \end{aligned}$$

These zeros correspond to the intersection points of the circle with midpoint $(10,1)^{\mathrm{T}}$ and radius 1 and the circle with midpoint $(11,1)^{\mathrm{T}}$ and radius 1. They lie within the box $([10,12],[0,3])^{\mathrm{T}}$ which can be used as starting box.

Exercise 13.2 Use the procedure *AllNLSS* to verify that $y \approx (1.6641, 1.6641)^{\mathrm{T}}$ is a zero of the function $f : I\!R^2 \to I\!R^2$ with

$$\begin{aligned} f_1(x) &= \left(\frac{x_1}{2}\right)^2 + \left(\frac{x_2}{3}\right)^2 - 1, \\ f_2(x) &= \left(\frac{x_1}{3}\right)^2 + \left(\frac{x_2}{2}\right)^2 - 1, \end{aligned}$$

by simply taking a small interval vector $[y]$ with $y \in [y]$ and $d_{\mathrm{rel},\infty}([y]) \approx 10^{-4}$ as input data for the procedure.

13.5 References and Further Reading

The method we discussed in this chapter is an *a priori* method because the iteration starts with a (possibly large) interval vector enclosing all the solutions which have to be found. Here, the iterates of the method are subintervals of the previous iterates. There are also methods for finding (and bounding) a single solution of a system of nonlinear equations called *a posteriori* methods. These methods start with an approximation of a zero and apply a test procedure for a neighborhood interval of the approximation to verify a zero within that interval. Our method presented in this chapter can also be applied to verify such an approximation, if we start the process with a small interval containing the approximation.

A large number of authors (Alefeld, Hansen, Moore, Neumaier, Kearfott, Rump, and many others) have considered the problem of computing enclosures for the solutions of a system of nonlinear equations. For further references in the field of *a priori* methods and for improvements for our method, see [3], [8], [28], [62], [63], [64], [71], or [72]. For *a posteriori* methods, see [11], [37], [38], [50], [53], [55], [60], [64], [77], [78], or [79]. In the field of preconditioning of the interval Gauss-Seidel step, interesting work and improvements can be found in [40] or [41].

Chapter 14

Global Optimization

In Chapter 7, we considered the problem of finding the global minimizers of one-dimensional nonlinear functions. Now, we consider its generalization, the problem of finding the global minimizers of multi-dimensional nonlinear functions. Our algorithm is based on the method of Hansen [26]. The algorithm computes enclosures for all global minimizers and for the global minimum value of a twice continuously differentiable function in the interior of a given interval vector.

Classical numerical global optimization methods for the multi-dimensional case start from some approximate trial points and iterate. Thus, classical optimization methods sample the objective function at only a finite number of points. There is no way to guarantee that the function does not have some unexpectedly small values between these trial points.

Hansen's algorithm uses interval arithmetic to evaluate the objective function and its first- and second-order partial derivatives over a continuum of points, including those points that are not finitely representable on a computer. Interval analysis supplies the prerequisite for solving the global optimization problem with *automatic result verification*, i.e. with the guarantee that the global minimum points and the global minimum values have been found.

14.1 Theoretical Background

Let $f : I\!R^n \to I\!R$ be a twice continuously differentiable function, and let $[x] \in II\!R^n$. We address the problem of finding all points x^* in the interior of $[x]$ such that

$$f(x^*) = \min_{x \in [x]} f(x). \tag{14.1}$$

We are interested in both the global minimizers x^* and the minimum value $f^* = f(x^*)$.

We use the method of Hansen [26, 28] with the modifications of Ratz [73]. Our algorithm does not find non-stationary minima on the boundary of $[x]$. A sophisticated method for handling this case can be found in [73]. Starting from the initial interval vector (box) $[x]$, our algorithm subdivides $[x]$ and stores the subboxes $[y] \subset [x]$ in a list L. Subintervals which are guaranteed not to contain a global minimizer of f are discarded from that list, while the remaining subintervals get subdivided again until the desired accuracy (relative diameter) of the intervals in the list is achieved. The power and speed of self-validating methods for global optimization comes not so much from the ability to find the answer as from the ability to discard from consideration regions where the answer is not. The tests we use to discard pending subboxes are multi-dimensional analogues of those we used for one-dimensional global optimization in Chapter 7. We use the four tests

- midpoint test,
- monotonicity test,
- concavity test, and
- interval Newton Gauss-Seidel step.

In the following sections, we consider each of these methods in detail. We use the notation $\underline{f_y}$ as abbreviation for the lower interval bound of the interval function evaluation $[f_y] := f_{[]}([y])$ for $[y] \in I\!R^n$.

14.1.1 Midpoint Test

If we are able to determine an upper bound $\widetilde{f}$ for the global minimum value f^*, then we can delete all subintervals $[y]$ for which

$$\underline{f_y} > \widetilde{f} \geq f^*. \tag{14.2}$$

The midpoint test first determines or improves such an upper bound for f^*. Initially, let $\widetilde{f} = +\infty$. We choose a box $[y]$ from the list L which satisfies $\underline{f_y} \leq \underline{f_z}$ for all intervals $[z]$ in the list L. That is, $[y]$ has the smallest lower bound for the range of f. Hence, it is a likely candidate to contain a minimizer. Let $c = m([y])$ (or any other point in $[y]$), and compute $\widetilde{f} = \min\{f(c), \widetilde{f}\}$. Such an upper bound can also be computed on a computer when rounding errors occur: we compute $f_{\Diamond}(c)$ and use the upper bound of the resulting interval as the possibly new value $\widetilde{f}$.

Now, with a possibly improved (decreased) value of $\widetilde{f}$, we can discard all intervals $[z]$ from the list L for which $\widetilde{f} < \underline{f_z}$. The midpoint test is relatively inexpensive, and it often allows us to discard from consideration large portions of the original interval $[x]$. Figure 7.1 in Chapter 7 illustrates this procedure, which deletes the intervals $[y]_2$, $[y]_3$, $[y]_8$, and $[y]_9$ in this special case. The midpoint test remains valid if an arbitrary $c' \in [y]$ is used instead of $c = m([y])$. Our algorithm could be extended by using a local approximate search to find a $c' \in [y]$ that is likely to give a smaller upper bound for f than c gives.

The value $\widetilde{f}$ is also used for function value checks when entering newly subdivided intervals $[w]$ in our list L. If we know that $[w]$ satisfies $\underline{f_w} > \widetilde{f}$, then $[w]$ cannot contain a global minimizer. Thus, we must only enter intervals $[w]$ in the list L if $\underline{f_w} \leq \widetilde{f}$ holds.

14.1.2 Monotonicity Test

The monotonicity test determines whether the function f is *strictly monotone* in an entire subbox $[y] \subset [x]$, in which case $[y]$ cannot contain a global minimizer (stationary point). Therefore, if $[g] := \nabla f([y])$ satisfies

$$0 \notin [g]_i \quad \text{for some } i = 1, \ldots, n, \tag{14.3}$$

then f is strictly monotone over the subbox $[y]$ with respect to the ith coordinate. Thus, $[y]$ can be deleted. Figure 7.2 in Chapter 7 demonstrates the monotonicity test in the one-dimensional case.

We emphasize that if f is monotone *in a single coordinate direction*, that subbox may be discarded. Therefore, a small number (relative to n) of interval-valued evaluations of partial derivatives are often enough to discard large portions of the original box $[x]$. We use the differentiation arithmetic of Chapter 12 to compute ∇f.

14.1.3 Concavity Test

The concavity test (non-convexity test) detects whether the function f is *not convex* in a subbox $[y] \subset [x]$, in which case $[y]$ cannot contain a global minimizer. The Hessian matrix of a convex function f must be positive semidefinite. We know that a necessary condition for this is that all diagonal elements of the Hessian are non-negative. If $[H] := \nabla^2 f([y])$ satisfies

$$\overline{H}_{ii} < 0 \quad \text{for some } i = 1, \ldots, n, \tag{14.4}$$

then $H_{ii} < 0$ for $H = \nabla^2 f(y)$ and all $y \in [y]$. Thus, f cannot be convex over $[y]$, the subbox $[y]$ cannot contain a stationary minimum, and $[y]$ can be deleted. Figure 7.3 in Chapter 7 demonstrates the concavity test in the one-dimensional case. The Hessian for the concavity test is computed by automatic differentiation described in Chapter 12.

14.1.4 Interval Newton Step

In our global optimization method, we apply one step of the extended interval Newton Gauss-Seidel method described in Chapter 13 to the nonlinear system

$$\nabla f(y) = 0, \quad y \in [y]. \tag{14.5}$$

The subbox $[y]$ is a candidate for containing a minimizer x^*, which we have assumed must satisfy $\nabla f(x^*) = 0$. When we apply the algorithm of Chapter 13, three things may happen. First, we may validate that $[y]$ contains no stationary point, in which case we may discard $[y]$. Second, the Newton step may contract $[y]$ significantly. Subsequently, f can be evaluated on the narrower box $[y]$ with less overestimation, so the midpoint, monotonicity, and concavity tests are likely to be more effective. Third, we may get splittings of the box $[y]$ due to gaps produced by the extended interval divisions applied in the Newton step. We only apply one Newton step because this test is relatively expensive, and the other test (with bisection) may subsequently discard even subboxes containing local minimizers. In addition, a stationary point is not necessarily even a local minimizer.

To apply the extended interval Newton Gauss-Seidel method, we first compute $[A] \in I\mathbb{R}^{n\times n}$ and $b \in \mathbb{R}^n$ by

$$[A] := R \cdot \nabla^2 f([y]) \quad \text{and} \quad b := R \cdot \nabla f(m([y])), \tag{14.6}$$

where $R \approx (m(\nabla^2 f([y])))^{-1}$. Then, we compute $N'_{\mathrm{GS}}([y])$ according to

$$\left.\begin{aligned} [z] &:= [y] \\ [z]_i &:= \Big(c_i - \Big(b_i + \sum_{\substack{j=1 \\ j\neq i}}^{n} [A]_{ij} \cdot ([z]_j - c_j) \Big) \Big/ [A]_{ii} \Big) \cap [z]_i \\ & \qquad i = 1, \ldots, n \\ N'_{\mathrm{GS}}([y]) &:= [z] \end{aligned}\right\} . \tag{14.7}$$

If the extended interval division results in a splitting of a component $[z]_i$, we will store one part of the actual box $[z]$, and we use the other part to continue the computing in (14.7). Thus, our interval Newton step results in at most $n+1$ subboxes to place on the pending list L.

14.1.5 Verification

When all subboxes in the list L have been reduced to have widths less than a specified tolerance, we have a list of small subboxes we could not discard by any of our tests. We attempt to verify the existence and uniqueness of a local minimizer within each subbox by checking two conditions. The first condition is

$$N'_{\mathrm{GS}}([y]) \overset{\circ}{\subset} [y], \tag{14.8}$$

which guarantees the existence and uniqueness of a stationary point of f, i.e. a zero of ∇f in $[y]$ (cf. Theorem 13.1). The second condition (14.9) guarantees that $\nabla^2 f([y])$ is positive definite, i.e. all real symmetric matrices $A \in \nabla^2 f([y])$ are positive definite. If the real matrix $B := I - \|A\|^{-1} \cdot A$ has only eigenvalues with absolute values less than 1, i.e. the spectral radius $\rho(B) < 1$, then the symmetric matrix A is positive definite. Thus, we use the following theorem:

Theorem 14.1 *Let $[H] \in I\mathbb{R}^{n\times n}$ and $[S]$ be defined by $[S] := I - \frac{1}{\kappa}[H]$ with $\|[H]\|_\infty \leq \kappa \in \mathbb{R}$. If $[S]$ satisfies*

$$[S] \cdot [z] \overset{\circ}{\subset} [z] \tag{14.9}$$

for an interval vector $[z] \in I\mathbb{R}^n$, then we have $\rho(B) < 1$ for all $B \in [S]$, and all symmetric matrices $A \in [H]$ are positive definite.

For a proof of this theorem, see [73].

To check the condition of Theorem 14.1, we first compute $[H] = \nabla^2 f([y])$, κ with $\|[H]\|_\infty \leq \kappa$, and $[S] := I - \frac{1}{\kappa}[H]$. Starting with an interval vector $[z]^{(0)}$ with every component $[z]_i^{(0)} = [-1, 1]$, we iterate

$$[z]^{(k+1)} := [S] \cdot [z]^{(k)}, \quad k = 0, 1, \ldots$$

until the condition $[z]^{(k+1)} \overset{\circ}{\subset} [z]^{(k)}$ is satisfied. We stop the iteration if a certain number of steps could not satisfy this condition.

We know of no way to verify the uniqueness of a global minimizer in general. The global minimizer can even be one or more continua of points. Hence, we settle for attempting to verify that intervals we compute as candidates for containing a global minimizer contain unique local minimizers. Failure to verify the uniqueness of a local minimizer in a subinterval is not grounds for discarding that subinterval from the list of candidates.

In fact, our method produces a final list containing enclosures for *locally unique candidates for global minimizers*. If we have in the final list exactly one subinterval $[y]$ in which we can validate a local minimizer, then we have validated a unique global minimizer in the starting interval $[x]$. If we have two or more subintervals validated to contain unique local minimizers, then the best we can say is that each contains a candidate for a global minimizer, which need not be unique (but in this case, the global minimizer could not be a continuum of points).

14.2 Algorithmic Description

The main algorithm AllGOp (Algorithm 14.7) consists of two parts. The first part is the subdivision method including the tests and the extended interval Newton steps described in Section 14.1. The second part is a verification step which tries to verify the local uniqueness of a minimizer within each of the remaining intervals of the pending list.

The algorithm for the monotonicity test (Algorithm 14.1) uses the interval gradient $[g] = \nabla f([y])$ and returns *true* if no further processing of the actual box $[y]$ is necessary because f is monotonic in $[y]$, or *false*, otherwise.

Algorithm 14.1: MonotonicityTest ($[g]$) {Function}

1. **for** $i := 1$ **to** n **do**

   if $0 \notin [g]_i$ **then return** MonotonicityTest := *true*;

2. **return** MonotonicityTest := *false*;

Algorithm 14.2 executes the concavity test using the interval Hessian $[H] = \nabla^2 f([y])$ returning *true* if no further processing of the actual box $[y]$ is necessary because f is not convex in $[y]$, or *false*, otherwise.

Algorithm 14.2: ConcavityTest $([H])$ {Function}

1. **for** $i := 1$ **to** n **do**
   if $\overline{H}_{ii} < 0$ **then return** ConcavityTest := *true*;
2. **return** ConcavityTest := *false*;

Next, we present the algorithm for the execution of the extended interval Newton Gauss-Seidel step (Algorithm 14.3), where (14.7) is applied to the box $[y] \in I\mathbb{R}^n$ resulting in at most $n+1$ boxes $[V]_i$. After some initializations, we use Algorithm 10.1 (**MatInv**) of Chapter 10 to compute an approximate inverse of the midpoint matrix of the interval Hessian matrix $[H] = \nabla^2 f([y])$. In Step 4, we compute $[A]$, $[b]$, and $[y_c]$ followed by some initializations for the loop. Interval arithmetic must be used to compute $\nabla f(c)$ for bounding all rounding errors.

We perform the single component steps of the Gauss-Seidel step for all i with $0 \notin [A]_{ii}$ (Step 6) and then for the remaining indices with $0 \in [A]_{ii}$ (Step 7). Using this strategy, it is possible that the intervals $[y]_i$ become smaller by the intersections with the old values $[y]_i$ in Step 6(b), before the first splitting is produced in Step 7(b). If a gap is produced in this case, we store one part of the actual box $[y]$ in the pth row of the interval matrix $[V]$, and we continue the iteration for the other part of $[y]$. The flag *NoSolution* signals that an empty intersection has occurred resulting in $p = 0$.

Algorithm 14.3: NewtonStep $(f, [y], [H], [V], p)$ {Procedure}

1. $[y_{\text{in}}] := [y]$; $c := m([y])$; {Initializations}
2. MatInv $(m([H]), R, \mathit{InvErr})$; {Invert the midpoint matrix}
3. **if** $(\mathit{InvErr} \neq$ "No Error") **then** $R := I$;
4. $[A] := R \cdot [H]$; $[b] := R \cdot \nabla f_\diamond(c)$; $[y_c] := [y] - c$;
5. $p := 0$; *NoSolution* := *false*; {Initializations for loop}
6. **for** $i := 1$ **to** n **do** {Interval Gauss-Seidel step for $0 \notin [A]_{ii}$}
   (a) **if** $(0 \in [A]_{ii})$ **then next**$_i$;
   (b) $[y]_i := \left(c_i - \left([b]_i + \sum\limits_{\substack{j=1 \\ j \neq i}}^{n} [A]_{ij} \cdot [y_c]_j \right) / [A]_{ii} \right) \cap [y]_i$;
   (c) **if** $[y]_i = \emptyset$ **then** *NoSolution* := *true*; **exit**$_{i\text{-loop}}$;
   (d) $[y_c]_i := [y]_i - c_i$;
7. **for** $i := 1$ **to** n **do** {Interval Gauss-Seidel step for $0 \in [A]_{ii}$}
   (a) **if** $(0 \notin [A]_{ii})$ **then next**$_i$;
   (b) $[z] := \left(c_i - \left([b]_i + \sum\limits_{\substack{j=1 \\ j \neq i}}^{n} [A]_{ij} \cdot [y_c]_j \right) / [A]_{ii} \right) \cap [y]_i$; $\{[z] = [z]_1 \cup [z]_2\}$

```
        if ([z] = ∅) then NoSolution := true;  exit_{i-loop};
      (d) [y]_i := [z]_1;  [y_c]_i := [y]_i − c_i;
      (e) if ([z]_2 ≠ ∅) then
              p := p + 1;  [V]_p := [y];  [V]_{pi} := [z]_2;  {Store part of [y] in [V]_p}
 8. if NoSolution then return p := 0;
 9. p := p + 1;  [V]_p := [y];
10. return [V], p;
```

Algorithm 14.4 manages the bisection of subboxes and their insertion in the pending list L. The subdivided boxes $[y]$ are stored together with the lower bound of the interval function evaluation $[f_y] := f_{[]}([y])$ as pairs $([y], \underline{f_y})$. Pairs are stored in the list sorted in nondecreasing order of lower bounds $\underline{f_y}$. Therefore, a newly computed pair is stored in the list L according to the ordering rule (cf. [73]):

$$\left.\begin{array}{ll} \bullet \text{ either} & \underline{f_w} \le \underline{f_y} < \underline{f_z} \text{ holds,} \\ \bullet \text{ or} & \underline{f_y} < \underline{f_z} \text{ holds, and } ([y], \underline{f_y}) \text{ is the first element of the list,} \\ \bullet \text{ or} & \underline{f_w} \le \underline{f_y} \text{ holds, and } ([y], \underline{f_y}) \text{ is the last element of the list,} \\ \bullet \text{ or} & ([y], \underline{f_y}) \text{ is the only element of the list,} \end{array}\right\} \quad (14.10)$$

where $([w], \underline{f_w})$ is the predecessor and $([z], \underline{f_z})$ is the successor of $([y], \underline{f_y})$ in L.

That is, the second components of the list elements may not decrease, and a new pair is entered behind all other pairs with the same second component. Thus, the first element of the list has the smallest second component. We can directly use the corresponding box to compute $f(c)$ for the improvement of $\tilde{f}$ in performing the midpoint test. We can also save some work when deleting elements in the midpoint test, by deleting the rest of the list when we have found the first element to be deleted. At each stage of our algorithm, $\tilde{f}$ is the best known upper bound for the global minimum value f^*. The second component $\underline{f_y}$ from the first element of the list is the best known lower bound for f^*. Therefore, we might choose to terminate the algorithm on the basis of the interval $[\underline{f_y}, \tilde{f}]$.

Given the list L and a list element E, we use the following notation in our algorithms:

Notation	Meaning
$L := \{\}$	Initialization by an empty list
$L := E$	Initialization by a single element
$L := L + E$	Enter element E in L according to condition (14.10)
$L := L - E$	Discard element E from L
$E :=$ **Head** (L)	Set E to the first element of L
MultiDelete $(L, \tilde{f})$	Discard all elements from L satisfying condition (14.2)
Length (L)	Delivers the number of elements in L

In **GlobalOptimize**, we first compute an upper bound for the global minimum value, and we do some initializations. Step 3 is the main iteration. Here, we first do a bisection of the actual interval $[y]$. Then in Step 3(c), we apply the monotonicity test, a function value check using the centered form, the concavity test, and the interval Newton step to the bisected boxes $[u]_1$ and $[u]_2$. The interval Newton step may result in at most p intervals. We have to handle them all in Step 3(c)viii, where we again apply a monotonicity test and a function value check with centered forms. If the actual box $[V]_j$ has not been discarded, then it is still a candidate for a minimizer, and we store it in L.

In Step 3(e), we remove the first element from the list L, i.e. the element of L with the smallest lower bound of the interval function evaluation, and we perform the midpoint test. Then, we check the tolerance criterion for the new actual interval. If the desired accuracy is achieved, we store this interval in the result list L_{res}. Otherwise, we go to the bisection step.

When the iteration stops because the pending list L is empty, we compute a final enclosure $[f^*]$ for the global minimum value in Step 4, and we return L_{res} and $[f^*]$. Procedure **GlobalOptimize** terminates because the elements of L move to L_{res} if $(d_{\text{rel},\infty}([y]) < \varepsilon)$ or if $(d_{\text{rel}}([f^*]) < \varepsilon)$. The bisection step (Step 3(b)) guarantees that the first condition is fulfilled at some stage of the iteration.

Algorithm 14.4: GlobalOptimize $(f, [x], \varepsilon, L_{\text{res}}, [f^*])$ (Procedure)

1. $c := m([x])$; $[f_c] := f_\diamond(c)$; $\tilde{f} := \overline{f_c}$; {Compute upper bound for f^*}
2. $[y] := [x]$; $L := \{\}$; $L_{\text{res}} := \{\}$;
3. **repeat** {Start iteration}
   (a) Find k with $d([y]_k) = d_\infty([y])$; {Component with maximum diameter}
   (b) $[u]_1 := [y]$; $[u]_2 := [y]$; $[u]_{1k} := [\underline{y}_k, c_k]$; $[u]_{2k} := [c_k, \overline{y}_k]$; {Bisect $[y]$}
   (c) **for** $i := 1$ **to** 2 **do**
      i. $[g] := \nabla f([u]_i)$;
      ii. **if** MonotonicityTest $([g])$ **then next**$_i$;
      iii. $[f_u] := (f(c) + [g] \cdot ([u]_i - c)) \cap f([u]_i)$; {Centered form}
      iv. **if** $\tilde{f} < \underline{f_u}$ **then next**$_i$;
      v. $[H] := \nabla^2 f([u]_i)$;
      vi. **if** ConcavityTest $([H])$ **then next**$_i$;
      vii. NewtonStep $(f, [u]_i, [H], [V], p)$; {Interval Newton Gauss-Seidel step}
      viii. **for** $j := 1$ **to** p **do**
         A. $[g] := \nabla f([V]_j)$;
         B. **if** MonotonicityTest $([g])$ **then next**$_j$;
         C. $c_V := m([V]_j)$;
         D. $[f_V] := (f(c_V) + [g] \cdot ([V]_j - c_V)) \cap f([V]_j)$; {Centered form}
         E. **if** $\tilde{f} \geq \underline{f_V}$ **then** $L := L + ([V]_j, \underline{f_V})$; {Store $[V]_j$}

Bisect := *false*;

(e) **while** ($L \neq \{\}$) **and** (**not** *Bisect*) **do**

i. $([y], \underline{f_y})$:= Head (L); $L := L - ([y], \underline{f_y})$; $c := m([y])$;

ii. $\tilde{f} := \min\{\tilde{f}, f(c)\}$; MultiDelete ($L, \tilde{f}$); {Midpoint test}

iii. $[f^*] := [\underline{f_y}, \tilde{f}]$;

iv. **if** ($d_{\text{rel}}([f^*]) < \varepsilon$) **or** ($d_{\text{rel},\infty}([y]) < \varepsilon$) **then** $L_{\text{res}} := L_{\text{res}} + ([y], \underline{f_y})$;
else *Bisect* := *true*;

until (**not** *Bisect*);

4. $([y], \underline{f_y})$:= Head (L_{res}); $[f^*] := [\underline{f_y}, \tilde{f}]$;

5. **return** L_{res}, $[f^*]$;

Algorithm 14.5 applies the test given by Theorem 14.1 to check whether the interval Hessian matrix over the current enclosure $[y]$ is positive definite. We use $k_{\mathbf{max}} = 10$ as the maximum number of iterations, $\varepsilon = 0.25$ as the starting value for the epsilon inflation, and a factor of 8 to increase ε within the iterations. It turned out that these are good values for minimizing the effort if no verification is possible (see also [46]).

Algorithm 14.5: PosDef ([H]) {Function}

1. Compute $\kappa \geq \|[H]\|_\infty$; $[S] := \Diamond(I - \frac{1}{\kappa}[H])$;
2. **for** $k := 1$ **to** n **do** $[z]_k := [-1, 1]$;
3. $k_{\text{max}} := 10$; $k := 0$; $\varepsilon := 0.25$;
4. **repeat**

   $[u] := [u] \bowtie \varepsilon$; $[z] := [S] \cdot [u]$; $k := k + 1$; $\varepsilon := 8 \cdot \varepsilon$;

   until ($[z] \overset{\circ}{\subset} [u]$) **or** ($k = k_{\text{max}}$);
5. **return** PosDef := ($[z] \overset{\circ}{\subset} [u]$);

Algorithm 14.6 describes the verification step checking the local uniqueness of a minimizer enclosed in the interval $[y]$. The procedure tries to do a "zero check" for the gradient ∇f according to condition 3 of Theorem 13.1 by applying interval Newton steps including an epsilon inflation of the iterates $[y]$ (Step 4). We also check the condition of Theorem 14.1 by applying function **PosDef** in Step 5. We use the same values for k_{max} and ε as in **PosDef**.

Algorithm 14.6: VerificationStep (f, [y], *Unique*) {Procedure}

1. **if not** *Unique* **then** *Unique* := ($\underline{y} = \overline{y}$); {Point interval $\Rightarrow$ uniqueness}
2. **if** *Unique* **then return** $[y]$, *Unique*;

$k_{\max} := 10$; $k := 0$; $[y_{\text{in}}] := [y]$; $\varepsilon := 0.25$; {Initializations}
4. **while** (**not** *Unique*) **and** ($k < k_{\max}$) **do** {Do $k_{\max}$ loops to achieve inclusion}
   (a) $[y_{\text{old}}] := [y] \bowtie \varepsilon$; {Epsilon inflation of $[y]$}
   (b) $[H] := \nabla^2 f([y_{\text{old}}])$;
   (c) $k := k + 1$; $c := m([y_{\text{old}}])$;
   (d) NewtonStep $(f, [y_{\text{old}}], [H], [Y_p], p)$; {Interval Newton step}
   (e) **if** $p \neq 1$ **then exit**$_{\text{while-loop}}$; {No verification possible}
   (f) $[y] := [Y_p]_1$;
   (g) *Unique* $:= ([y] \overset{\circ}{\subset} [y_{\text{old}}])$; {Inner inclusion $\Rightarrow$ uniqueness}
   (h) **if** $[y] = [y_{\text{old}}]$ **then** $\varepsilon := \varepsilon \cdot 8$; {Increase ε}
5. **if** *Unique* **then** *Unique* := PosDef $([H])$;
6. **if not** *Unique* **then** $[y] := [y_{\text{in}}]$;
7. **return** $[y]$, *Unique*;

Algorithm 14.7 now combines these procedures to compute enclosures for all global minimizers x^* of the function f and for the global minimum value f^* within the input interval vector $[x]$ and tries to prove the local uniqueness of the minimizers within the computed enclosures. The desired accuracy (relative diameter) of the interval enclosures is specified by the input parameter ε. 1 ulp accuracy is chosen if the specified value of ε is too small (for example 0). The enclosures for the global minimizers of f are returned in the interval matrix $[Opt]$ row by row, the corresponding information on the local uniqueness of the optimizer is returned in the Boolean vector *Info*. The number of enclosures computed is returned in the integer variable N.

We use a function called **CheckParameters** as an abbreviation for the error checks for the parameters of **AllGOp** which are necessary in an implementation. If no error occurs, **AllGOp** delivers the N enclosures $[Opt]_i$, $i = 1, 2, \ldots, N$, satisfying

if $Info_i$ = *true*, then $[Opt]_i$ encloses a locally unique minimizer of f,
if $Info_i$ = *false*, then $[Opt]_i$ may enclose a local or global minimizer of f.

If $N = 0$, then it is guaranteed that there is *no* stationary global minimizer of f in the starting interval $[x]$.

Algorithm 14.7: AllGOp $(f, [x], \varepsilon, [Opt], Info, N, [f^*], Err)$ {Procedure}

1. Err := CheckParameters;
2. **if** $Err \neq$ "No Error" **then return** Err;
3. GlobalOptimize $(f, [x], \varepsilon, L_{\text{res}}, [f^*])$;
4. N := Length (L_{res});
5. **for** $i := 1$ **to** N **do**

$[Opt]_i :=$ Head (L_{res});

VerificationStep ($f, [Opt]_i, Info_i$);

6. **return** $[Opt], Info, N, [f^*], Err$;

Applicability of the Algorithm

We have assumed that f is twice continuously differentiable. However, we can apply our algorithm to functions that are only once continuously differentiable or to functions that are not differentiable if we leave out some parts of the algorithm. If we do not use the interval Newton step (Step 3(c)vii in Algorithm 14.4), and replace it by the sequence

$$[V]_1 := [u]_i; \quad p := 1;$$

then our method works for functions that are only once differentiable. Algorithm 14.6 cannot be applied in this case.

If we replace Step 3(c) of Algorithm 14.4 by the sequence

for $i := 1$ **to** 2 **do**
  $[f_u] := f([u]_i)$;
  if $\widetilde{f} \geq \underline{f_u}$ **then** $L := L + ([u]_i, \underline{f_u})$;

then we can apply our method, consisting of subdividing and midpoint test, to functions that are *not* differentiable.

The closer the upper bound $\widetilde{f}$ is to the global minimum value f^*, the more intervals we can delete in the midpoint test (Step 3(e)ii of Algorithm 14.4). Thus, the method can be improved by incorporating an approximate local search procedure, to try to decrease the value $\widetilde{f}$. See [28] or [72] for the description of such local search procedures.

For a multiple zero x^* of ∇f, the algorithm cannot verify the existence and the local uniqueness of x^* in the enclosing result box. Nevertheless, the zero of ∇f, which is possibly a global minimizer, is bounded to the desired accuracy specified by ε. In this case, the corresponding component of the *Info*-vector has the value *false*.

As a consequence of the bisecting of the boxes, it may happen that a minimizer lying exactly at the splitting boundary is enclosed in several intervals. Furthermore, because of the known overestimation effects of interval arithmetic, the algorithm may also find "near-global" minima when rounding prevents determination of the true minimum of several candidates. Sophisticated supplements to our method avoiding these effects can be found in [28] or [73].

14.3 Implementation and Examples

14.3.1 PASCAL–XSC Program Code

We begin by describing our implementation of the operations needed for handling lists and list elements. Then, we describe the implementation of Algorithm 14.7 and its subalgorithms.

14.3.1.1 Module lst_ari

The dynamic concept of PASCAL–XSC does not allow to use dynamic arrays as record components. Thus, we use an indirect list indexing for ordering the vectors which have to be stored in our pending list. The list only stores an index value k, whereas the pair $([y], f_y)$, represented by an interval vector with an additional 0-component, is stored in the kth row of an interval matrix provided by the user of module *lst_ari*.

The module supplies the type *PairPtr* representing a list of pairs. The local variable *FreeList* and the procedures *NewPP*, *Free*, and *FreeAll* generate and free list elements and prevent memory garbage. *MakePair*, *Int*, and *Fyi* are transfer and access functions for pairs.

The global function *EmptyList* represents an empty list. The function *Enter* enters a new list element *Pair* in the list *List* (by storing the pair vector in the matrix *VecMat*) according to condition (14.10). The procedure *MultiDelete* deletes all elements *P* in *List* for which *Fyi*(*P*) > *fmax*. This procedure assumes that the list elements are ordered according to condition (14.10). Function *Next* sets the list pointer *List* to the next list element. *Head* delivers the first pair of *List*, whereas *DelHead* deletes the first pair of *List*. Function *Length* delivers the number of elements in *List*.

```
{----------------------------------------------------------------------------}
{ Purpose: Definition of a list arithmetic used in connection with an        }
{    interval bisection method in global optimization for storing pairs of   }
{    an interval vector and a real value.                                    }
{ Method: Representing pairs of an interval vector and a real value by a     }
{    interval vector with additional 0-component storing the real value.     }
{    Overloading of functions and procedures for the data type 'PairPtr'.    }
{ Global types, functions, and procedures:                                   }
{    types      PairPtr, PairElmt: list of pairs                             }
{    functions  MakePair         : transfer function for pairs               }
{               Int, Fyi         : access functions for pairs                }
{               Next, Head       : access functions for lists                }
{               Length           : access function to length of list         }
{               EmptyList        : delivers an empty list                    }
{    procedures FreeAll          : free complete list                        }
{               MultiDelete      : deletes several elements in a list        }
{               DelHead          : deletes first element of a list           }
{----------------------------------------------------------------------------}
module lst_ari;

use
  i_ari,     { Interval arithmetic              }
  mvi_ari;   { Interval matrix/vector arithmetic }
```

```
{----------------------------------------------------------------------------}
{ Global type definitions                                                    }
{----------------------------------------------------------------------------}
global type                    { List of pairs of an interval vector 'y' and   }
  PairPtr  = ↑PairElmt;        { the real value 'inf(f(y))'.  The pair is re-  }
  PairElmt = record            { presented by an interval vector with index    }
               P : integer;    { bounds 0..n and the 0-component contains the  }
               N : PairPtr;    { real value. Within the list, each pair-vector }
             end;              { is represented by the index 'P' and indirect- }
                               { ly stored in 'VecMat[P]', where 'VecMat' is a }
                               { storage matrix provided by the routine using  }
                               { this module. 'N' means NEXT, the pointer to   }
                               { the next element of the list.                 }
                               {-----------------------------------------------}
{----------------------------------------------------------------------------}
{ Local variable storing list of free elements (automatic garbage recycling) }
{----------------------------------------------------------------------------}
var
  FreeList : PairPtr;
  IndexNo  : integer;

{----------------------------------------------------------------------------}
{ Procedures for generating and freeing of list elements (pairs)             }
{----------------------------------------------------------------------------}
procedure NewPP (var pp: PairPtr);     { 'NewPP' generates a new list element }
begin                                  { or gets one from 'FreeList'.         }
  if FreeList = nil then               {--------------------------------------}
    begin
      new(pp);  pp↑.N:= nil;  IndexNo:= IndexNo+1;  pp↑.P:= IndexNo;
    end
  else
    begin
      pp:= FreeList;  FreeList:= FreeList↑.N;  pp↑.N:= nil;
    end;
end;

procedure Free (var pp: PairPtr);            { 'Free' enters one element of a }
begin                                        { list in the 'FreeList'.        }
  if pp <> nil then                          {--------------------------------}
  begin
      pp↑.N:= FreeList;  FreeList := pp;  pp:= nil;
  end;
end;

global procedure FreeAll (var List: PairPtr);{ 'FreeAll' enters all elements }
var  H : PairPtr;                             { of 'List' in the 'FreeList'.  }
begin                                         {-------------------------------}
  if List <> nil then
  begin
    H:= List;
    while H↑.N <> nil do H:= H↑.N;
    H↑.N:= FreeList;  FreeList:= List;  List:= nil;
  end;
end;

{----------------------------------------------------------------------------}
{ Transfer and access functions for pairs                                    }
{----------------------------------------------------------------------------}
global function MakePair (var int: ivector; fyi: real) : ivector[0..ub(int)];
var i : integer;                                            { Generate pair }
begin                                                       {---------------}
  MakePair[0]:= fyi;
  for i:=1 to ub(int) do MakePair[i]:= int[i];
```

```
end;

global function Int (var Pair: ivector) : ivector[1..ub(Pair)];
var i : integer;                        { Get interval vector component of pair }
begin                                   {--------------------------------------}
  for i:=1 to ub(Pair) do Int[i]:= Pair[i];
end;

global function Fyi (var Pair: ivector) : real; { Get real component of pair }
begin                                           {----------------------------}
  Fyi:= Pair[0].inf;
end;

{----------------------------------------------------------------------------}
{ Functions, and procedures for lists of pairs                               }
{----------------------------------------------------------------------------}
{ Global function 'EmptyList' representing an empty list of pairs.           }
{----------------------------------------------------------------------------}
global function EmptyList : PairPtr;
begin
  EmptyList := nil;
end;

{----------------------------------------------------------------------------}
{ Function 'Enter' enters 'Pair' the list 'List' (by storing the vector in   }
{ the matrix 'VecMat') in such a way that after entering, one of the four    }
{ following condition holds:                                                 }
{ 1) Fyi(O) <= Fyi(Pair) < Fyi(Q),                                           }
{ 2)           Fyi(Pair) < Fyi(Q) and 'Pair' is the first element of 'List', }
{ 3) Fyi(O) <= Fyi(Pair)          and 'Pair' is the last  element of 'List', }
{ 4)                                  'Pair' is the only  element of 'List', }
{ where 'O' is the preceding and 'Q' is the succeeding element of 'Pair' in  }
{ the resulting list.  If the list (ie. the storage matrix 'VecMat') is full,}
{ the parameter 'Full' returns true.                                         }
{----------------------------------------------------------------------------}
global procedure Enter (var List: PairPtr; var VecMat: imatrix;
                        var Pair: ivector; var Full   : boolean);
var
  H, HN             : PairPtr;
  ready, alreadyIn : boolean;
begin
  Full:= false;
  if (List = nil) then                        { List is empty, so new list }
    begin                                     { only contains 'Pair'.      }
      NewPP(H);                               {----------------------------}
      if H↑.P > ub(VecMat) then
        Full:= true
      else
        VecMat[H↑.P]:= Pair;
      H↑.N:= nil;  List:= H;
    end
  else if (Fyi(VecMat[List↑.P]) > Fyi(Pair)) then { 'Pair' becomes new first }
    begin                                         { element of the list.     }
      NewPP(H);                                   {--------------------------}
      if H↑.P > ub(VecMat) then
        Full:= true
      else
        VecMat[H↑.P]:= Pair;
      H↑.N:= List;  List:= H;
    end
  else
    begin
      H:= List;  HN:= H↑.N;  ready:= false;
      alreadyIn:= (Int(VecMat[H↑.P]) = Int(Pair));
```

```
      while not (ready or alreadyIn) do          { Search for the right       }
      begin                                      { position to enter 'Pair'.  }
        if (HN = nil) then                       {----------------------------}
          ready:= true
        else if (Fyi(VecMat[HN↑.P]) > Fyi(Pair)) then
          ready:= true
        else
          begin
            H:= HN;  HN:= H↑.N;  alreadyIn:= (Int(VecMat[H↑.P]) = Int(Pair));
          end;
      end;

      if not alreadyIn then
      begin
        NewPP(H↑.N);  H:= H↑.N;                     { Enter 'Pair' between H }
        if H↑.P > ub(VecMat) then                   { and HN. Return List.   }
          Full:= true                               {------------------------}
        else
          VecMat[H↑.P]:= Pair;
        H↑.N:= HN;
      end;
    end;
end;

{----------------------------------------------------------------------------}
{ 'MultiDelete' deletes all elements 'P' in 'List' for which the condition   }
{ 'Fyi(P) > fmax' holds.  This procedure assumes that the 'fyi' components   }
{ of the list elements are sorted in increasing order (see function 'Enter').}
{----------------------------------------------------------------------------}
global procedure MultiDelete (var List: PairPtr; var MV: imatrix; fmax: real);
var
  DelPrev, Del : PairPtr;
  ready        : boolean;
begin
  if (List <> nil) then
  begin
    if (Fyi(MV[List↑.P]) > fmax) then             { All list elements fulfill }
      begin                                       { 'Fyi(P) > fmax'.          }
        Del:= List;  List:= nil;                  {---------------------------}
      end
    else
      begin
        DelPrev:= List;  Del:= DelPrev↑.N;  ready:= (Del=nil);

        while not ready do
        begin
          if (Del = nil) then
            ready := true
          else if (Fyi(MV[Del↑.P]) > fmax) then
            begin
              ready:= true;  DelPrev↑.N:= nil;
            end
          else
            begin
              DelPrev:= Del;  Del:= Del↑.N;
            end;
        end;
      end;
    FreeAll(Del);
  end;
end;

global function Next (List: PairPtr) : PairPtr;   { Sets list pointer to the }
```

```
begin                                                { next list element       }
  Next:= List↑.N;                                    {-------------------------}
end;

global function Head (List: PairPtr; var MV: imatrix): ivector[0..ub(MV,2)];
                                                { Delivers first pair of the }
begin                                           { list, i.e. the pair P with }
  Head:= MV[List↑.P];                           { the smallest value P[0].   }
end;                                            {----------------------------}

global procedure DelHead (var List: PairPtr);   { Deletes the first pair of  }
var                                             { the List.                  }
  Del : PairPtr;                                {----------------------------}
begin
  Del := List;  List:= List↑.N;  Free(Del);
end;

global function Length (List: PairPtr) : integer;  { 'Length' delivers the   }
var  i : integer;                                  { number of elements in   }
begin                                              { list 'List'.            }
  i:= 0;                                           {-------------------------}
  while List <> nil do
  begin
    i:= succ(i); List:= List↑.N;
  end;
  Length := i;
end;

{----------------------------------------------------------------------------}
{ Module initialization                                                      }
{----------------------------------------------------------------------------}
begin
  FreeList := nil;         { List of freed elements which can be used again  }
  IndexNo  := 0;           { Index of last row of storage matrix used        }
end.                       {-------------------------------------------------}
```

14.3.1.2 Module gop

The module *gop* supplies the global routines *AllGOp* (the implementation of Algorithm 14.7) and the corresponding function *AllGOpErrMsg* to get an error message for the error code returned by *AllGOp*. The functions *MaxDiamComp*, *MaxNorm*, *PosDef*, *MonotonicityTest*, and *ConcavityTest*, and the procedures *NewtonStep*, *GlobalOptimize*, and *VerificationStep* are defined locally. All derivative evaluations use the differentiation arithmetic *hess_ari* described in Chapter 12.

The procedure *AllGOp* uses the *HessType* function *f* and the starting interval *Start* as input parameters and stores all computed enclosures in the interval matrix *OptiVector*, a vector of interval vectors). If this matrix is not big enough to store all result interval vectors, the corresponding error code is returned together with the *OptiVector* containing all solutions it is able to store. If this error occurs, the user must increase the upper index bound of *OptiVector* to compute *all* optimizers.

The same applies to the information about the uniqueness, stored in the Boolean vector *InfoVector*. The user must declare both vectors *OptiVector* and *InfoVector* with lower index bound equal to 1 and with upper index bounds which are equal. These conditions as well as the condition *Epsilon* $\geq$ *MinEpsilon* are checked at the beginning of procedure *AllGOp*. *Epsilon* in our program corresponds to the parameter ε in the algorithms, and *MinEpsilon* corresponds to 1 ulp accuracy.

```
{----------------------------------------------------------------------------}
{ Purpose: Computing enclosures for all global minimizers and for the global }
{    minimum value of a twice continuously differentiable multi-dimensional, }
{    scalar valued function, assuming that the global minimum is a stationa- }
{    ry point. If it is a boundary point of the search area with gradient of }
{    the function being different from zero, the method fails in its form    }
{    presented here.                                                         }
{    Note that some routines are implemented using the "modified call by     }
{    reference" of PASCAL-XSC (see Language Reference for details) to avoid  }
{    the very inefficient allocation of local memory for the copies of the   }
{    actual parameters.                                                      }
{ Method: Bisection method combined with midpoint, monotonicity, concavity   }
{    test and extended interval Newton step.                                 }
{ Global procedures and functions:                                           }
{    procedure AllGOp(...)       : computes enclosures for all zeros         }
{    function  AllGOpErrMsg(...) : delivers an error message text            }
{----------------------------------------------------------------------------}
module gop;

use
  i_ari,     { Interval arithmetic               }
  xi_ari,    { Extended interval arithmetic      }
  lst_ari,   { List arithmetic                   }
  i_util,    { Interval utilities                }
  b_util,    { Boolean utilities                 }
  hess_ari,  { Differentiation arithmetic        }
  mv_ari,    { Real matrix/vector arithmetic     }
  mvi_ari,   { Interval matrix/vector arithmetic }
  matinv,    { Inversion of real matrices        }
  mvi_util;  { Interval matrix/vector utilities  }

const
  ListSize = 1000;    { List size, i.e. number of vectors which can be stored }

{----------------------------------------------------------------------------}
{ Error codes used in this module.                                           }
{----------------------------------------------------------------------------}
const
  NoError       = 0;  { No error occurred.                                   }
  lbOptiVecNot1 = 1;  { Lower bound of variable OptiVector is not equal to 1.}
  lbInfoVecNot1 = 2;  { Lower bound of variable InfoVector is not equal to 1.}
  VecsDiffer    = 3;  { Bounds of OptiVector and InfoVector do not match.    }
  VecTooSmall   = 4;  { OptiVector too small. Not all zeros can be stored.   }
  NoStatOpti    = 5;  { No stationary point is a global optimizer.           }
  ListTooSmall  = 6;  { Internal List is too small.                          }

{----------------------------------------------------------------------------}
{ Error messages depending on the error code.                                }
{----------------------------------------------------------------------------}
global function AllGOpErrMsg ( Err : integer ) : string;
var
  Msg : string;
begin
  case Err of
    NoError      : Msg := '';
    lbOptiVecNot1: Msg := 'Lower bound of OptiVector is not equal to 1';
    lbInfoVecNot1: Msg := 'Lower bound of InfoVector is not equal to 1';
    VecsDiffer   : Msg := 'Bounds of OptiVector and InfoVector do not match';
    VecTooSmall  : Msg := 'Not all optimizers found. OptiVector is too small';
    NoStatOpti   : Msg := 'No global optimizer (stationary point!) found';
```

```
    ListTooSmall : Msg := 'Internal list too small. ' +
                          'Increase "ListSize" and recompile "gop.p"!';
    else         : Msg := 'Code not defined';
  end;
  if (Err <> NoError) then Msg := 'Error: ' + Msg + '!';
  AllGOpErrMsg := Msg;
end;

{----------------------------------------------------------------------------}
{ Determine the component 'mc' with maximum interval diameter of 'iv[mc]'.   }
{----------------------------------------------------------------------------}
function MaxDiamComp (var iv: ivector) : integer;
var
  mc, k : integer;
  d     : rvector[lb(iv)..ub(iv)];
begin
  d:= diam(iv);  mc:= lb(iv);
  for k:=lb(iv)+1 to ub(iv) do
    if d[k] > d[mc] then mc:= k;
  MaxDiamComp:= mc;
end;

function MonotonicityTest (var gradY: ivector) : boolean; { returns true if  }
var                                                       { 'f' is monotone  }
  i, n   : integer;                                       {------------------}
  Delete : boolean;
begin
  Delete:= false; i:= 1;  n:= ub(gradY);
  while (i <= n) and (not Delete) do
  begin
    if (0 < gradY[i].inf) or (0 > gradY[i].sup) then    { 'f' is monotone }
      Delete:= true;                                     {-----------------}
    i:= i+1;
  end;
  MonotonicityTest:= Delete;
end;

function ConcavityTest (var HessY: imatrix) : boolean;  { returns true if    }
var                                                     { 'f' is not convex  }
  i, n   : integer;                                     {--------------------}
  Delete : boolean;
begin
  Delete:= false;  i:= 1;  n:= ub(HessY);
  while (i <= n) and (not Delete) do
  begin
    if (HessY[i,i].sup < 0) then  { 'f' is not convex }
      Delete:= true;              {-------------------}
    i:= i+1;
  end;
  ConcavityTest:= Delete;
end;

{----------------------------------------------------------------------------}
{ Purpose: Execution of one single interval Newton Gauss-Seidel step for the }
{    interval vector 'Y' and the gradient of 'f'.                            }
{ Parameters:                                                                }
{    In     : 'f'         : must be declared for the type 'HessType' to      }
{                           enable the internal use of the differentiation  }
{                           arithmetic 'hess_ari'.                           }
{             'Y'         : specifies the starting interval.                 }
{             'HessY'     : Hessian matrix of 'f(Y)', already computed       }
{                           outside of 'NewtonStep'.                         }
{    Out    : 'V'         : stores the enclosures 'V[i]' for the splittings }
{                           generated by the Newton step.                    }
```

```
{           'p'          : number of non-empty interval vectors 'V[i]'.     }
{ Description:                                                              }
{    'NewtonStep' executes the extended interval Newton Gauss-Seidel step   }
{    for 'Y' with result interval vector(s) 'V[i]' which can be empty.      }
{    'p' gives the number of non-empty interval vectors stored as vectors   }
{    'V[1]',...,'V[p]'.                                                     }
{---------------------------------------------------------------------------}
procedure NewtonStep (function f (x: HTvector) : HessType[0..ub(x,1),
                                                          0..ub(x,2)];
                                                  Y : ivector;
                          var                 HessY : imatrix;
                          var                     V : imatrix;
                          var                     p : integer);
var
  c                           : rvector[1..ub(Y)];
  GradC, b, Yin, Y_minus_c    : ivector[1..ub(Y)];
  R                           : rmatrix[1..ub(Y),1..ub(Y)];
  A                           : imatrix[1..ub(Y),1..ub(Y)];
  i, i0, n, InvErr, j         : integer;
  NoSolution                  : boolean;
  h, fC                       : interval;
  z                           : ivector[1..2];
begin
  Yin:= Y;  c:= mid(Y);
  fgEvalH(f, intval(c), fC, GradC);          { Midpoint gradient evaluation   }
                                             {--------------------------------}
  MatInv(mid(HessY),R,InvErr);               { Invert the midpoint matrix     }
  if InvErr <> 0 then R:= id(R);             {--------------------------------}

  A:= R * HessY;  b:= R * GradC;         { Compute data for Gauss-Seidel step }
  Y_minus_c:= Y - c;                     {------------------------------------}

  p:= 0; i:= 0; i0:= 0; n:= ub(Y); { Initializations, A[i0,i0] contains zero }
                                   {----------------------------------------}
  NoSolution:= false;
  while (i < n) and (not NoSolution) do      { Interval Gauss-Seidel step for }
  begin                                      { non-zero A[i,i] elements       }
    i:= succ (i);                            {--------------------------------}
    if not (0 in A[i,i]) then
      begin
        h:= ## (b[i] + (for j:=1 to i-1 sum (A[i,j] * Y_minus_c[j]) )
                     + (for j:=i+1 to n sum (A[i,j] * Y_minus_c[j]) ) );
        h:= c[i] - h / A[i,i];

        if (Y[i] >< h) then
          NoSolution:= true
        else
          begin
            Y[i]:= Y[i] ** h;  Y_minus_c[i]:= Y[i] - c[i];
          end;
      end
    else
      i0:= i;                                    { Largest i with 0 in A[i,i] }
  end;                                           {----------------------------}

  i:= 0;
  while (not NoSolution) and (i < i0) do     { Interval Gauss-Seidel step for }
  begin                                      { zero A[i,i] elements           }
    i:= succ (i);                            {--------------------------------}
    if (0 in A[i,i]) then
    begin
      h:= ## ( b[i] + ( for j:=1 to i-1 sum (A[i,j] * Y_minus_c[j]) )
                    + ( for j:=i+1 to n sum (A[i,j] * Y_minus_c[j]) ) );
      z:= Y[i] ** (c[i] - h div A[i,i]);         { Extended interval division }
```

```
                                                    {---------------------------}
      if (z[1] = EmptyIntval) then                  { z[1] = z[2] = EmptyIntval }
        NoSolution:= true                           {---------------------------}
      else
        begin                                                { Compute new 'Y' }
          Y[i]:= z[1];  Y_minus_c[i]:= Y[i] - c[i];          {-----------------}
          if z[2] <> EmptyIntval then
          begin                                    { Store further bisections }
            p:= p+1; V[p]:= Y; V[p][i]:= z[2];     {--------------------------}
          end;
        end;
    end; { if 0 in A[i,i] ... }
  end; { while (not NoSolution) ... }

  if NoSolution then
    p := 0
  else
    begin
      p:= p+1;  V[p]:= Y;
    end;
end; { NewtonStep }


{----------------------------------------------------------------------------}
{ Purpose: Execution of the global optimization method including a bisection }
{    method, midpoint test, monotonicity test, concavity test, and extended  }
{    interval Newton steps.                                                  }
{ Parameters:                                                                }
{    In    : 'f'           : must be declared for the 'DerivType' to enable  }
{                            the internal use of the differentiation         }
{                            arithmetic 'ddf_ari'.                           }
{            'Start:       : specifies the starting interval.                }
{            'Epsilon'     : specifies the desired relative accuracy.        }
{            'VecMat'      : matrix for storing the vectors of the list.     }
{    Out   : 'ResultList'  : stores the candidates for enclosure of a global }
{                            minimizer.                                      }
{            'Minimum'     : stores the enclosure of the global minimum      }
{                            value.                                          }
{            'ListFull'    : signals that list (matrix VecMat) is full       }
{ Description:                                                               }
{   The procedure manages the list 'L' of pending subintervals that may      }
{   contain global minimizers. Subintervals are removed from the list and    }
{   placed in the accepted list 'ResultList' when they satisfy relative      }
{   error acceptance criteria. Subintervals are also removed from the list   }
{   by the midpoint, monotonicity, concavity tests, or by the interval       }
{   Newton steps. Subintervals are added to the pending list when an element }
{   from the list is bisected or when the extended interval Newton step      }
{   yields two candidate intervals.                                          }
{   'ResultList' returns the list of enclosures of the global minimizers,    }
{   'Minimum' returns the enclosure of the global minimum value.             }
{----------------------------------------------------------------------------}
procedure GlobalOptimize (function f (x: HTvector) : HessType[0..ub(x,1),
                                                             0..ub(x,2)];
                          var         Start : ivector;
                                    Epsilon : real;
                          var    ResultList : PairPtr;
                          var       Minimum : interval;
                          var        VecMat : imatrix;
                          var      ListFull : boolean);

var
  PairY                  : ivector[0..ub(Start)];  { Pair ( Y, inf(f(Y) )       }
  Y                      : ivector[1..ub(Start)];
  U                      : imatrix[1..2,           { Subboxes of Y              }
```

```
                           1..ub(Start)];
  V                    : imatrix[1..ub(Start)+1, { Subboxes of U                }
                           1..ub(Start)];
  fY, fU, fV, fC, fCV  : interval;              { Function evaluations of f   }
  gradU, gradV         : ivector[1..ub(Start)]; { Gradient evaluations of f   }
  HessU                : imatrix[1..ub(Start),  { Hessian evaluation of f     }
                           1..ub(Start)];
  fmax                 : real;                  { Upper bound for minimum     }
  c, cV                : rvector[1..ub(Start)]; { Midpoints of Y and V        }
  WorkList             : PairPtr;               { List of pairs               }
  i, j, p, k           : integer;               { Control variables           }
  Bisect               : boolean;               { Flag for iteration          }

begin
  c:= mid(Start);
  fEvalH(f, intval(c), fC);                  { Compute upper bound for minimum }
  fmax:= sup(fC);                            {---------------------------------}

  ResultList:= EmptyList;  ListFull:=false;

  if not UlpAcc(Start,1) then                                { Start iteration }
    begin                                                    {-----------------}
      Y:= Start; WorkList:= EmptyList;
      repeat
        k:= MaxDiamComp(Y);                            { d(Y[k] = max_i d(Y[i])    }
        U[1]:= Y;  U[2]:= Y;                           { Bisect 'Y' with respect   }
        U[1][k].sup:= c[k];  U[2][k].inf:= c[k];       { to component 'k'.         }
        for i:= 1 to 2 do                              {---------------------------}
        begin
          fgEvalH(f, U[i], fU, GradU);                { Compute interval gradient }
                                                      {---------------------------}
          if not MonotonicityTest(GradU) then
          begin                                       { Try centered form to get  }
            fU:= (fC + GradU*(U[i] - c)) ** fU;       { better enclosure of 'f(U)'}
                                                      {---------------------------}
            if (fmax >= inf(fU)) then
            begin
              fghEvalH(f, U[i], fU, GradU, HessU);{ Compute interval Hessian }
                                                  {--------------------------}
              if not ConcavityTest(HessU) then
              begin
                NewtonStep(f,U[i],HessU,V,p);{ Extended interval Newton step }
                                             {-------------------------------}
                for j:=1 to p do
                begin
                  fgEvalH(f, V[j], fV, GradV);  { Compute interval gradient }
                                                {---------------------------}
                  if not MonotonicityTest(GradV) then
                  begin
                    cV:= mid(V[j]);                       { Try centered form}
                    fEvalH(f,intval(cV),fCV);             { to get better en-}
                    fV:= (fCV + GradV*(V[j] - cV)) ** fV; { closure of 'f(U)'}
                                                          {------------------}
                    if (fmax >= inf(fV)) then                      { Store V}
                    begin                                          {--------}
                      PairY:= MakePair(V[j],inf(fV));
                      Enter(WorkList, VecMat, PairY, ListFull);
                    end;
                  end;
                end; { for j ... }
              end;  { if not ConcavityTest(HessU) }
            end; { if fmax >= ... }
          end; { if not MonotonicityTest(gradU) ... }
        end; { for i ... }
```

```
                                                   { Get next 'Y' of }
                                                   { the work list   }
      Bisect:= false;                              {-----------------}
      while (WorkList <> EmptyList) and (not Bisect) and (not ListFull) do
      begin
        PairY:= Head(WorkList,VecMat);  DelHead(WorkList);

        Y:= Int(PairY);  c:= mid(Y);
        fEvalH(f, intval(c), fC);                    { Compute f(c) }
                                                     {--------------}
        if sup(fC) < fmax then fmax:= sup(fC);
        MultiDelete(WorkList,VecMat,fmax);

        Minimum:= intval(Fyi(PairY),fmax);
                                          { Check termination criteria }
        if (RelDiam(Minimum)<Epsilon) or (MaxRelDiam(Y)<Epsilon) then
          Enter(ResultList, VecMat, PairY, ListFull)      { Store 'PairY' }
        else                                              {---------------}
          Bisect:= true;
      end;

    until (not Bisect);

  end { if not UlpAcc(Start,1) }
 else
  begin                                      { Store starting interval }
    fEvalH(f, Start, fY);                    { and interval evaluation }
    Enter(ResultList,VecMat,MakePair(Start,inf(fY)),ListFull);
  end;
                                          { Compute good enclosure of  }
 if ResultList <> EmptyList then          { the global minimum value   }
  Minimum:= intval(Fyi(Head(ResultList,VecMat)),fmax);{--------------------}
end;

{---------------------------------------------------------------------------}
{ 'MaxNorm' delivers an upper bound of the maximum norm of a symmetric in-  }
{ terval matrix 'H', i.e. the row sum norm or infinity norm.                }
{---------------------------------------------------------------------------}
function MaxNorm (var H: imatrix) : real;
var
  Nm, MaxNm : real;
  i, j      : integer;
begin
  MaxNm:= 0;
  for i:=lb(H) to ub(H) do
  begin
    Nm:= 0;
    for j:=lb(H) to ub(H) do Nm := Nm +> sup (abs (H[i,j]));
    if Nm > MaxNm then MaxNm:= Nm;
  end;
  MaxNorm:= MaxNm;
end;

{---------------------------------------------------------------------------}
{ 'PosDef' delivers true if it is guaranteed that all real symmetric matri- }
{ ces enclosed in 'H' are positive definite.                                }
{---------------------------------------------------------------------------}
function PosDef (var H: imatrix) : boolean;
const
  kmax = 10;  { Maximum number of iterations }
var
  pd          : boolean;
  kappa, eps  : real;
  S           : imatrix[lb(H,1)..ub(H,1),lb(H,2)..ub(H,2)];
```

```
  Z, U        : ivector[lb(H,1)..ub(H,1)];
  k           : integer;
begin
  kappa:= MaxNorm(H);   S:= Id(H) - H/kappa;
  for k:=lb(Z) to ub(Z) do Z[k]:= intval(-1,1);
  k := 0;  eps:= 0.25;
  repeat
    U:= blow(Z,eps);  Z:= S * U;  pd:= Z in U;
    k:= k+1;  eps:= 8*eps;
  until pd or (k = kmax);
  PosDef:= pd;
end;

{----------------------------------------------------------------------------}
{ Purpose: Execution of a verification step including the use of an epsilon  }
{    inflation.                                                               }
{ Parameters:                                                                 }
{    In     : 'f'       : function of 'HessType'.                             }
{    Out    : 'yUnique' : returns 'true' if the verification is successful.   }
{    In/Out : 'y'       : interval enclosure to be verified.                  }
{ Description: This procedure checks the uniqueness of the local minimizer   }
{    enclosed in the interval variable 'y' by a verification step including  }
{    the use of an epsilon inflation of the iterates.                         }
{----------------------------------------------------------------------------}
procedure VerificationStep (function f (x: HTvector) : HessType[0..ub(x,1),
                                                                 0..ub(x,2)];
                                   var               y : ivector;
                                   var         yUnique : boolean);
const
  kmax = 10;  { Maximum number of iterations }
var
  fY                 : interval;
  yIn, yOld, GradY   : ivector[1..ub(y)];
  HessY              : imatrix[1..ub(y),1..ub(y)];
  yp                 : imatrix[1..ub(y)+1,1..ub(y)];
  k, p               : integer;
  eps                : real;
begin
  yUnique := (inf(y) = sup(y));                   { y is a point interval vector }
  if not yUnique then                             {------------------------------}
  begin
    yIn := y;  yOld := y;  k := 0;  eps:= 0.25;              { Initializations }
                                                             {-----------------}
    while (not yUnique) and (k < kmax) do{ Do kmax loops to achieve inclus.  }
    begin                                {-----------------------------------}
      yOld := blow(y,eps);               { Epsilon inflation of 'y'          }
                                         {-----------------------------------}
      { Perform interval Newton step }
      {------------------------------}
      k := k+1;
      fghEvalH(f, yOld, fY, GradY, HessY);    { Compute gradient and Hessian }
                                              {------------------------------}
      NewtonStep(f, yOld, HessY, yp, p);

      if (p <> 1) then                             { No verification possible }
        k := kmax                                  {--------------------------}
      else if (yp[1] = yOld) then
        eps := eps * 8                             { Increase 'eps'           }
      else                                         {--------------------------}
        begin
          y:= yp[1];  yUnique:= y in yOld;{ Inner inclusion ===> uniqueness  }
        end;                              { of a stationary point            }
    end;                                  {----------------------------------}
    if yUnique then yUnique:= PosDef(HessY);  { Positive definite ===> local }
```

```
  end;                                      { minimizer is verified       }
  if not yUnique then y:= yIn;              {-----------------------------}
end;

{---------------------------------------------------------------------------}
{ Purpose: Computation of enclosures for all global minimizers and for the  }
{    global minimum value of a twice continuously differentiable multi-di-  }
{    mensional, scalar-valued function.                                     }
{ Parameters:                                                               }
{    In     : 'f'             : objective function, must be declared for the }
{                               'HessType' to enable the internal use of    }
{                               the differentiation arithmetic 'hess_ari'.  }
{             'Start',        : specifies the starting interval vector.     }
{             'Epsilon'.      : specifies the desired relative accuracy     }
{                               (interval diameter) of the result boxes.    }
{    Out    : 'OptiVector'    : stores (row by row) and returns the boxes   }
{                               (enclosures) for the global optimizers of   }
{                               'f'. 'OptiVector' is a vector of interval   }
{                               vectors (ie. an interval matrix).           }
{             'InfoVector'    : stores the corresponding information on the }
{                               uniqueness of the local optimizers in these }
{                               enclosures.                                 }
{             'NumberOfOptis' : number of enclosures computed.              }
{             'Minimum'     ' : enclosure for the minimum value.            }
{             'Err'           : error code.                                 }
{ Description:                                                              }
{    The enclosures for the global minimizers of 'f' are computed by calling }
{    procedure 'GlobalOptimize'. Then a verification step is applied.       }
{    The enclosures (boxes) for the global minimizers of 'f' are stored in  }
{    the interval matrix 'OptiVector' row by row, the corresponding         }
{    information on the uniqueness of the local minimizers in these         }
{    enclosures is stored in the Boolean vector 'InfoVector'. The number of }
{    enclosures computed is returned in the integer variable 'NumberOfOptis'.}
{    The enclosure for the global minimum value is returned in the variable }
{    'Minimum'. If an error occurs, the value of 'Err' is different from 0. }
{---------------------------------------------------------------------------}
global procedure AllGOp (function f (x: HTvector) : HessType[0..ub(x,1),
                                                             0..ub(x,2)];
                               var          Start : ivector;
                                          Epsilon : real;
                               var     OptiVector : imatrix;
                               var     InfoVector : bvector;
                               var  NumberOfOptis : integer;
                               var        Minimum : interval;
                               var            Err : integer);
var
  i, k           : integer;
  ResultList, L  : PairPtr;
  Start1n        : ivector[1..ub(Start)-lb(Start)+1];
  ResMat         : imatrix[1..ListSize,0..ub(Start)-lb(Start)+1];
  ListFull       : boolean;
  MinEpsilon     : real;
begin
  NumberOfOptis:= 0;
  if (lb(OptiVector) <> 1) then        { Check index bounds of result vectors }
    Err:= lbOptiVecNot1                {-------------------------------------}
  else if (lb(InfoVector) <> 1) then
    Err:= lbInfoVecNot1
  else if (ub(InfoVector) <> ub(OptiVector)) then
    Err:= VecsDiffer
  else
    begin                              { Start global optimization method    }
      Err:= NoError;                   {-------------------------------------}
      MinEpsilon:= succ(1.0) - 1.0;    { Relative machine accuracy (1 ulp)   }
```

```
    Start1n:= Start;                   { Resize to standard bounds 1..n      }
                                       {-------------------------------------}
    if (Epsilon < MinEpsilon) then Epsilon := MinEpsilon;  { Set 'Epsilon' }
                                                           { to 1 ulp acc. }
    GlobalOptimize(f,Start1n,Epsilon,ResultList,Minimum,ResMat,ListFull);

    NumberOfOptis:= Length(ResultList);

    if NumberOfOptis = 0 then  Err:= NoStatOpti;

    if ListFull then Err:= ListTooSmall;
                                            { Check if there are more opti- }
    if ub(OptiVector) < NumberOfOptis then  { mizers than storage space     }
    begin                                   {-------------------------------}
      Err:= VecTooSmall;  NumberOfOptis:= ub(OptiVector);
    end;

    L:= ResultList;                            { Verification step for the }
    for i:=1 to NumberOfOptis do               { enclosure intervals       }
    begin                                      {---------------------------}
      OptiVector[i]:= Int(Head(L, ResMat));  L := Next(L);
      VerificationStep(f, OptiVector[i], InfoVector[i]);
    end;

    FreeAll(ResultList);
  end;
end;

{----------------------------------------------------------------------------}
{ Module initialization part                                                 }
{----------------------------------------------------------------------------}
begin
  { Nothing to initialize }
end.
```

14.3.2 Examples

We illustrate the use of *AllGOp* to compute all global minimizers and the global minimum of the function of Branin [85]

$$f_B(x) = \left(\frac{5}{\pi}x_1 - \frac{5.1}{4\pi^2}x_1^2 + x_2 - 6\right)^2 + 10\left(1 - \frac{1}{8\pi}\right)\cos x_1 + 10$$

in the starting interval vector $([-5,10],\ [0,15])^{\mathrm{T}}$ and of the function of Levy (see [58])

$$\begin{aligned} f_L(x) &= \sum_{i=1}^{5} i\cos((i-1)x_1 + i)\sum_{j=1}^{5} j\cos((j+1)x_2 + j) \\ &\quad + (x_1 + 1.42513)^2 + (x_2 + 0.80032)^2 \end{aligned}$$

in the starting interval vector $([-10,10],\ [-10,10])^{\mathrm{T}}$.

Figures 14.1 and 14.2 show the plots of these two functions. Branin's function has the global minimizers $(-\pi, \frac{491}{40})^{\mathrm{T}}$, $(\pi, \frac{91}{40})^{\mathrm{T}}$, and $(3\pi, \frac{99}{40})^{\mathrm{T}}$, and the global minimum value $\frac{5}{4}\pi$. The large number of local optimizers of f_L, make it extremely difficult

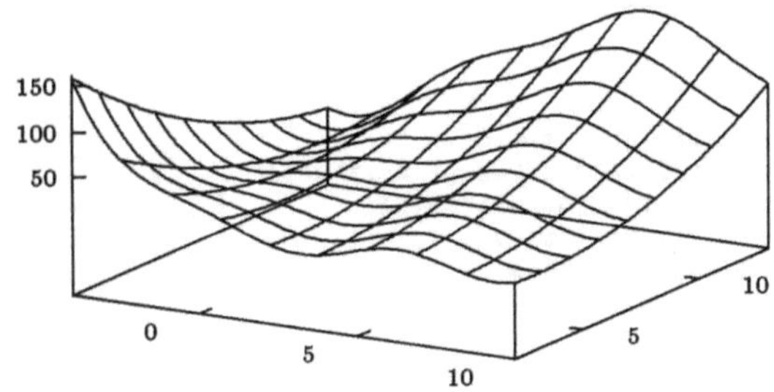

Figure 14.1: Function of Branin

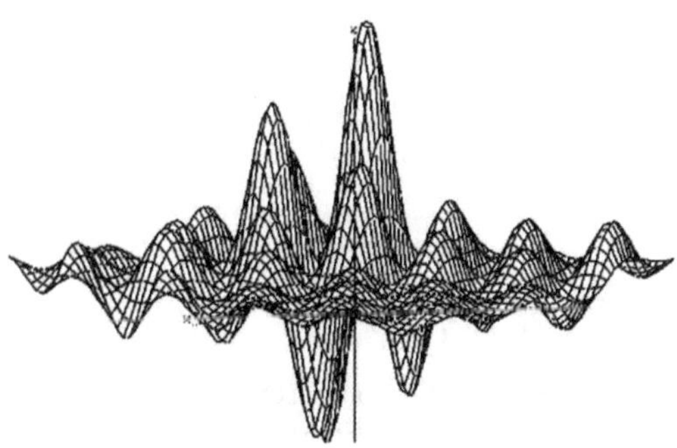

Figure 14.2: Function of Levy with about 700 local minima

for an approximation method to find the global minimizer $(-1.3068..., -1.424...)^T$. The global minimum value is $-1.7613....$

We first define the functions f_B and f_L for the type *HessType*. Then, we use a procedure *compute* with a function and a textual description of the function as parameters. This procedure reads the necessary input data for the call of procedure *AllGOp*. If we add additional function definitions in our program, we can apply this procedure in the same way.

```
{---------------------------------------------------------------------------}
{ This program uses module 'gop' to compute the global optimizers of the    }
{ function of Branin                                                          }
{                                                                             }
{    fB(x) = sqr(5/pi*x[1] - 51/(40*sqr(pi))*sqr(x[1]) + x[2] - 6)           }
{                                                                             }
{             + 10*(1-1/8/pi)*cos(x[1]) + 10                                  }
{                                                                             }
{ and the function of Levy                                                    }
{             5                       5                                       }
{    fL(x) = sum i cos((i-1)x[1] + i)  sum j cos((j+1)x[2] + j)               }
{            i=1                      j=1                                     }
{                                                                             }
{             + sqr(x[1] + 1.42513) + sqr(x[2] + 0.80032)                     }
{                                                                             }
```

```
{ A starting interval and a tolerance must be entered.                         }
{------------------------------------------------------------------------------}
program gop_ex;

use
  i_ari,     { Interval arithmetic                 }
  mvi_ari,   { Interval matrix/vector arithmetic }
  hess_ari,  { Differentiation arithmetic         }
  i_util,    { Interval utilities                  }
  b_util,    { Boolean utilities                   }
  gop;       { Global optimization                 }
const
  n    = 20;   { Maximum number of optimizers to be computed }
  fDim = 2;    { Dimension of sample functions               }

function fBranin (x: HTvector) : HessType[0..fDim,0..fDim];
begin
  fBranin := sqr(5/Pi*x[1] - 51/(40*sqr(Pi))*sqr(x[1]) + x[2] - 6)
             + 10*(1-1/8/Pi)*cos(x[1]) + 10;
end;

function fLevy (x: HTvector) : HessType[0..fDim,0..fDim];
var
  isum, jsum : HessType[0..fDim,0..fDim];
  i          : integer;
begin
  isum := 0; jsum := 0;
  for i:=1 to 5 do
  begin
    isum := isum + i*cos((i-1)*x[1] + i);
    jsum := jsum + i*cos((i+1)*x[2] + i);
  end;
  fLevy := isum * jsum + sqr(x[1] + intval(142513)/100000) { Avoid real con- }
                       + sqr(x[2] + intval(80032)/100000); { version error   }
end;

{------------------------------------------------------------------------------}
{ Procedure for printing and reading informations to call the procedure       }
{ 'AllZeros'. This procedure must be called with the function 'f', a string  }
{ 'Name' containing textual description of that function, and an integer     }
{ 'dim' specifying the dimension of the problem.                              }
{------------------------------------------------------------------------------}
procedure compute (function f(x:HTvector): HessType[0..ub(x),0..ub(x)];
                                    Name: string;
                                     dim: integer);
var
  SearchInterval             : ivector[1..dim];
  Minimum                    : interval;
  Tolerance                  : real;
  Opti                       : imatrix[1..n,1..dim];
  Unique                     : bvector[1..n];
  NumberOfOptis, i, ErrCode : integer;
begin
  writeln('Computing all global minimizers of the ', Name);
  write('Search interval      : '); read(SearchInterval);
  write('Tolerance (relative) : '); read(Tolerance);
  writeln;
  AllGOp(f, SearchInterval, Tolerance,
            Opti, Unique, NumberOfOptis, Minimum, ErrCode);
  for i:=1 to NumberOfOptis do
  begin
    write(Opti[i]);
    if unique[i] then
      writeln('encloses a locally unique candidate for a global minimizer!')
```

```
    else
      writeln('may contain a local or global minimizer!')
  end;
  writeln;
  if (NumberOfOptis <> 0) then
  begin
    writeln(Minimum);  writeln('encloses the global minimum value!');
  end;
  if ErrCode <> 0 then writeln(AllGOpErrMsg(ErrCode));
  writeln;  writeln(NumberOfOptis:1, ' interval enclosure(s)');  writeln;
  if (NumberOfOptis = 1) and (unique[1]) then
    writeln('We have validated that there is a unique global optimizer!');
end;

begin { Main program }
  compute(fBranin, 'Function of Branin', fDim);  writeln;  writeln;
  compute(fLevy,   'Function of Levy',   fDim);
end.
```

If we execute this program, we get the following runtime output:

```
Computing all global minimizers of the Function of Branin
Search interval      : [-5, 10]  [0, 15]
Tolerance (relative) : 1e-8

[     9.424777960768E+000,     9.424777960771E+000 ]
[     2.474999999999E+000,     2.475000000001E+000 ]
encloses a locally unique candidate for a global minimizer!

[  3.141592653589792E+000,  3.141592653589795E+000 ]
[   2.274999999999999E+000,   2.275000000000001E+000 ]
encloses a locally unique candidate for a global minimizer!

[  -3.14159265358980E+000,  -3.14159265358979E+000 ]
[   1.22749999999999E+001,   1.22750000000001E+001 ]
encloses a locally unique candidate for a global minimizer!

[    3.9788735772973E-001,    3.9788735772975E-001 ]
encloses the global minimum value!

3 interval enclosure(s)


Computing all global minimizers of the Function of Levy
Search interval      : [-10, 10]  [-10, 10]
Tolerance (relative) : 1e-8

[   -1.3068530097537E+000,   -1.3068530097535E+000 ]
[   -1.4248450415608E+000,   -1.4248450415606E+000 ]
encloses a locally unique candidate for a global minimizer!

[     -1.76137578003E+002,     -1.76137578001E+002 ]
encloses the global minimum value!

1 interval enclosure(s)

We have validated that there is a unique global optimizer!
```

Thus, we know that there are three locally unique minimizers of f_B which are good candidates for global minimizers and one global minimizer of f_L within the specified starting boxes.

14.3.3 Restrictions and Hints

The objective function f must be expressible in PASCAL–XSC code as a finite sequence of arithmetic operations and elementary functions supported by the differentiation arithmetic module *hess_ari*.

The procedure *AllGOp* stores all enclosures for candidates for global minimizers in the interval matrix *OptiVector*, which must be of sufficient size. If the first run of *AllGOp* is not able to compute all optimizers because *OptiVector* is not big enough, then the routine must be called again with an increased index range for *OptiVector*.

The method is not very fast if a very small value of ε (*Epsilon*) is used, if the interval Newton step does not improve the actual iterates, *and* if the different tests do not discard intervals any more because of rounding and overestimation effects of the machine interval arithmetic. Under these circumstances, the method is equivalent with a bisection method.

In *GlobalOptimize*, the evaluation of the function with differentiation arithmetic can cause a runtime error if the interval argument of an elementary function does not lie in the domain specified for this interval function (see [65]) or if a division by an interval containing zero occurs. This also may be due to the known overestimation effects of interval arithmetic (see Section 3.1). To get rid of these errors, the user may try to split the starting interval in several parts and call *AllGOp* for these parts.

The rules for getting true enclosures in connection with conversion errors (see Section 3.7) also apply here. That is why we expressed 1.42513 as `intval(142513)/100000` in the code for the Levy function.

14.4 Exercises

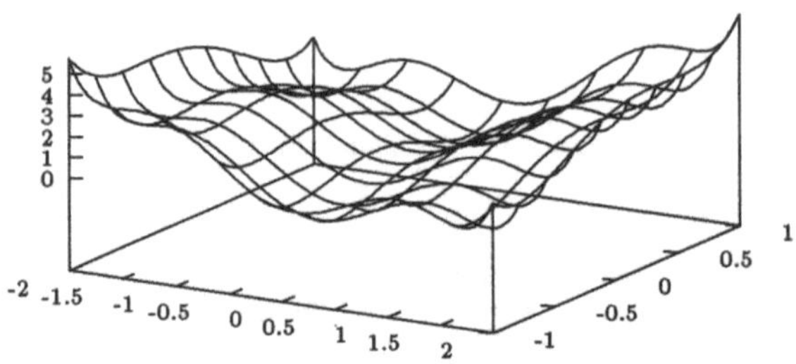

Figure 14.3: Six-Hump Camel-Back function

Exercise 14.1 Use our procedure *AllGOp* to compute the global minimizers and the global minimum value of the *Six-Hump Camel-Back* function [72]

$$f(x) = 4x_1^2 - 2.1x_1^4 + \frac{1}{3}x_1^6 + x_1x_2 - 4x_2^2 + 4x_2^4$$

shown in Figure 14.3 within the box $([-2.5, 2.5], [-2.5, 2.5])^{\mathrm{T}}$. The global optimizers are $\pm(0.089842..., -0.712656...)^{\mathrm{T}}$ and the global minimum value is -1.03162.

Exercise 14.2 Use our procedure *AllGOp* to compute the global minimizer and the global minimum value of the function $f : I\!R^n \to I\!R$ with

$$f(x) = \sum_{i=1}^{n} \left(\sum_{j=1}^{i} x_j \right)^2 .$$

Study the behavior of CPU time for our algorithm for increasing values of n.

Exercise 14.3 Use our procedure *AllGOp* to compute the global minimizers (a continuum of points) and the global minimum value of the function $f : I\!R^2 \to I\!R$ with

$$f(x) = (x_1 + x_2)^2.$$

Exercise 14.4 Use our procedure *AllGOp* to compute the global minimizer $(1, \ldots, 1)^{\mathrm{T}}$ and the global minimum value 0 of $f : I\!R^n \to I\!R$ with

$$f(x) = \sum_{i=2}^{n} \left(100(x_i - x_{i-1}^2)^2 + (1 - x_{i-1})^2 \right).$$

f is a generalization of Rosenbrock's two-dimensional function [75], which has become accepted as a difficult test for new optimization routines. Study the behavior of CPU time for our algorithm for increasing values of $n = 5, 10, 15, \ldots$.

14.5 References and Further Reading

The method we discussed in this chapter is an *a priori* method because the iteration starts with a (possibly large) interval enclosing all the solutions which have to be found. Here, the iterates of the method are subintervals of the previous iterates.

There are also methods for finding (and bounding) one single *local* optimizer called *a posteriori* methods. These methods start with an approximation of a local minimizer of f and apply a test procedure for a neighborhood interval of the approximation to verify that a zero of ∇f lies within that interval. There are a huge number of approximate optimization methods for local and global optimization available without any verification of the result. For an overview on such approximation methods, see [85] or [88].

The method presented in this chapter can be extended to become more efficient and faster. For more sophisticated extensions, see [26], [28], [72], [73], and [74].

Appendix A

Utility Modules

The data types and routines described in the following sections are used by several problem solving routines in this book. They are grouped according to the basic data types of PASCAL-XSC.

A.1 Module b_util

The module b_util gives the definition of a dynamic vector whose components are of type *boolean*. Variables of this type are used as tag fields to indicate whether the components of a corresponding interval vector or matrix are verified.

```
{----------------------------------------------------------------------------}
{ Purpose: Utilities of type 'boolean'.                                      }
{ Global type:                                                               }
{    type bvector : Used as tag field to indicate whether the components of  }
{                   a corresponding interval vector or matrix are verified.  }
{----------------------------------------------------------------------------}
module b_util;       { Boolean Utilities }
                     { -       ----      }
global type
  bvector = global dynamic array [*] of boolean;

{----------------------------------------------------------------------------}
{ Module initialization part                                                 }
{----------------------------------------------------------------------------}
begin
  { Nothing to initialize }
end.
```

A.2 Module r_util

The module r_util provides the function *Max* to find the maximum of two real numbers.

```
{----------------------------------------------------------------------------}
{ Purpose: Utilities of type 'real'.                                         }
{ Global function:                                                           }
{    function Max(...) : Delivers the maximum of two real numbers.           }
{----------------------------------------------------------------------------}
module r_util;       { Real Utilities }
                     { -    ----      }

global function Max ( a, b : real ) : real;       { Maximum function for reals }
```

```
begin
  if (a >= b) then  Max := a  else  Max := b;
end;

{----------------------------------------------------------------------}
{ Module initialization part                                           }
{----------------------------------------------------------------------}
begin
  { Nothing to initialize }
end.
```

A.3 Module i_util

The module i_util provides procedures and functions based on the type *interval*. The real-valued functions *AbsMin*, *AbsMax*, and *RelDiam* may be used to compute the smallest absolute value $\langle[x]\rangle$, the greatest absolute value $|[x]|$, and the relative diameter $d_{\text{rel}}([x])$ of a real interval $[x]$, according to the Definitions (3.1) and (3.2). The Boolean function *UlpAcc* is used to check for a desired accuracy specified in ulp (see Section 3.6). The interval function *Pi* delivers an enclosure of π.

Finally, the module gives an implementation of the interval power function $[x]^n$ with integer exponent n as defined in Example 3.5. The implementation makes use of the local function *Power* for real argument x, positive (!) integer exponent n, and a parameter indicating the rounding mode. Depending on the rounding parameter, a lower or an upper bound of x^n is computed using the binary shift method. The binary shift method is used to reduce an exponentiation by an integer to a sequence of multiplications. Therefore in general, it does not yield a result of maximum accuracy. The number of multiplications depends logarithmically on n. An algorithmic description of the method is given by Algorithm A.1.

Algorithm A.1: Power (x, n) {(Function)}

1. {Initialization}
   $p := 1;\quad z := x;$
2. {Binary shift method}
   while $(n > 0)$ **do**
     if Odd (n) **then** $p := p \cdot z;$
     ShiftRight (n); {Equivalent to $n := n/2$}
     if $(n > 0)$ **then** $z := z \cdot z;$
   end;
3. {Return function result}
   return Power $:= p;$

The method works by binary shifting the exponent n. Depending on the digit which was shifted out, the factor p is updated, while z is updated as long as n does not vanish.

Example A.1 To compute x^{13}, the following intermediate results are generated. The binary representation of the exponent is $13_{10} = 1101_2$.

Digit shifted out	p	z
1	x	x^2
0	x	x^4
1	x^5	x^8
1	x^{13}	–

To get lower and upper bounds of x^n, respectively, Algorithm A.1 is modified by introducing directed-rounding multiplications (see the listing below). For negative x and odd n, we have $\nabla(x^n) = -\Delta(-x^n)$ and $\Delta(x^n) = -\nabla(-x^n)$. The complete listing of the module *i_util* is given below.

```
{----------------------------------------------------------------------------}
{ Purpose: Utilities of type 'interval'.                                      }
{ Global functions:                                                           }
{    function AbsMin(...)  : Smallest absolute value of an interval.          }
{    function AbsMax(...)  : Greatest absolute value of an interval.          }
{    function RelDiam(...) : Relative diameter of an interval.                }
{    function UlpAcc(...)  : To check whether the width of an interval is     }
{                            less than a certain number of ulps (ulp = unit   }
{                            in the last place of the mantissa).              }
{    function Power(...)   : Exponentiation by an integer for intervals.      }
{    function Pi           : Returns an enclosure of pi.                      }
{----------------------------------------------------------------------------}
module i_util;          { Interval Utilities }
                        { -        ----      }
use
  i_ari;                { Interval arithmetic }

var
  LocalPi : interval;   { Local variable to store an enclosure of pi }

global function AbsMin ( x : interval ) : real;          { Absolute minimum of }
begin                                                    { an interval         }
  if (0 in x) then                                       {---------------------}
    AbsMin := 0
  else if (inf(x) > 0) then
    AbsMin := inf(x)
  else
    AbsMin := -sup(x);
end;

global function AbsMax ( x : interval ) : real;          { Absolute maximum of }
var                                                      { an interval         }
  a, b : real;                                           {---------------------}
begin
  a := abs(inf(x));  b := abs(sup(x));
  if (a > b) then  AbsMax := a  else  AbsMax := b;
end;

global function RelDiam ( x : interval ) : real;         { Relative diameter   }
begin                                                    { of an interval      }
  if (0 in x) then                                       {---------------------}
    RelDiam := diam(x)
  else
    RelDiam := diam(x) /> AbsMin(x);
end;
```

```
{----------------------------------------------------------------------------}
{ Checks whether the width of the interval 'x' is less or equal to 'n' ulp.  }
{ An ulp is an abbreviation for: units in the last place of the mantissa.    }
{----------------------------------------------------------------------------}
global function UlpAcc ( x : interval; n : integer ) : boolean;
var
  i       : integer;
  Infimum : real;
begin
  Infimum := inf(x);
  for i := 1 to n do Infimum := succ(Infimum);
  UlpAcc := (Infimum >= sup(x));
end;

{----------------------------------------------------------------------------}
{ Purpose: The local function 'Power()' is used to compute a lower or an     }
{    upper bound for the power function with real argument and integer       }
{    exponent, respectively.                                                 }
{ Parameters:                                                                }
{    In  : 'x'       : real argument.                                        }
{          'n'       : integer exponent.                                     }
{          'RndMode' : rounding mode,                                        }
{                      (-1 = downwardly directed, +1 = upwardly directed)    }
{ Description:                                                               }
{    This function is used to speed up the interval power function defined   }
{    below. The exponentiation is reduced to multiplications using the       }
{    binary shift method. Depending on 'n', this function is up to 40 times  }
{    as fast as the standard power function for real argument and real       }
{    exponent. However, its accuracy is less than one ulp (unit in the last  }
{    place of the mantissa) since about log2(n) multiplications are executed }
{    during computation. Since directed roundings are antisymmetric, one     }
{    gets                                                                    }
{                                                                            }
{       down(x↑n) = -up((-x)↑n)   and   up(x↑n) = -down((-x)↑n)              }
{                                                                            }
{    for x < 0 and odd n, where 'down' and 'up' denote the downwardly and    }
{    upwardly directed roundings, respectively.                              }
{----------------------------------------------------------------------------}
function Power ( x : real; n, RndMode : integer ) : real;
var                          { Signals change of the rounding mode }
  ChangeRndMode : boolean;   { for x < 0 and odd n                 }
  p, z          : real;
begin
  ChangeRndMode := ((x < 0) and odd(n));
  if ChangeRndMode then
    begin z := -x; RndMode := -RndMode; end
  else
    z := x;

  p := 1;                                    { Note: Seperate while-loops    }
  case RndMode of                            { used to gain speed at runtime }
    -1 : while (n > 0) do                    {-------------------------------}
         begin
           if odd(n) then p := p *< z;
           n := n div 2;
           if (n > 0) then z := z *< z;
         end;
    +1 : while (n > 0) do
         begin
           if odd(n) then p := p *> z;
           n := n div 2;
           if (n > 0) then z := z *> z;
         end;
```

```
  end;
  if ChangeRndMode then  Power := -p  else  Power := p;
end;

{----------------------------------------------------------------------------}
{ Purpose: This version of the function 'Power()' is used to compute an      }
{    enclosure for the power function with interval argument and integer     }
{    exponent.                                                               }
{ Parameters:                                                                }
{    In  : 'x' : interval argument.                                          }
{          'n' : integer exponent.                                           }
{ Description:                                                               }
{    In general, this implementation does not deliver a result of maximum   }
{    accuracy, but it is about 30-40 times faster than the standard power    }
{    function for interval arguments and interval exponents. The resulting   }
{    interval has a width of approximately 2*log2(n) ulps. Since x↑n is      }
{    considered as a monomial, we define x↑0 := 1. For negative exponents    }
{    and 0 in 'x', the division at the end of the function will cause a      }
{    runtime error (division by zero).                                       }
{----------------------------------------------------------------------------}
global function Power ( x : interval; n : integer ) : interval;
var
  m            : integer;
  Lower, Upper : real;
begin
  if (n = 0) then
    Power := 1
  else
    begin
      if (n > 0) then  m := n  else  m := -n;

      if (0 < x.inf) or odd(m) then
        begin  Lower := Power(x.inf,m,-1);  Upper := Power(x.sup,m,+1);  end
      else if (0 > x.sup) then
        begin  Lower := Power(x.sup,m,-1);  Upper := Power(x.inf,m,+1);  end
      else
        begin  Lower := 0;  Upper := Power(AbsMax(x),m,+1);  end;

      if (n > 0) then
        Power := intval(Lower,Upper)
      else                                     { Propagates a 'division by }
        Power := 1 / intval(Lower,Upper);      { zero' error if 0 in 'x'.  }
    end;
end;

{----------------------------------------------------------------------------}
{ The function 'Pi' returns an enclosure of pi. To save computation time, pi }
{ is computed only once in the module initialization part by evaluating      }
{ 4*arctan(intval(1)). Thus, this function only returns the value of the     }
{ local variable 'LocalPi'.                                                  }
{----------------------------------------------------------------------------}
global function Pi : interval;
begin
  Pi := LocalPi;
end;

{----------------------------------------------------------------------------}
{ Module initialization part                                                 }
{----------------------------------------------------------------------------}
begin
  LocalPi := 4*arctan(intval(1));    { Computes an enclosure of pi }
end.
```

A.4 Module mvi_util

The module mvi_util provides procedures and functions based on the types *ivector* and *imatrix*. The function *MaxRelDiam* may be used to get the maximum relative diameter of an interval vector as defined in Section 3.4. The Boolean function *UlpAcc* is used to check any component of an interval vector for a desired accuracy specified in ulp. The complete listing of the module is given below.

```
{----------------------------------------------------------------------------}
{ Purpose: Utilities of type 'ivector' and 'imatrix'.                        }
{ Global functions:                                                          }
{    MaxRelDiam(...) : To get the maximum of the relative diameters of the   }
{                      components of an interval vector.                     }
{    UlpAcc(...)     : To check whether all components of an interval vector }
{                      have a width less than a certain number of ulps       }
{                      (ulp = unit in the last place of the mantissa).       }
{----------------------------------------------------------------------------}
module mvi_util;        { Matrix Vector Interval Utilities }
                        { -      -      -        ----      }
use
  i_ari,                { Interval arithmetic        }
  i_util;               { Utilities of type interval }

{----------------------------------------------------------------------------}
{ Compute maximum of relative diameters of the components of ivector 'v'.    }
{----------------------------------------------------------------------------}
global function MaxRelDiam ( var v: ivector ) : real;
var
  k               : integer;
  Max, RelDiam_k  : real;
begin
  Max := 0;
  for k := lb(v) to ub(v) do
  begin
    RelDiam_k := RelDiam(v[k]);
    if (RelDiam_k > Max) then Max := RelDiam_k;
  end;
  MaxRelDiam := Max;
end;

{----------------------------------------------------------------------------}
{ Checks if the diameter of the interval vector 'x' is less or equal to 'n'  }
{ ulps. An ulp is an abbreviation for: units in the last place of the        }
{ mantissa.                                                                  }
{----------------------------------------------------------------------------}
global function UlpAcc ( v : ivector; n : integer ) : boolean;
var
  k : integer;
begin
  k := lb(v);
  while (k < ub(v)) and (UlpAcc(v[k],n)) do k := k+1;
  UlpAcc := UlpAcc(v[k],n);
end;

{----------------------------------------------------------------------------}
{ Module initialization part                                                 }
{----------------------------------------------------------------------------}
begin
  { Nothing to initialize }
end.
```

Bibliography

[1] Adams, E., Kulisch, U. (Eds.): *Scientific Computing with Automatic Result Verification*. Academic Press, New York, 1993.

[2] Alefeld, G., Herzberger, J.: *Einführung in die Intervallrechnung*. Bibliographisches Institut, Mannheim, 1974.

[3] Alefeld, G., Herzberger, J.: *Introduction to Interval Computations*. Academic Press, New York, 1983.

[4] American National Standards Institute / Institute of Electrical and Electronics Engineers: *A Standard for Binary Floating-Point Arithmetic*. ANSI/IEEE Std. 754-1985, New York, 1985.

[5] American National Standards Institute / Institute of Electrical and Electronics Engineers: *A Standard for Radix-Independent Floating-Point Arithmetic*. ANSI/IEEE Std. 854-1987, New York, 1987.

[6] Atanassova, L., Herzberger, J. (Eds.): *Computer Arithmetic and Enclosure Methods*. North-Holland, Elsevier, Amsterdam, 1992.

[7] Auzinger, W., Stetter, H. J.: *Accurate Arithmetic Results for Decimal Data on Non-Decimal Computers*. Computing **35**, pp 141–151, 1985.

[8] Bauch, H., Jahn, K.-U., Oelschlägel, D., Süsse, H., Wiebigke, V.: *Intervallmathematik*. Teubner, Leipzig, 1987.

[9] Berz, M.: *Forward Algorithms for High Orders and Many Variables with Application to Beam Physics*. In [21], pp 147–156, 1991.

[10] Böhm, H.: *Berechnung von Polynomnullstellen und Auswertung arithmetischer Ausdrücke mit garantierter maximaler Genauigkeit*. Dissertation, Universität Karlsruhe, 1983.

[11] Buchberger, B., Kutzler, B., Feilmeier, M., Kratz, M., Kulisch, U., Rump, S. M.: *Rechnerorientierte Verfahren*. Teubner, Stuttgart, 1986.

[12] Corliss, G. F.: *Automatic Differentiation Bibliography*. In [21], pp 331–353, 1991.

[13] Dantzig, G. B.: *Lineare Programmierung und Erweiterung*. Springer-Verlag, Heidelberg, 1966.

[14] Dennis, J. E., Schnabel, R. B.: *Numerical Methods for Unconstrained Optimization and Nonlinear Equations*. Prentice-Hall, Englewood Cliffs, New Jersey, 1983.

[15] Falcó Korn, C.: *Die Erweiterung von Software-Bibliotheken zur effizienten Verifikation der Approximationslösung linearer Gleichungssysteme*. Dissertation, Universität Basel, 1993.

[16] Fischer, H.-C.: *Schnelle automatische Differentiation, Einschließungsmethoden und Anwendungen*. Dissertation, Universität Karlsruhe, 1990.

[17] Fischer, H.-C., Haggenmüller, R., Schumacher, G.: *Auswertung arithmetischer Ausdrücke mit garantierter hoher Genauigkeit*. Siemens Forschungs- und Entwicklungsberichte **16**, Nr. 5, pp 171–177, 1987.

[18] Fischer, H.-C., Schumacher, G., Haggenmüller, R.: *Evaluation of Arithmetic Expressions with Guaranteed High Accuracy*. In [56], pp 149–157, 1988.

[19] Geörg, S.: *Two Methods for the Verified Inclusion of Zeros of Complex Polynomials*. In [87], pp 229–244, 1990.

[20] Griewank, A.: *On Automatic Differentiation*. In: Iri, M., Tanabe, K. (Ed.): *Mathematical Programming: Recent Developments and Applications*, pp 83–108, Kluwer Academic Publishers, 1989.

[21] Griewank, A., Corliss, G. (Eds.): *Automatic Differentiation of Algorithms: Theory, Implementation, and Applications*. SIAM, Philadelphia, Pennsylvania, 1991.

[22] Grüner, K.: *Solving Complex Problems for Polynomials and Linear Systems with Verified High Accuracy*. In [37], pp 199–220, 1987.

[23] Hammer, R.: *Maximal genaue Berechnung von Skalarproduktausdrücken und hochgenaue Auswertung von Programmteilen*. Dissertation, Universität Karlsruhe, 1992.

[24] Hammer, R., Neaga, M., Ratz, D.: *PASCAL-XSC – New Concepts for Scientific Computation and Numerical Data Processsing*. In [1], pp 15–44, 1993.

[25] Hansen, E.: *Global Optimization Using Interval Analysis – The One-Dimensional Case*. Journal of Optimization Theory and Applications **29**, pp 331–344, 1979.

[26] Hansen, E.: *Global Optimization Using Interval Analysis – The Multi-Dimensional Case*. Numerische Mathematik **34**, pp 247–270, 1980.

[27] Hansen, E.: *An Overview of Global Optimization Using Interval Analysis*. In [63], pp 289–307, 1988.

[28] Hansen, E.: *Global Optimization Using Interval Analysis*. Marcel Dekker, New York, 1992.

[29] Hansen, E., Sengupta, S.: *Bounding Solutions of Systems of Equations Using Interval Analysis*. BIT **21**, pp 203–211, 1981.

[30] Heuser, H.: *Funktionalanalysis*. Teubner, Stuttgart, 1986.

[31] IBM *High-Accuracy Arithmetic Subroutine Library* (ACRITH). General Information Manual, GC 33-6163-02, 3rd Edition, 1986.

[32] IBM *High-Accuracy Arithmetic Subroutine Library* (ACRITH). Program Description and User's Guide, SC 33-6164-02, 3rd Edition, 1986.

[33] IBM *High Accuracy Arithmetic - Extended Scientific Computation* (ACRITH-XSC), Reference. SC 33-6462-00, IBM Corp., 1990.

[34] Jansson, C.: *Zur Linearen Optimierung mit unscharfen Daten*. Dissertation, Universität Kaiserslautern, 1985.

[35] Jansson, C.: *A Global Optimization Method Using Interval Arithmetic*. In: [6], pp 259–267, 1992.

[36] Karmarkar, N.: *A New Polynomial-Time Algorithm for Linear Programming*. Combinatorica **4**, pp 373–395, 1985.

[37] Kaucher, E., Kulisch, U., Ullrich, Ch. (Eds.): *Computer Arithmetic - Scientific Computation and Programming Languages*. Teubner, Stuttgart, 1987.

[38] Kaucher, E., Rump, S. M.: *E-Methods for Fixed Point Equations $f(x) = x$*. Computing **28**, pp 31–42, 1982.

[39] Kaucher, E., Markov, S. M., Mayer, G. (Eds.): *Computer Arithmetic, Scientific Computation and Mathematical Modelling*. IMACS Annals on Computing and Applied Mathematics **12**, J. C. Baltzer AG, Scientific Publishing Co., Basel, 1991.

[40] Kearfott, R. B.: *Preconditioners for the Interval Gauss-Seidel Method*. SIAM Journal of Numerical Analysis **27**, 1990.

[41] Kearfott, R. B., Hu, C., Novoa, M.: *A Review of Preconditioners for the Interval Gauss-Seidel Method*. Interval Computations **1**, pp 59–85, Institute for New Technologies, St. Petersburg, 1991.

[42] Khachiyan, L. G.: *A Polynomial Algorithm in Linear Programming*. Doklady Akademiia Nauk SSSR **244**, pp 1093–1096 (in Russian), English translation in Sov. Math. Dokl. **20**, pp 191–194, 1979.

[43] Klatte, R., Kulisch, U., Lawo, C., Rauch, M., Wiethoff, A.: *C-XSC - A C++ Class Library for Extended Scientific Computing*. Springer-Verlag, Heidelberg, 1993.

[44] Klatte, R., Kulisch, U., Neaga, M., Ratz, D., Ullrich, Ch.: *PASCAL-XSC - Sprachbeschreibung mit Beispielen*. Springer-Verlag, Heidelberg, 1991.

[45] Klatte, R., Kulisch, U., Neaga, M., Ratz, D., Ullrich, Ch.: *PASCAL-XSC - Language Reference with Examples*. Springer-Verlag, New York, 1992.

[46] König, S.: *On the Inflation Parameter Used in Self-validating Methods*. In [87], pp 127–132, 1990.

[47] Krämer, W.: *Inverse Standardfunktionen für reelle und komplexe Intervallargumente mit a priori Fehlerabschätzungen für beliebige Datenformate*. Dissertation, Universität Karlsruhe, 1987.

[48] Krämer, W.: *Einschluß eines Paares konjugiert komplexer Nullstellen eines reellen Polynoms*. ZAMM **71**, pp T820–T824, 1991.

[49] Krämer, W.: *Evaluation of Polynomials in Several Variables with High Accuracy.* In [39], pp 239–249, 1991.

[50] Krawczyk, R.: *Newton-Algorithmen zur Bestimmung von Nullstellen mit Fehlerschranken.* Computing **4**, pp 187–220, 1969.

[51] Krawzcyk, R.: *Fehlerabschätzungen bei linearer Optimierung.* In: Nickel, K. (Ed.): *Interval Mathematics.* Lecture Notes in Computer Science, No. 29, pp 215–227, Springer-Verlag, Vienna, 1975.

[52] Kulisch, U.: *Grundlagen des Numerischen Rechnens – Mathematische Begründung der Rechnerarithmetik.* Reihe Informatik, No. 19, Bibliographisches Institut, Mannheim, 1976.

[53] Kulisch, U. (Ed.): *Wissenschaftliches Rechnen mit Ergebnisverifikation — Eine Einführung.* Ausgearbeitet von S. Geörg, R. Hammer und D. Ratz. Akademie Verlag, Berlin, and Vieweg, Wiesbaden, 1989.

[54] Kulisch, U., Miranker, W. L.: *Computer Arithmetic in Theory and Practice.* Academic Press, New York, 1981.

[55] Kulisch, U., Miranker, W. L. (Eds.): *A New Approach to Scientific Computation.* Academic Press, New York, 1983.

[56] Kulisch, U., Stetter, H. J. (Eds.): *Scientific Computation with Automatic Result Verification.* Computing Suppl. **6**, Springer-Verlag, Vienna, 1988.

[57] Kulisch, U., Ullrich, Ch. (Eds.): *Wissenschaftliches Rechnen und Programmiersprachen.* Berichte des German Chapter of the ACM, No. 10, Teubner, Stuttgart, 1982.

[58] Levy, A. V., Montalvo, A., Gomez, S., Claderon, A.: *Topics in Global Optimization.* Lecture Notes in Mathematics, No. 909, Springer-Verlag, New York, 1981.

[59] Mayer, G.: *Grundbegriffe der Intervallrechnung.* In [53], pp 101–118, 1989.

[60] Miranker, W. L., Toupin, R. A.: *Accurate Scientific Computations.* Lecture Notes in Computer Science, No. 235, Springer-Verlag, Berlin, 1986.

[61] Moore, R. E.: *Interval Analysis.* Prentice-Hall, Englewood Cliffs, New Jersey, 1966.

[62] Moore, R. E.: *Methods and Applications of Interval Analysis.* SIAM, Philadelphia, Pennsylvania, 1979.

[63] Moore, R. E. (Ed.): *Reliability in Computing: The Role of Interval Methods in Scientific Computations.* Academic Press, New York, 1988.

[64] Neumaier, A.: *Interval Methods for Systems of Equations.* Cambridge University Press, Cambridge, 1990.

[65] Numerik Software GmbH: *PASCAL–XSC: A PASCAL Extension for Scientific Computation. User's Guide.* Numerik Software GmbH, P.O.Box 2232, D-76492 Baden-Baden, Germany, 1991.

[66] Rall, L. B.: *Automatic Differentiation, Techniques and Applications.* Lecture Notes in Computer Science, No. 120, Springer-Verlag, Berlin, 1981.

[67] Rall, L. B.: *Differentiation in PASCAL—SC: Type GRADIENT.* ACM Trans. Math. Softw. **10**, pp 161–184, 1984.

[68] Rall, L. B.: *Optimal Implementation of Differentiation Arithmetic.* In [37], pp 287–295, 1987.

[69] Rall, L. B.: *Differentiation Arithmetics.* In [86], pp 73–90, 1990.

[70] Ratschek, H.: *Die Subdistributivität in der Intervallarithmetik.* ZAMM **51**, pp 189–192, 1971.

[71] Ratschek, H., Rokne, J.: *Computer Methods for the Range of Functions.* Ellis Horwood Limited, Chichester, 1984.

[72] Ratschek, H., Rokne, J.: *New Computer Methods for Global Optimization.* Ellis Horwood Limited, Chichester, 1988.

[73] Ratz, D.: *Automatische Ergebnisverifikation bei globalen Optimierungsproblemen.* Dissertation, Universität Karlsruhe, 1992.

[74] Ratz, D.: *An Inclusion Algorithm for Global Optimization in a Portable PASCAL-XSC Implementation.* In: [6], pp 329–338, 1992.

[75] Rosenbrock, H.: *An Automatic Method for Finding the Greatest and the Least Value of a Function.* Computer Journal **3**, pp 175–184, 1960.

[76] Rump, S. M.: *Kleine Fehlerschranken bei Matrixproblemen.* Dissertation, Universität Karlsruhe, 1980.

[77] Rump, S. M.: *Lösung linearer und nichtlinearer Gleichungssysteme mit maximaler Genauigkeit.* In: [57], pp 147–174, Teubner, Stuttgart, 1982.

[78] Rump, S. M.: *Solving Algebraic Problems with High Accuracy.* In: [55], pp 51–120, 1983.

[79] Rump, S. M.: *New Results on Verified Inclusions.* In [60], pp 31–69, 1986.

[80] SIEMENS: *ARITHMOS (BS 2000) Unterprogrammbibliothek für Hochpräzisionsarithmetik. Kurzbeschreibung, Tabellenheft, Benutzerhandbuch.* BNr.: U2900-J-Z87-1, 1986.

[81] Spaniol, O.: *Die Distributivität in der Intervallarithmetik.* Computing **5**, pp 6–16, 1970.

[82] Stetter, H. J.: *Sequential Defect Correction for High-Accuracy Floating-Point Algorithms.* In: *Numerical Analysis*, Lecture Notes in Mathematics, Vol. 1066, pp 186–202, 1984.

[83] Stoer, J., Bulirsch, R.: *Introduction to Numerical Analysis.* Springer-Verlag, New York, 1980.

[84] Syslo, M., Deo, N., Kowalik, J.: *Discrete Optimization Algorithms with Pascal Programs.* Prentice-Hall, Englewood Cliffs, New Jersey, 1983.

[85] Törn, A., Žilinskas, A.: *Global Optimization*. Lecture Notes in Computer Science, No. 350, Springer-Verlag, Berlin, 1989.

[86] Ullrich, Ch. (Ed.): *Computer Arithmetic and Self-Validating Numerical Methods*. Academic Press, San Diego, 1990.

[87] Ullrich, Ch. (Ed.): *Contributions to Computer Arithmetic and Self-Validating Numerical Methods*. J. C. Baltzer AG, Scientific Publishing Co., IMACS, 1990.

[88] Wolfe, M. A.: *Numerical Methods for Unconstrained Optimization – An Introduction*. Van Nostrand Reinhold, New York, 1978.

[89] Zurmühl, R., Falk, S.: *Matrizen und ihre Anwendungen. Teil 2: Numerische Methoden*. Springer-Verlag, Heidelberg, 1984.

Index of Special Symbols

Index

Order Form

All computer programs presented in this volume are available on diskettes and via ftp (see Page vii). To order the programs on diskettes in IBM-PC compatible format, complete the information on the order form and mail it (or a photocopy) to:

Numerik Software GmbH
P.O.Box 2232
D-76492 Baden-Baden
Germany
Fax: +49 721 694418

All orders must be prepaid. Prices are not guaranteed. Ordinary postage is paid by Numerik Software GmbH.

ORDER FORM

We would like to order

... Toolbox Program Diskette(s) – Volume 1 (in PASCAL–XSC)
US$ 19.95 (outside Germany)
DM 19.95 (inside Germany)

Address

Name ....
Street ....
ZIP, City ....

Mode of Payment

☐ Check ☐ Visa ☐ Eurocard/Mastercard ☐ American Express
Card No. .... Exp. Date ....
Date/Signature ....

SPRINGER SERIES IN COMPUTATIONAL MATHEMATICS

Volume 1: **R. Piessens, E. de Doncker-Kapenga, C. W. Überhuber, D. K. Kahaner**

QUADPACK

A Subroutine Package for Automatic Integration

1983. ISBN 3-540-12553-1

Volume 2: **J. R. Rice, R. F. Boisvert**

Solving Elliptic Problems Using ELLPACK

1985. ISBN 3-540-90910-9

Volume 3: **N. Z. Shor**

Minimization Methods for Non-Differentiable Functions

1985. ISBN 3-540-12763-1

Volume 4: **W. Hackbusch**

Multi-Grid Methods and Applications

1985. ISBN 3-540-12761-5

Volume 5: **V. Girault, P-A. Raviart**

Finite Element Methods for Navier-Stokes Equations

Theory and Algorithms

1986. ISBN 3-540-15796-4

Volume 6: **F. Robert**

Discrete Iterations

A Metric Study

1986. ISBN 3-540-13623-1

Volume 7: **D. Braess**

Nonlinear Approximation Theory

1986. ISBN 3-540-13625-8

Volume 8: **E. Hairer, S. P. Nørsett, G. Wanner**

Solving Ordinary Differential Equations I

Nonstiff Problems

1993. 2nd revised edition.
ISBN 3-540-56670-8

Volume 9: **Z. Ditzian, V. Totik**

Moduli of Smoothness

1987. ISBN 3-540-96536-X

Springer

B3.10.130

Volume 10: **Yu. Ermoliev, R. J.-B. Wets** (Eds.)

Numerical Techniques for Stochastic Optimization

1988. ISBN 3-540-18677-8

Volume 11: **J-P. Delahaye**

Sequence Transformations

With an Introduction by C. Brezinski

1988. ISBN 3-540-15283-0

Volume 12: **C. Brezinski**

History of Continued Fractions and Pade Approximants

1990. ISBN 3-540-15286-5

Volume 13: **E. L. Allgower, K. Georg**

Numerical Continuation Methods

An Introduction

1990. ISBN 3-540-12760-7

Volume 14: **E. Hairer, G. Wanner**

Solving Ordinary Differential Equations II

Stiff and Differential-Algebraic Problems

1991. ISBN 3-540-53775-9

Volume 15: **F. Brezzi, M. Fortin**

Mixed and Hybrid Finite Element Methods

1991. ISBN 3-540-97582-9

Volume 16: **J. Sokolowski, J. P. Zolesio**

Introduction to Shape Optimization

Shape Sensitivity Analysis

1992. ISBN 3-540-54177-2

Volume 17: **A. R. Conn, N. I. M. Gould, Ph. R. Toint**

LANCELOT

A Fortran Package for Large-Scale Nonlinear Optimization (Release A)

1992. ISBN 3-540-55470-X

Volume 18: **W. Hackbusch**

Elliptic Differential Equations

Theory and Numerical Treatment

1992. ISBN 3-540-54822-X

Volume 19: **A. A. Gonchar, E. B. Saff** (Eds.)

Progress in Approximation Theory

An International Perspective

1992. ISBN 3-540-97901-8

Volume 20: **F. Stenger**

Numerical Methods Based on Sinc and Analytic Functions

1993. ISBN 3-540-94008-1

Springer

B3.10.130

MIX
Papier aus verantwortungsvollen Quellen
Paper from responsible sources
FSC® C105338
FSC
www.fsc.org

If you have any concerns about our products,
you can contact us on
ProductSafety@springernature.com

In case Publisher is established outside the EU,
the EU authorized representative is:
Springer Nature Customer Service Center GmbH
Europaplatz 3, 69115 Heidelberg, Germany

Printed by Libri Plureos GmbH
in Hamburg, Germany